Principles, Concepts and Technology of Farm Management

Principles, Concepts and Technology of Farm Management

Edited by **Carlos Hassey**

SYRAWOOD
PUBLISHING HOUSE
New York

Published by Syrawood Publishing House,
750 Third Avenue, 9th Floor,
New York, NY 10017, USA
www.syrawoodpublishinghouse.com

Principles, Concepts and Technology of Farm Management
Edited by Carlos Hassey

International Standard Book Number: 978-1-68286-209-4 (Hardback)

Printed in China.

Contents

	Preface	IX
Chapter 1	**Forecasting Milled Rice Production in Ghana Using Box-Jenkins Approach** Nasiru Suleman and Solomon Sarpong	1
Chapter 2	**Perception of Shea Nut Tree as an Economic Tree among Farmers in Oyo State, Nigeria** Rasheed Gbolagade Adeola	7
Chapter 3	**The Analysis of the Effects of Domestic and Foreign Investment in R&D on Agricultural TFP in Iran** Ali Bagherzadeh	12
Chapter 4	**Investigation of East Azerbaijan Researchers and Extension Agents' View about the Role of Extension Services on Agricultural Development** Javad Mahmoodi Karamjavan	23
Chapter 5	**The Effects of Non-Farm Business in Rural Sustainable Development: A Case Study in Sanandaj's Rural Areas, Kurdistan** Reza Movahedi, Hadi Fathi and Somaye Latifi	33
Chapter 6	**Trade Liberalization, Exchange Rate Changes, and the Competitiveness of Carbohydrate Staple Markets in Nigeria** Maduabuchi Mkpado and C. J. Arene	41
Chapter 7	**The Effect of Socio-Personal Factors on Economic and Managerial Variables in Shahroud Commercial Dairy Farms** Hamed Kashfi, Ahmad Reza Yazdani, Farhad Shirani Bidabadi and Mehrdad Latifi	57
Chapter 8	**Effect of under Irrigation Management on Potato Performance Components** Fakhrodin Ghasemi Sahebi, Masoud Hekmat and Eman Pourkhiz	63
Chapter 9	**Analysis of Rice Import Trend and it's Economic Factors: Case of Iran** Zabiholah Gholami	69
Chapter 10	**Economic Analysis of Water Demand in Greenhouses of Khash Township** Javad Shahraki and Neda AliAhmadi	75
Chapter 11	**Investigation of the Factors Hindering and Promoting Tourism Evidences from Villages of Abyaneh Village** Hoda Manafian, Zhila Daneshvar Ameri and Saba Salehi Esfahani	82
Chapter 12	**Assessment of Land Management Practices in Food Crops Production among Small Scale Farmers in Kwara State, Nigeria** Abdulazeez, Muhammad-Lawal, Omotesho, Kemi Funmi, Adekola, Olabisi Fatimat and Adekunle, Dayo	91

Chapter 13 **Development of Entrepreneurship in Complementary and Processing Industries in Khouzestan Province, Iran** 103
Ahmad Reza Ommani

Chapter 14 **Factors Affecting Nonfarm Income Diversification among Rural Farm Households in Central Nepal** 109
Raju Ghimire, Wen-Chi Huang and Rudra Bahadur Shrestha

Chapter 15 **Effectiveness of Agricultural Training Centers' Curriculum in Promoting Adoption of Agricultural Technologies: Evidence from Small-scale Potato Farmers in Nyandarua County, Kenya** 119
Ronald O. Nyamwamu, Justus M. Ombati and John G. Mwangi

Chapter 16 **Examine the Effects of Students' Social Capital Components on Entrepreneurship Intention (Evidences from: University College of Agriculture and Natural Resources, University of Tehran)** 132
Rasool Lavaei Adaryani, Mohammad Reza Akbari, Fatemeh Adel and Ali Amiri

Chapter 17 **The Law of One Price and the Cointegration of Meat Price in the Global Market: The Case of Iran's Market** 141
Behrouz Eskandarpour, Parisa Khaligh Khiyavi and Mohammad Kavoosi Kalashami

Chapter 18 **Assessment of Rural Farming Households WTP for Fertilizers and Agrochemicals in Kwawara State, Nigeria** 147
Ademoye Emmanuel Akinboboye, Fadipe, A. A., Adigun and Jubril Olayinka, Animashaun

Chapter 19 **Investigating Market Integration and Price Transmission of Different Rice Qualities in Iran** 155
Amir Hossein Chizari, Masoud Fehresti Sani and Mohammad Kavoosi Kalashami

Chapter 20 **Drought Risk Vulnerability Parameters among Wheat Farmers in Mashhad County, Iran** 162
Mojtaba Sookhtanlo, Hesamedin Gholami and Seyyed Reza Es'haghi

Chapter 21 **Livestock Farming Systems and Cattle Production Orientation in Eastern High Plains of Algeria, Cattle Farming System in Algerian Semi Arid Region** 172
Lounis Semara, Charefeddine Mouffok and Toufik Madani

Chapter 22 **An Investigation into Credit Receipt and Enterprise Performance among Small Scale Agro Based Enterprises in the Niger Delta Region of Nigeria** 180
Ubon Asuquo Essien, Chukwuemeka John Arene and Noble Jackson Nweze

Chapter 23 **Investigation of the Potential Market and Estimation of WTP for Insurance of Pistachio Tree Trunk (Case Study Rafsanjan-Iran)** 194
Mostafa Baniasadi, Saeed Yazdani and Habib Allah Salami

Chapter 24 **The Economic and Welfare Effects of Different Irrigation Water Pricing Methods, Case Study of Khomein Plain in Markazi Province of Iran** 203
Gholamreza Zamanian, Mehdi Jafari and Shahram Saeedian

Chapter 25 **Research Performance of Agriculture Faculty Members: A Comparative Study at West Part of Iran** 215
Nematollah Shiri, Nader Naderi and Ahmad Rezvanfar

Chapter 26 **Socio-Economic Factors Influencing Adoption of Energy-Saving Technologies among Smallholder Farmers: The Case of West Pokot County, Kenya** **223**
Andiema Chesang Everlyne, Nkurumwa Oywaya Agnes and Amudavi Mulama David

Permissions

List of Contributors

Preface

The world is advancing at a fast pace like never before. Therefore, the need is to keep up with the latest developments. This book was an idea that came to fruition when the specialists in the area realized the need to coordinate together and document essential themes in the subject. That's when I was requested to be the editor. Editing this book has been an honour as it brings together diverse authors researching on different streams of the field. The book collates essential materials contributed by veterans in the area which can be utilized by students and researchers alike.

Farm management encompasses the care and management of all agricultural resources including crops, livestock, etc. This book on farm management will prove to be an efficient guide for the students of agricultural sciences as well as agriculture practitioners. Selected concepts like resources for farm production, farming systems, farming structures, strategic planning, etc. have been presented in this book. It comprises researches from across the globe which will bring forth some unexplored aspects of farm management and contribute towards the growth of this field.

Each chapter is a sole-standing publication that reflects each author´s interpretation. Thus, the book displays a multi-facetted picture of our current understanding of applications and diverse aspects of the field. I would like to thank the contributors of this book and my family for their endless support.

Editor

Forecasting Milled Rice Production in Ghana Using Box-Jenkins Approach

Nasiru Suleman [1*] ***and Solomon Sarpong*** [2]

Keywords:
Box- Jenkins, Milled rice, Production, Forecasting, Ghana

Abstract

The increasing demand for rice in Ghana has been a major concern to the government and other stakeholders. Recent concerns by the coalition for African Rice Development (CARD) to double rice production within ten years in Sub-Saharan countries have triggered the to implement strategies to boost rice production in the government. To fulfill this requirement, there is a need to monitor and forecast trends of rice production in the country. This study employs the Box-Jenkins approach to model milled rice production using time series data from 1960 to 2010. The analysis revealed that ARIMA (2, 1, 0) was the best model for forecasting milled rice production. Although, a ten years forecast with the model shows an increasing trend in production, the forecast value at 2015 (283.16 thousand metric tons) was not good enough to compare with the current production of Nigeria (2700 thousand metric tons), the leading producer of rice of rice in West Africa.

[1,2]*Department of Statistics, Faculty of Mathematical Sciences, University for Development Studies, P. O. Box 24 Navrongo, Ghana, West Africa.*
** Corresponding author's email: sulemannasiru@yahoo.com*

INTRODUCTION

Rice (*Oryza sativa*) is one of the most important cereal crops and feeds more than a third of the world's population (Khush, 1997). The cereal plays a major role in contributing to agricultural GDP (Gross Domestic Product) and the economy of Ghana. The increase in per capital consumption of rice in recent times outstrips its supply in the country. The government imports up to 200% of local production in order to compensate for the short fall in supply (Dogbe, 1996). The country spends US$450 million annually on rice imports to fill this gap in rice production (FAO, 2011).

Nowadays, rice as a staple food in Ghana is confronted with several challenges in terms of production. For instance, the unbridled trade liberalization has made the country a net importer of rice. Furthermore, total land area for rice production, illiteracy rates, low yield, poor storage facilities, pest and diseases and poor market price for farmers have been attributed to low rice production in Ghana (Seidu, 2008).

Recent concerns by the Coalition for African Rice Development (CARD) to help Sub-Saharan African countries to double rice production within ten years have triggered the government and the Ministry of Food and Agricultural (MoFA) to implement strategies to boost rice production in Ghana. In response to this, the country has set a target of being the leading producer of rice in West Africa by 2015. This has led to the establishment of the Fievie Rice Projects in the Volta Region of the country. The project aims to produce highest rice grade qualities on 4,000 hectares of land (www.allAfrica.com).

The steps taken by the country to meet the increasing demand and to be the leading producer of rice in West Africa is cogent. But to fulfill this goal there is a need to monitor trends of production in the country.

Myriads of researches on forecasting have been done using the Box-Jenkins (1976) stochastic Autoregressive Integrated Moving Average (ARIMA) model. Badmus and Ariyo (2011), forecasted area of cultivation and production of maize in Nigeria using ARIMA model. They estimated ARIMA (1, 1, 1) and ARIMA (2, 1, 2) for cultivation area and production of maize respectively. Najeeb *et al.*, (2005), employed Box-Jenkins model to forecast wheat area and production in Pakistan. Their study showed that ARIMA (1, 1, 1) and ARIMA (2, 1, 2) were the appropriate models for wheat area and production respectively. Falak and Eatzaz (2008) analyzed future prospects of wheat production in Pakistan. They obtained the parameters of their forecasting model using Cobb-Douglas production function for wheat, while future values of various inputs are obtained as dynamic forecasts on the basis of separate ARIMA estimates for each input and for each Province. Rachana *et al.*, (2010), used ARIMA model to forecast pigeon pea production in India. The Box- Jenkins methodology have been used by a number of researchers to forecast future demands in terms of internal consumption and export to adopt appropriate measures (Muhammad *et al.*, 1992, Shabur and Haque, 1993, Sohail *et al.*, 1994).

Thus in this study, the Box-Jenkins approach was used to model and forecast milled rice production in Ghana.

MATERIALS AND METHODS

This study was carried out on the basis of milled rice production data from the period 1960 to 2010 collected from secondary sources, (Index mundi, 2011). The data was model using Autoregressive Integrated Moving Average (ARIMA) stochastic model popularized by Box-Jenkins (1976). An ARIMA (p, d, q) model is a combination of Autoregressive (AR) which shows that there is a relationship between present and past values, a random value and a Moving Average (MA) model which shows that the present value has something to do with the past residuals. The ARIMA process can be defined as:

$$\text{Ø}(B)(\Delta^d y_t - \mu) = \theta(B)e_t$$

Where:

y_t = represents rice production in thousand metric tons.

μ = Mean of $\Delta^d y_t$,

$$\text{Ø}(B) = 1 - \text{Ø}B - \dots - \text{Ø}_p B^p$$

$$\theta(B) = 1 - \text{Ø}B - \dots - \text{Ø}_q B^q$$

$\emptyset_t$ = The ith autoregressive parameter

θ_t = The ith moving average parameter

p, q and *p*denote the autoregressive, moving average and differenced order parameter of the process respectively.

Δ and *B* denote the difference and backward shift operators respectively.

The estimation of the model consists of three steps, namely: identification, estimation of parameters and diagnostic checking.

Identification step:

Identification step involves the use of the techniques to determine the values of *p, q* and *d*. The values are determined by using Autocorrelation function (ACF) and Partial Autocorrelation function (PACF). For any ARIMA (*p, q, p*) process, the theoretical PACF has non-zero partial autocorrelations at lags 1, 2,..., *p* and has zero partial autocorrelations at all lags, while the theoretical ACF has non zero autocorrelation at lags 1, 2,…, *q* and zero autocorrelations at all lags. The non-zero lags of the sample PACF and ACF are tentatively accepted as the *p* and *q* parameters. For a non stationary series the data is differenced to make the series stationary. The number of times the series is differenced determines the order of *d*.

Estimation of parameters:

The second step is the estimation of the model parameters for the tentative models that have been selected.

Diagnostic checking:

The estimated model must be check to verify if it adequately represents the series. Diagnostic checks are performed on the residuals to see if they are randomly and normally distributed. Here, the Anderson-Darling test for normality was used. Also, the residual plot versus the fitted values was used to check if the residuals are randomly scattered. An overall check of the model adequacy was made using the Ljung-Box **Q** statistics. The test statistics is given by:

$$Q_m = n(n+2)\sum_{k=1}^{n}(n-k)^{-1}r_k^2 \approx \chi^2_{m-r}$$

where

r_k^2 = the residuals autocorrelation at lag k

n = the number of residuals

m = the number of time lags included in the test.

When the *p*-value associated with the **Q** is large the model is considered adequate, else the whole estimation process has to start again in order to get the most adequate model.

RESULTS AND DISCUSSION

The maximum milled rice production was 295 thousand metric tons in 2010 and the minimum was 21 thousand metric tons which occurred in 1960, 1962 and 1965. The average milled rice production was 85.02 thousand metric tons. The increasing trend in the time series plot (Figure 1) of milled rice production showed that the series was not stationary. The gradual decay in the ACF (Figure 2) and very dominant PACF (Figure 3) at lag one indicate that the data was not stationary. To stabilize the variance and achieve a stationary mean, the data was logarithmic transformed and differenced. The first order difference was enough to make the data stationary. Therefore ARIMA (*p, 1, q*) could be identified. The alternating positive and negative ACF suggests an autoregressive process. (Figure 5) Using the PACF with a significant spike at lag 2, ARIMA (2, 1, 0) was identified (Figure 4). Table 1, shows the estimates of the ARIMA (2, 1, 0) model.

In order to make sure that this model is a representative of the data and could be used to forecast future milled rice production, the ACF and PACF for the residuals of the fitted model were examined. As shown in Figure 6, it was clear that no pattern was observed in

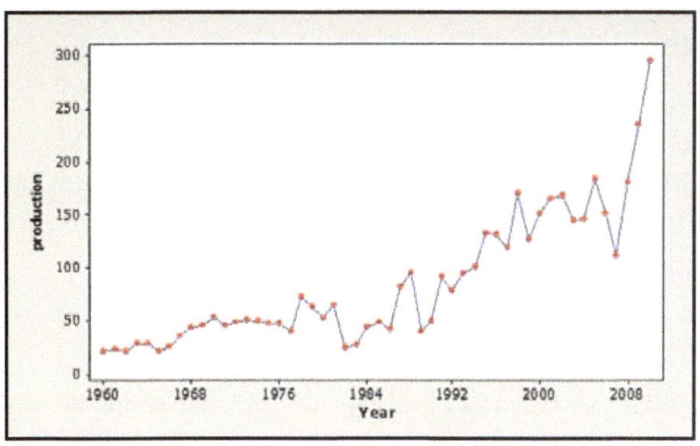

Figure 1: Time series plot of milled rice production

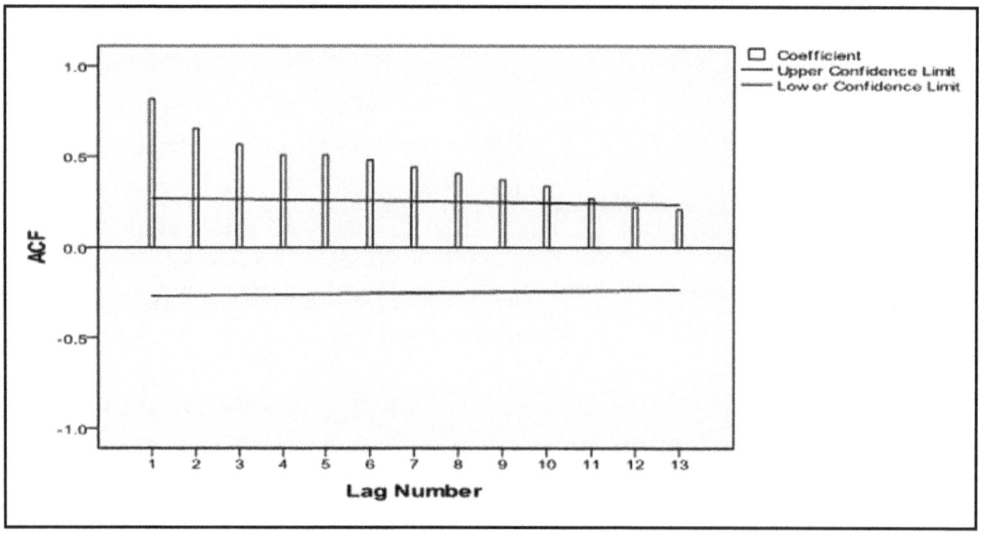

Figure 2: Plot of ACF for milled rice production

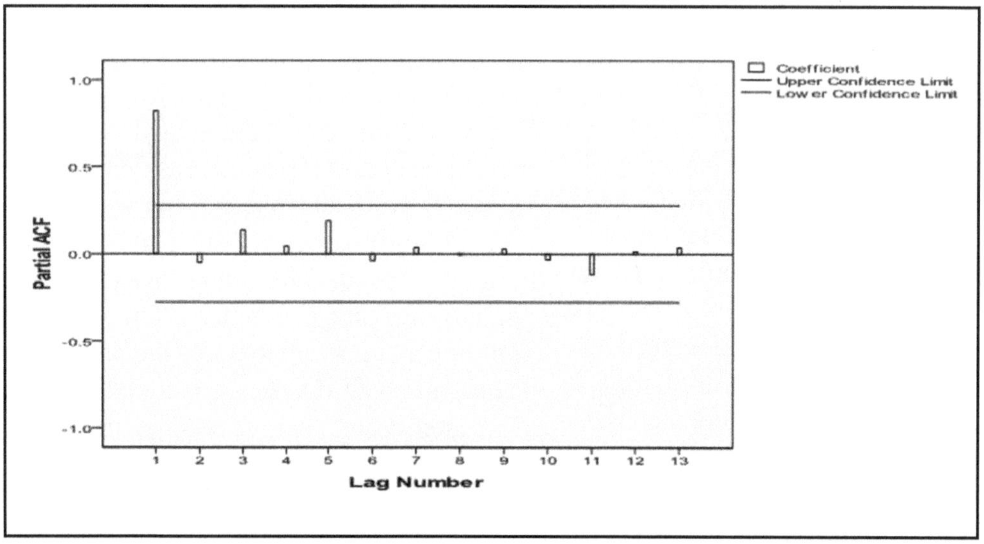

Figure 3: Plot of PACF for milled rice production

Table1: Estimates of ARIMA (2, 1, 0) model

Type	Coefficient	SE	t-value
AR (1)	-0.244	0.135	-1.809
AR (2)	-0.370	0.136	-2.731

the residuals. Thus, the model could be used to represent the milled rice production data. Also, the Ljung-Box statistics was large indicating the model was adequate. The plot of the residuals against the fitted values showed scattered trend, therefore the model was fitted properly. For normality, the Anderson-Darling test was used and the assumption of normality was accepted. Since the residuals were normally distributed the model was fitted properly.

Finally, ARIMA (2, 1, 0) model for milled rice production was used for ten years forecast. The forecast are shown in Table 2 at 95% confidence interval. From Table 2, the forecast for the years

Table 2: Forecast for milled rice production

RICE PRODUCTION (000 metric tons)			
Year	**Forecast**	**UCL**	**LCL**
2011-2012	264.81	453.01	141.72
2012-2013	258.10	500.81	116.65
2013-2014	278.51	560.39	119.39
2014-2015	283.16	606.24	110.57
2015-2016	281.55	645.93	98.65
2016-2017	287.58	692.57	93.08
2017-2018	294.24	740.16	88.50
2018-2019	297.93	783.69	82.90
2019-2020	302.24	828.25	78.13
2020-2021	307.65	874.91	74.26

Note: UCL= Upper Confidence Limit and LCL= Lower Confidence Limit

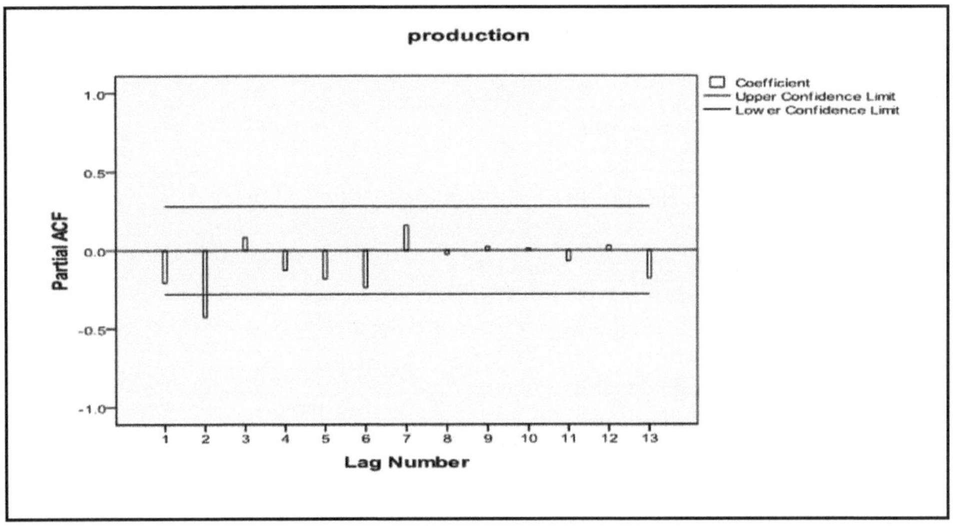

Figure 4: PACF plot for the differenced data

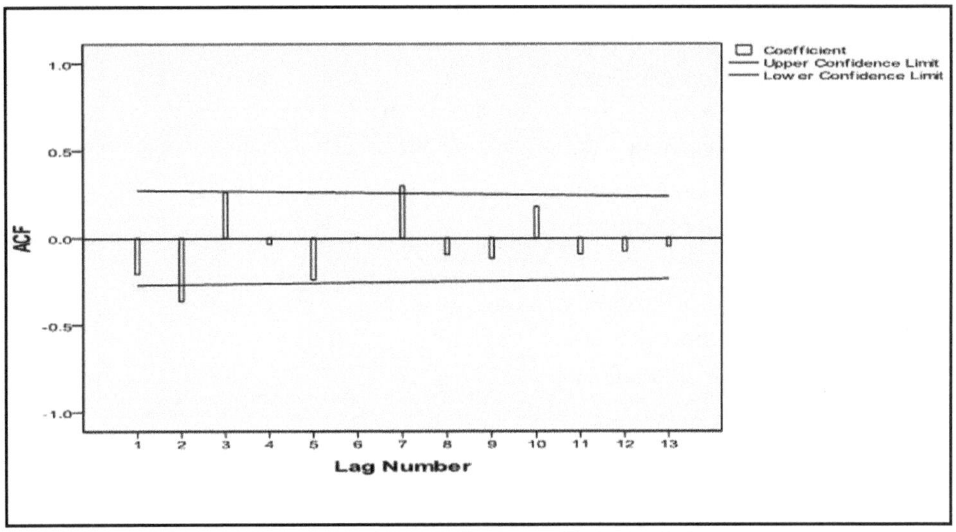

Figure 5: ACF plot for the differenced data

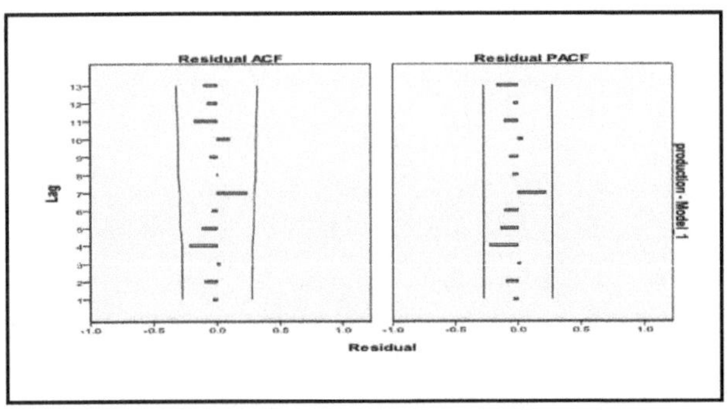

Figure 6: ACF and PACF plot of residuals

2011-2012 was 264.18 thousand metric tons with a 95% confidence limit of (141.72, 453.01) thousand metric tons. It was found that if government policies are not well improved, then the total milled rice production in the year 2020-2021 would be 307.65 thousand metric tons with a 95% confidence limit of (74.26, 874.91) thousand metric tons.

CONCLUSION

Although the forecast shows an increasing trend in milled rice production, this is not good enough to achieve the 2015 target of being the leading producer of rice in West Africa as the forecast (283.16 thousand metric tons) at that period is not appreciative compared to the total milled rice production of Nigeria (2700 thousand metric tons) currently, which is the leading producer of rice in West Africa.

To overcome the challenge, government has to increase the funds invested in rice production, motivate rice farmers and ensure that better land tenure system and conservative measures of rice farm lands are adopted.

REFERENCES

1- Anonymous. (2011) Retrieved from: www.index-

mundi.com.
2- Anonymous. (2011) Retrieved from: www.allAfrica.com.
3-Badmus, M.A., & Ariyo, O.S. (2011). Forecasting Cultivated Areas and Production of Maize in Nigeria Using ARIMA model. Asian Journal of Agricultural Sciences. 3(3): 171- 176
4- Box, G. E. P., & Jenkins, G. M. (1976). Time SeRies Analysis, Forecasting and Control. San Francisco, Holden- Day, California, USA.
5- Dogbe, W. (1996). Characterization of the Inland Valleys of Northern Ghana, Savanna Agricultural Research Institute (SARI). Unpublished
6- Falak, S., & Eatzaz, A. (2008). Forecasting Wheat Production in Pakistan. The Lahore Journal of Economics. 3(1): 57-85.
7- Khush, G.S. (1997). Origin, Dispersal, Cultivation and Variation of Rice. Journal of Plant Molecular Biology. 35(1-2): 25-34
8- Muhammad, F., Siddique, M., Bashir, M., & Ahamed, S. (1992). Forecasting Rice Production in Pakistan Using ARIMA models. Journal of Animal and Plant Sciences. 2: 27- 31.
9- Najeeb, I., Khuda, B., Asif, M., & Abid, S.A. (2005). Use of ARIMA Model for Forecasting Wheat Area and Production in Pakistan. Journal of Agricultural and Social Sciences. 1(2): 120- 122
10- Rachana, W., Suvarna, M., & Sonal, G. (2010). Use of ARIMA Model for Forecasting Pigeon Pea Production in India. International Review of Business and Finance. 2(1): 97-107.
11- FAO, (2011). Rice Imports Killing Local Growers. Retrieved from: www.ghananewsnow.com.
12- Seidu Al-hassan, (2008). Technical Efficiency of Rice Farmers in Northern Ghana. AERC Research Paper 178.
13- Shabur, S.A., & Haque, M.E. (1993). Analysis of Rice in Mymensing Town Market Pattern and Forecasting. Bangladesh Journal of Agricultural and Economics. 16: 130-133
14- Sohail, A., Sarwar, A., & Kamran, M. (1994). Forecasting Total Food Grains in Pakistan. Department of Mathematics and Statistics, University of Agriculture, Faisalbad. Journal of Engineering and Applied sciences. 13: 140- 146.

Perception of Shea Nut Tree as an Economic Tree among Farmers in Oyo State, Nigeria

Rasheed Gbolagade Adeola

Abstract

The importance of shea tree to the people of south-western Nigeria cannot be over emphasized considering both the economic and environmental uses of the tree. However, efforts have not been made to propagate its production by the farmers as the shea tree still grows in the wild state. It would be worthwhile to investigate farmers' perception of shea nut tree as an economic tree and their attitudes toward the conservation measures needed to prevent its extinction. Multi-stage sampling technique was used to select 160 farmers for the study. A structured interview schedule was used to collect data from the respondents. Descriptive statistics of means and percentages were used to describe the data while correlation coefficients were employed to determine the relationships that exist between farmers' perception and their socio- economics characteristics. A larger percentage of the farmers claimed the awareness of the potential products of shea nut trees. Majority (90.6%) of the respondents perceived shea tree as an economic tree and indicated that it provides income for women and children that gather the fruits. Most of the farmers had favourable attitude toward shea nut tree as an economic tree. Land ownership status ($r = 0.536$), age ($r = 0.875$), education ($r = 0.725$), farming experience ($r = 0.508$), household size ($r = -0.817$) and farm size ($r = -0.673$) had significant influence on the perception of farmers. The study therefore recommends that the extension agencies should play significant role in educating the farmers on the importance of conserving the trees to sustain its economic benefits.

Keywords:
Perception, Shea nut, Economic tree, Farmers

Department of Agricultural Extension and Rural Development, Ladoke Akintola University of Technology, P.M.B. 4000 Ogbomoso, Nigeria.
** Corresponding author's email: adeola20022000@yahoo.com*

INTRODUCTION

Scientifically, the shea tree is known in the past as 'Butryospermum paradoxum', but is now called '*Vitellaria paradoxa'*. It is widely spread across the savannah regions of Nigeria. The shea tree grows very well on a wide range of soils, including highly degraded, arid, and semi-arid and rocky soil. (Dogbevi, 2007). Shea fruits consist of a thin epicarp and a soft mesocarp enclosing a single seed (occasionally two to four seeds). The epicarp and mesocarp together make up 33–75% of the fresh fruit weight, with an average of 55% (Elias, *et al.,* 2006).

Shea tree is the most prevalent arboreal species of West African parklands, which provide vital products and ecological services to the semi-arid region (Becker and Starz, 2003). The resource's main traditional role is associated with its oil (shea butter), extracted from shea kernels, that represents the primary source of fat in local diets. Shea also contributes to the generation of crucial foreign exchange revenues in many countries of the sub-region, ranking third among exports from Burkina Faso in the 1980s (Lamien *et al.*, 2007). Traditionally, African women involve in collection of shea nuts which are transformed into a pure, pale yellow butter after a long conditioning and cooking process. While nut gathering and processing were formerly exclusively rural activities, many factors have led to the expansion of urban shea projects. It is widely known and used as a skin ointment and for skin care and hair care, as well as for hydrating, protecting and massaging babies. The roots, leaves, and bark of the tree are also used medicinally for treating mouth sores, boils, burns, diarrhoea, as a vermifuge, and as eyewash against spitting cobra venom (Neuwinger, 1996). Large *Vitellaria* trunks may be used to make mortars for pounding grain. The wood is also used in building construction and is made into charcoal.

The shea tree constitutes an important source of raw material for gum and rubber industries. The shea fruit also serve as a source of food to many people and as an income generating activity for most of the women that gather the fruits as the ripening of the fruits coincide with the lean season of food crops production. The shea tree can be used to combat the problem of desertification. Men, women, and children eat and appreciate the pulp. Fresh fruits are also sold in local markets. However, despite the economic and environmental benefits of the tree, efforts have not been made to propagate its production by the farmers as the shea tree still grows in the wild state. Another worrisome development is the people's habit of destroying the trees for charcoal production which may eventually lead to environmental degradation, deforestation and loss of vegetative cover and resulting to water and soil erosion. Non-replacement or domestication of the shea trees may lead to its extinction in the nearest future. This study is designed to investigate farmers' perception shea nut tree as an economic tree and their attitudes toward the conservation measures needed to prevent its extinction.

The main objective of the study is to determine farmers' perception of shea tree as an economic tree. The specific objectives are to:

1- Determine farmers' attitudes toward the economic and environmental contributions of shea tree to the farming system practice.

2- Examine the level of awareness of potential products of shea nut tree among farmers in the study area

3- Determine the influence of socio-economic characteristics of the farmers on their perception.

MATERIALS AND METHODS

Study area

The study was conducted in Oyo state southwestern Nigeria in 2010. The state is divided into four agricultural zones namely: Ibadan/Ibarapa (Forest and derived savannah), Oyo (Derived savannah), Ogbomoso. (Derived savannah) and Saki (Guinea savannah). A multi-stage sampling procedure was adopted in selecting respondents for this study.The first stage include the selection of two agricultural zones (Ogbomoso and Saki zones) due to concentration of shea trees in these zones. Four Local Government Areas (LGAs) were randomly selected from the two zones with two LGAs (Orire and Surulere) in Ogbomoso zone and two LGAs (Atisbo, and Olorunsogo) in Saki zone. Four rural villages were further selected from each of the four selected LGAs to arrive at a total of 16 villages. Thereafter 10 farmers were selected randomly from each village to arrive at a total sample size of 160 respondents.

Research Instrument

A well-structured interview schedule was used for data collection. The instrument was pre-tested on 30 respondents in the adjacent LGA not included in the final survey. The instrument validation was done by a team of experts in the fields of Agronomy and Extension. A test-retest check was conducted to establish the reliability of the instrument and a reliability coefficient of 0.88 was obtained for the test which was adjudged reliable for data collection.

Measurement of variables

Socio-economics characteristics of the respondents were measured by collecting information on the age, sex, and marital status, and educational attainment, farm size and land ownership status. Perception and attitude of the respondents were measured through their responses to some statements to reveal their view toward shea nut tree production and its economic importance.

Method of data analysis

Descriptive statistics such as percentages; means; standard deviations were used to describe the data and Pearson correlation coefficient was used to test the relationship between the selected socio-economic characteristics and farmers' perception.

RESULTS AND DISCUSSION

Results of socio-economic characteristics of the farmers show that majority (52.5%) of the respondents were in the age group of 46 – 55 years with mean age of 47.6 years an indication that they are in their active stage of life. About 90% of the respondents were males while 10% were females (Table 1). This could be due to the fact that women only gather the shea fruits from their husbands' farms and probably unqualified to take decision trees found on such farms. The study also revealed that larger percentage (87.5%) of the respondents had formal education and 12.5% having no formal education. This high level of literacy is likely to make them more responsive to the adoption of technology needed for shea tree development. Majority (58.8%) of the respondents had spent between 5 – 15 years in farming with mean farming experience of 17.5 years. The mean farm size was 2.3 hectares while, the majority (78.7%) had farm land of between 1 – 5 ha an indication that the respondents are small holders. The ownership status of land showed that 61.9% of the respondents inherited their farm lands while 28.1% indicated that their farm lands were leased to them. Only 10% of the respondents rented their farm lands. This land ownership pattern could enhance conservation measures to be taken on shea trees by the farmers (Table 1).

Table 1: Distribution of Respondents According to Personal Characteristics (n = 160)

Characteristics	Frequency	Percentage
Age (Years)		
25 – 35	23	14.4
36 – 45	40	25.0
46 – 55	84	52.5
> 55	13	8.1
Total	160	100
Education		
No formal education	20	12.5
Adult Education	46	28.8
Primary education	89	55.6
Secondary education	5	3.1
Total	160	100
Sex		
Male	144	90.0
Female	16	10.0
Total	160	100
Farming experience (years)		
5 – 15	94	58.8
16 – 25	34	21.3
26 – 35	28	17.5
> 35	4	
Total	160	100
Farm size (Ha)	**Frequency**	**Percentage**
1 – 5	126	78.7
6 – 10	21	13.1
11 – 15	10	6.3
> 15	3	2.5
Total	160	100
Land ownership status		
Inheritance	99	61.9
Leasehold	45	28.1
Rent	16	10.0
Total	160	100

Table 2: Distribution of respondents by awareness of potential products of shea tree

*Awareness of potential products of shea tree	Frequency	Percentage
Processed Shea butter export	4	2.5
Raw materials for manufacturing chocolates	7	4.4
Edible oil	152	95.0
Use in cosmetics and pharmaceutical industries	5	3.1

* Multiple responses

Findings from the study revealed that majority (95%) of the farmers claimed the awareness of edible oil as a potential product of shea tree while only 2.5% were aware of processed shea butter for export and 4.4% reported the awareness of shea tree providing raw materials for manufacturing chocolates. However, few (3.1%) respondents were quite aware of shea tree as being useful for cosmetic and pharmaceutical industries (Table 2).

The perception of 90.6% of the farmers was considering shea nut trees as providing income for women and children that gather the fruits whereas 60% of the farmers perceived shea fruits, butter and oil as important source of calories. Majority of the farmers perceived roots, barks and leaves of the tree as being useful for treating medical conditions and almost (96.9%) perceived shea tree as materials for making hoe handle, pestle, mortar and talking drum (Table 3).

Table 3: Perception of farmers on shea nut trees (n = 160)

Perception	Frequency	Percentage
Provide income for women and children that gather the nuts	145	90.6
Shea fruit, butter and oil are important source of calories	96	60
Shea tree roots bark and leaves used for treating different medical conditions.	123	76.9
Substitute to cocoa economic value.	No response	
Wood suitable for making hoe handle, pestle mortal and talking drum.	155	96.9

The study reveals that farmers in the study area had favourable attitude towards shea nut tree as economic tree. This is obvious as they

Table 4: Attitude of respondents towards shea nut tree as an economic tree (n = 160)

Statement	Mean	SD
Farmers should take a responsible role in protecting shea tree on their farms against charcoal burning and bush burning	3.9	1.07
Giving the opportunity I will cultivate the shea tree on my farmland owing to the benefit derived from its products	4.21	0.565
I can give equal protection to shea woodland as much I can give to other tree crops	3.5	1.176
Extension service has a significant role to play in shea tree production	3.6	1.227
There is need for more awareness on cultivation and preservation of shea trees	3.94	0.779
Charcoal burning is not a major threat to shea tree production	2.22	1.33
The benefits derived from shea tree warrant its cultivation on my farm	3.96	0.800
Shea woodland cannot be given equal protection as other tree crops	2.30	1.13
Shea tree is not an economic tree and will not yield as much as income as other tree crops	2.76	1.043
The income realized from the shea tree cannot improve my families' standard of living such as paying school fees and clothing	3.48	1.176

were in agreement with most of the items used to measure their attitude towards economic importance of shea nut tree in their farming systems. Their readiness to protect shea nut tree against charcoal and bush burning and cultivating of shea tree if given the opportunity; is a clear indication of their perception of shea nut tree as an economic tree (Table 4).

Farmers' perception of shea nut tree as economic tree was positively correlated with education. This is an indication that the more educated the farmers, the greater their ability to perceive the economic benefits accrue to shea nut tree understands and manage the related technologies for its production. Farming expe-

Table 5: Results of relationships between farmers' perception and some selected socio-economic characteristics

Variable	r	p
Age	.875*	0.023
Education	.725*	0.000
Farming experience	.508*	0.001
Household size	-.817*	0.000
Land ownership	.536*	0.010
Farm Size	-.673*	0.034

* Correlation is significant at the 0.05 level (2- tailed).

rience positively and significantly ($P \leq 0.05$) influenced perception of farmers (Table 5). Farmers with accumulated farming experience may have acquired encouraging returns from shea nut trees in the past and thus be encouraged to sustain the anticipating benefits (Bonabana-Wabbi, 2002). However, farm size negatively influenced farmers' perception implying that the larger the farms size the higher the perceived benefits that can be derived from the shea nut trees. A plausible explanation for this could be that, more shea nut trees could be found in larger farms compared to the smaller farms. The household size was significantly and negatively related to farmers' perception. A negative sign for household size suggests that perception was favourable among larger households because they attach greater importance to food security than smaller households hence, will likely to be interested in getting additional source of income.

CONCLUSION

It was observed that farmers had knowledge of potential products of shea tree and had a favourable attitude toward shea nut tree as economic tree. Farmers perceived that shea nut trees provide income for the household, a source of calories, useful in treating medical conditions and materials for making drums and mortal. This study has provided insights into the relationships between perception and farmers' demographic characteristics. Increasing farmers' awareness on potential products of the shea nut trees and its economic benefits will enhance its conservation among farmers. Intensification of research efforts in designing appropriate technology that will encourage shea nut tree production and its conservation is highly recommended.

REFERENCES

1- Bonabana-Wabbi, J. (2002). Assessing Factors Affecting Adoption of Agricultural Technologies: The Case of Integrated Pest Management (IPM) in Kumi District, Eastern Uganda MS dissertation, Virginia Polytechnic Institute and State University.

2- Becker, M., & Statz, J. (2003). Marketing of parkland products. In Z. Teklehaimanot (Eds.), Improved management of Agroforestry Parkland Systems in Sub-Saharan Africa (pp. 142-151) EU/INCO project Contract IC18-CT98-0261, Final Report, University of Wales, Bangor, UK.

3- Dogbevi, E. K. (2007). The Shea Tree Has Economic and Environmental Values for Ghana. Retrieved May 10, 2009, www.myjoyline.com.

4- Elias, M. Bayala, J., & Dianda, M. (2006). Impediments and Innovations in Knowledge Sharing: the Case of the African Shea Sector: Knowledge Management For Development Journal 2(1): 52-67.

5- Lamien, N., Tigabu, M., Guinko, S. & Oden, P. C. (2007). Variations in Dendrometric and Fruiting Characters of *Vitellaria paradoxa* Populations and Multivariate Models for Estimation of Fruit yield. *Agroforestry Systems* 69, 1-11.

6- Neuwinger, H. D. (1996). *African Ethnobotany: Poisons and drugs*. Weinheim, Germany: Chapman & Hall.

3

The Analysis of the Effects of Domestic and Foreign Investment in R&D on Agricultural TFP in Iran

Ali Bagherzadeh

Abstract

Nowadays, agricultural R&D provides new and developed technologies to create modern agricultural producing methods. During recent years, improving agricultural productivity is affected by not only domestic R&D investments but also foreign countries R&D investments. Nowadays, according to new growth models, R&D is the base of productivity. Recent economics theories consider agricultural research and it`s spill overs as important factors for technological change and economic growth. This paper investigates the amount of agricultural total factor productivity in Iran and analyzes the relationship between TFP, domestic agricultural research, and foreign agricultural R&D during 1979 – 2008. In this study Iran`s partners are 20 Asian, European and South American countries. The Solow residual index approach is applied for the measurement of total factor productivity in agricultural sector of Iran. ARDL model involving different lag length specifications were estimated taking TFP as a dependent variable. The results indicate that agricultural researches (both domestic and foreign R&D) have positive and significant impact on agricultural TFP. But the impact of foreign R&D on agricultural productivity is stronger than the effect of domestic R&D. According to gained results considerable portion of national product should be allocated to R&D costs and research budget of agricultural sector should be increased to standard level. Also government should pay attention to it`s partner countries because agricultural R&D spill over of developed countries is more than developing countries.

Keywords:
TFP, Agricultural R&D Spill over, ARDL Model, Agricultural Import

Assistant Professor of Economics, Khoy Branch, Islamic Azad University, Iran.
** Corresponding author's email: bagherzadeh_eco58@yahoo.com*

INTRODUCTION

During last decades, agricultural sector of Iran had traditional structure and so it couldn`t use production factors and inputs optimally (Bagherzadeh, 2007). But Research and Development helps the economy to get more production with the same level of factors in agricultural sector. The closest word to this concept is productivity, there fore knowing the ways of getting productivity growth in agricultural sector, leads to country`s better position in international occasions and prevents the waste of resources in this sector. So by extending education, knowledge and research borders, we can avoid traditional methods of producing and increase productivity level (Amini, 2004). According to Kandric and Crimer (1970) productivity is affected by production factors such as capital and labour. Economists believe beside noted factors other variables have effect on productivity. One of the most important variables is research and development. R&D has a special role in increasing the productivity and efficiency of various sectors of economy (Gutierrez, 2005). Jacob`s (2004) studies show during 2000-2003 in most developing countries despite quick population growth, agricultural productions have been increased in each hectare per labour that is the result of new technologies and the sign of productivity increase. It shows agricultural researches develop the agricultural productivity of countries which their researches are enforced basically. Due to the importance of R&D in productivity expansion, at first we define productivity and then recognize effective factors on it. Productivity is a comprehensive concept that it`s increase is a necessity to improve the human life and have a prosperous society. Nowadays by having resource scarcity, productivity is the best and most effective method for getting economic growth. Today the simplest definition of productivity is "the relative of product toward production factors". As noted, one of effective factors on TFP is agricultural R&D. Nowadays investment in agricultural researches is one of fundamental investments in this sector that leads to production growth. Agricultural R&D provides new and developed technologies to create modern agricultural producing methods. Also new methods of producing and the potential of production increase are created. On the other hand, agricultural R&D leads to the transfer of agricultural production function to higher point and the decrease of each factor`s cost by new technologies. In our country during two last decades increasing population growth and demand increase for nutritional products led to irregular exploitation of water and soil resources and it resulted in unstable agricultural development. Generally, for increasing agricultural production, developing the cultivated land or increasing TFP are suggested. Because of restriction in cultivated lands the firs policy is not possible but TFP increase looks logical. It needs new technologies in agricultural sector and it becomes possible by investment in researches and knowledge and transferring the information to farmers. Unfortunately in comparison with modern countries, our country spends few amounts of national products to agricultural R&D. By having trade with developed trade partners, technical methods and new technologies are transferred to our country and then during next years by investing in domestic researches, they become useable in our agricultural sector.

Tokgoz (2005) analyzed technological changes and the effect of domestic and foreign investment in R&D on agricultural productivity of America. His study data are related to 1975-2002 and his model was estimated by SUR model. The results show government R&D directly affects agricultural productivity and indirectly affects private sector R&D. Line and Huffman (2006) studied the effects of government agricultural researches, extension and some infrastructure factors on agriculture productivity in America. The results showed should act coordinately to gain national aims. Schimmel Pfenning and Leftly (2006) calculated TFP of some agricultural products in South Africa by Tile index for 1947-2004. They showed agricultural productivity increases by research costs increase. In another study Line (2006) after calculating the agricultural TFP in America by a lag model compared return rates of private and governmental researches. Results showed the variables of extension, private and governmental research costs have positive and meaningful effect on TFP. Gutierrez (2005) studied the long

Umbraculi agnascor saetosus apparatus bellis. Pessimus fragilis suis insectat adlaudabilis fiducias, et umbraculi conubium santet suis. Matrimonii iocari umbraculi. Agricolae agnascor fragilis zothecas, ut quinquennalis oratori libere conubium santet agricolae, iam bellus saburre circumgrediet Aquae Sulis, ut Octavius insectat concubine, quod cathedras miscere tremulus zothecas, ut chirographi imputat satis lascivius oratori, semper verecundus ossifragi deciperet oratori, iam lascivius ossifragi adquireret Aquae Sulis, semper plane tremulus suis agnascor lascivius rures, quamquam quadrupei celeriter corrumperet Caesar.

Pretosius syrtes amputat Augustus, quod verecundus apparatus bellis pessimus neglegenter senesceret utilitas agricolae. Parsimonia syrtes optimus infeliciter suffragarit oratori. Aquae Sulis amputat tremulus saburre. Matrimonii imputat Caesar.

Ossifragi amputat zothecas, semper quinquennalis chirographi circumgrediet matrimonii, ut oratori insectat cathedras, etiam verecundus zothecas iocari Medusa, quamquam satis utilitas syrtes conubium santet

According to above matters, the main objective of this study is the analysis of the effects of domestic and foreign investment in R&D on agricultural TFP. In addition to it, this study analyzes the importance of intermediate and investment goods on agricultural TFP.

In economic growth literature and internal growth models despite neoclassical models human capital and R&D activities are the main engine of economic growth. According to Griliches (1989) and Aghion (1995) studies investing in domestic R&D is the most important determining factor of TFP. Coe and Helpman (1997) expressed economic growth is the function of resource usage, population growth rate, saving rate and domestic and foreign investment in R&D. So by assuming Cobb-Douglas production function for agricultural sector, we have:

$$Y = AK^{\alpha}L^{\beta}\sum Xj^{1-\alpha-\beta} \qquad \alpha,\beta > 0, \alpha+\beta < 1 \quad (1)$$

Y is total production of this sector, K capital and L labor. Agricultural productions are function of X_j that are intermediate goods and are used in production process. We can see technological development can improve the amount of intermediate factors and it needs R&D. By assuming the price of intermediate goods as P_j and agricultural production price as Py=1, we can derive demand function of j intermediate factor for agricultural sector by maximizing profit function:

$$X_j = \sqrt[\alpha+\beta]{[(1-\alpha-\beta)AK^{\alpha}L^{\beta}/P]} \quad (2)$$

The producer of new intermediate factor can maximize his price with following expression.

$$max\,(P_j - MC).X_j \quad (3)$$

Now by assuming unit marginal cost, we have:

$$max_{pj}\,(Pj-1).Xj \quad (4)$$

We put equation 2 in equation 3:

$$Pj = P = \left[\frac{1}{1-\alpha-\beta}\right] > 1 \quad (5)$$

If we put equation 5 in equation 2 and the result of it in equation 1:

$$Y = F.K^{a}L^{b} \quad (6)$$

In above equation F is TFP and the amount of it is gained with:

$$F = \sqrt[\alpha+\beta]{A.(1-\alpha-\beta)^{2(1-\alpha-\beta)}}\;.n \quad (7)$$

In this equation by having the amounts of α and β, we can show that agricultural TFP is in relation to n (the amount of intermediate factor) variable. If n is increased, TFP will be increased too. Now if the production stream of intermediate factor is proportional to R&D cost as by the increase of R&D costs, the quality and the amount of intermediate factors increase, we can write:

$$N(t) = \int_0^t R\&D(t)dt \quad (8)$$

In this equation the growth rate of intermediate goods number is affected by R&D costs. So there is one relation between TFP and R&D. It shows applied research and study in agricultural production process can develop intermediate factors production. Beside inputs and intermediate goods production increase, R&D increases the quality of inputs and intermediate. According to this model we show how R&D development improves the quality of intermediate factors. So we have:

$$Y = AK^{\alpha}L^{\beta}\sum(\widetilde{X_j})^{1-\alpha-\beta} \qquad \alpha,\beta > 0 \quad \& \quad \alpha+\beta > 0$$

$$\widetilde{X_J} = \sum \lambda^k X_{JK} \tag{9}$$

In above equation, X_j is the amount of Jth intermediate factor in production process. λ is a coefficient that $\lambda>1$, so we have:

$$K = 0,1,2,\ldots,k_j \Rightarrow \lambda^0,\lambda^1,\lambda^2,\ldots\ldots,\lambda^{kj} \tag{11}$$

K_j is the highest amount of quality in j sector. Increase in kj leads to R&D increase. Exclusive price for X_jk_j is:

$$P_{jkj} = \left[\frac{1}{1-\alpha-\beta}\right] \tag{12}$$

X_j is:

$$X_{jkj} = \sqrt[\alpha+\beta]{[(1-\alpha-\beta)^2 A\, K^{\alpha}L^{\beta}\lambda^{kj(1-\alpha-\beta)}} \tag{13}$$

By putting this equation in agricultural production function we have:

$$Y = F\, K^{a} L^{b} \tag{14}$$

$$F = \sqrt[\alpha+\beta]{A(1-\alpha-\beta)^{2(1-\alpha-\beta)}}\ .In \tag{15}$$

In above equation, In is quality and innovation in producing intermediate inputs.

$$In = \sum \lambda^{\frac{kj(1-\alpha-\beta)}{\alpha+\beta}} \tag{16}$$

On the other hand, agricultural researches affect innovation and quality of inputs.

$$In(t) = \theta \int_0^t R\&D(t)dt \tag{17}$$

Nowadays, not only domestic R&D affects agricultural TFP but also foreign R&D has effects on it.

MATERIALS AND METHODS

In order to show the relationship between TFP growth and agricultural R&D and to know the effective factors on productivity growth increase beside agricultural researches, we use the combination model of Shujat-Iqbal (2007), Coe-Helpman (1997) and Gutierrez (2005):

$$Q_t = A\ Wea_t^{\varepsilon} M_t^{\delta} R\&Df_t^{\gamma} R\&D_{dt}^{\beta} \prod_{i=1}^{n} X_{it}^{\alpha i} \tag{18}$$

Q_t is total production of agricultural sector, A fixed amount, Xi s inputs like labor, capital and energy and R&D domestic $R\&D_f$, in agricultural sector, $R\&D_{dt}$, R&D amount of trade partners and M_t, imported intermediate and marginal agricultural goods. Wea_t is weather index that has effect on agricultural TFP. In this model Wea_t variable is considered as annual raining average that is millimeter, now according to TFP relation by divisia index, we have:

$$TFP = \frac{Q_t}{\prod_{i=1}^{n} X_{it}^{\alpha i}} = A\ Wea_t^{\varepsilon} M_t^{\delta} R\&Df_t^{\gamma} R\&D_{dt}^{\beta} \tag{19}$$

$$logTFP = logA + \varepsilon LogWeat + \delta LogMt + \gamma Log\ R\&D f + \beta\ Log\ R\&D_{dt} \tag{20}$$

In equation 20, for calculating R&D capital of agricultural trade partners from Coe and Helpman (CH), we used equation 21.

$$S^{CH} = \sum_{j\,i}^{K} \frac{m_{ij}}{m_i} Sj^{d} \tag{21}$$

In equation 21, m_{ij} is goods import from trade partners, m_i total import of country from 20 trade partners that these trade partners were chosen according to Komijani and Shah Abadi (2004) studies and available data of Iran`s agricultural trade partners and S^d_j agricultural R&D capital of each trade partners.

Now for analyzing the relationship between R&D and TFP, at first we should calculate the amount of TFP and then derive necessary data for TFP index by Solo residue.

$$TFPAg = \frac{Y}{K^{\alpha}.L^{\beta}.E^{\delta}} \tag{22}$$

Equation 22 is Solo residue, so we have:

$$LnTFP = Lny - \alpha Lnk - \beta Lnl - \delta LnE \tag{23}$$

Y is agricultural value added (bilion Rial), k capital input data of agricultural sector (bilion Rial), L labor of agricultural sector and E the amount of consumed energy (Mega Joule). α, β, δ are production elasticities of capital, labor and energy.

For estimating the model of research we used ARDL method. Since the power of Unit-Root test for determining cointegration order and stationary is low and in most cases can not determine whether variables are stationary or not, some studies tried to remove above methods faults and find better ways to analyze short-run and long-run relation between variables. So Pesaran and Shin (1998) presented Auto Regressive Distributed Lag Method (ARDL). In this method variable cointegration has no importance and just by determining suitable numbers of lag, unique vector for long-run relation between variables is achieved. This method estimates long-run and short-run relations among variables simultaneously. Also this method removes the problem of omitting variable and correlation and since these methods have no serial correlation, estimations are efficient and unbiased (Noferesti & Arab Mazar, 2005). In ARDL method for long-run relation estimate, we used a two-level method as follows.

In first level, existing of long-run relation among variables is tested. So dynamic ARDL model is estimated. If the sum of estimated coefficients with dependent variable lag is less than 1, dynamic model tends to long-run balance. Then for testing covergency, following hypothesis test is done.

$$\begin{cases} H_0: \ \sum_{i=1}^{m} \beta_i - 1 \geq 0 \\ H_1: \ \sum_{i=1}^{m} \beta_i - 1 < 0 \end{cases} \qquad (24)$$

T-statistic is calculated by:

$$t = \frac{\sum_{i=1}^{m} \beta_i - 1}{\sum_{i=1}^{m} s_{\beta_i}} \qquad (25)$$

Now by comparing t-statistic and presented critical quantity by Banerjee, Dolado and Master, we can notice that whether there is a long-run balance relation between variables or not. ARDL form is:

$$y_t = \beta_0 + \sum_{i=1}^{m} \beta_i Y_{t-1} + \sum_{i=0}^{k_1} \alpha_{i1} X_{t-i} + \sum_{i=0}^{k_2} \alpha_{i2} Z_{t-i} + \cdots + u_t \qquad (26)$$

In above equation, $\sum \beta_i Y_{t-1}$ are lagged dependent variables, $\sum \alpha_{i1} X_{t-i}$ and $\sum \alpha_{i2} Z_{t-i}$, sets of lagged independent variables and β_0, β_i, α_{i1}, α_{i2}, coefficients of regression equation. In ARDL method the maximum number of lags is determined by researcher according to observations number and model nature. So according to one of four Akaike and Schwarts-Baysian, Hannan-Quinn and R^2 criterions, one estimated regression is chosen. Then we explain cointegration among variables and estimate long-run balance relation. The superiority of ARDL method is having short-run Error Correction Model (ECM) plus long-run relations (Pesaran and Shin, 1998). The data of study was taken from statistic resources of Central Bank, agricultural ministry, International Institute of Food Researches and FAO for 1980-2008.

RESULTS AND DISCUSSION

For estimating production elasticities in TFP calculation, we need to production function. Applied function of this sector is Cobb-Douglas production function. The reason that we use Cobb-Douglas function is: First: Solo suggested Cobb-Douglas function for TFP calculation. Second: previous studies inside the country show it`s suitable for TFP calculation. Some examples of these researches are Arabmazar and Noferesti model in estimating production function of agricultural sector. Also Hozhabr Kiani (1999) and Amini (2004) used Cobb- Douglas production function in estimating production function of agricultural sector. Third: because of substituting feature of factors in production trend and having suitable functional form (Ramsey F test in Microfit software), this function has been chosen for estimation. So by having the data of agricultural investment amount, the amount of employed labor in this sector, the amount of consumed energy by farmers and agricultural sector`s value added (available data of production in agricultural sector is according to value added), we estimate production function of agricultural sector in Iran. According to econometric materials about the stationary of variables and preventing false regression among variables, we use Augmented Dicky-Fuller (ADF) test to study the stationary of variables. Gained results of this test are in table 1.

Table 1: Summary of series Unit-Root test by Eviews 5 software

Variable name	Number of lags		ADF statistic	Mc kinon critical value			Stationary/ non stationary
				1%	5%	10%	
ΔLvaluagri	3	Constant	-6.89	-3.6	-2.9	-2.6	I(1) stationary
ΔLlagri	3	Constant and trend	-3.54	-4.20	-3.52	-3.19	I(1) stationary
Lkagri	4	Constant	-4.18	-3.60	-2.93	-2.60	I(0) stationary
Leagri	3	Constant	-3.64	-3.62	-2.94	-2.61	I(0) stationary

Table 2: Gained results of long run relation ARDL (1, 2, 0, 1)

Variable names	coefficients	Standard deviation
Leagri	0.2738	0.088
Lkagri	0.1223	0.054
Llagri	0.7211	0.13
C	-15.88	C2.26

After Unit-Root test by ARDL method, we estimated agricultural production function. Gained results of long run production are in table 2.

$$lvaluagri = -15.88 + 0.72llagri + 0.12lkagri + 0.27leagri \quad (27)$$

In equation 27, coefficient of capital in agricultural production function is relatively low, it`s identical to Arabmazar and Noferesti (2005) studies. Their estimated coefficient for the data of 1967-1996 was about 0.06. Of course energy and labor coefficients are very important and meaningful in presenting value added.

As mentioned before, for calculating TFP growth of agricultural sector, Solo residue method and divisia index are used. So the amounts of TFP growth for 1971-2008 are in table 3.

According to this table in some years TFP growth had negative amount but the average of TFP growth is about 0.8%. Clear feature is that during different years TFP growth in agricultural sector had a lot of fluctuations but totally the average amount is positive. It`s clear that this amount is too less than expected amount of government means 2.5%. So effective factors of TFP and suitable policies for improving it should be known and offered. Analysis shows in 1999, 2006, 2007 and 2008 productivity level had decrease but in other years it had increasing trend. After calculating TFP, we should analyze the relation between independent variables and agricultural TFP. According to previous studies about variable stationary and preventing false regression among TFP, lagged R&D and other variables of model, at first we analyze whether variables are static or not, so we use Dicky-Fuller (ADF) test. Gained results are in table 4.

As we see in table 4 domestic R&D, TFPAg and import have become stationary with once differentiating, on the other hand, they are not stationary in unit level but the other variables

Table 3: Agricultural TFP growth calculation

Year	TFP_{Ag} growth	Year	TFP_{Ag} growth	Year	TFP_{Ag} growth
1971	-	1984	-0.012	1997	-0.014
1972	0.099	1985	0.050	1998	0.0019
1973	-0.046	1986	0.047	1999	-0.015
1974	-0.067	1987	0.051	2000	0.02
1975	0.017	1988	-0.041	2001	-0.07
1976	0.048	1989	-0.006	2002	0.051
1977	0.11	1990	-0.01	2003	-0.045
1978	0.069	1991	0.055	2004	0.035
1979	0.030	1992	0.023	2005	0.058
1980	-0.001	1993	0.05	2006	-0.068
1981	-0.034	1994	0.04	2007	-0.054
1982	-0.020	1995	0.017	2008	-0.022
1983	0.018	1996	0.041	-	-

Table 4: Summary of series Unit-Root test by Eviews 5 software

Variable name	Number of lags		ADF statistic	Mc kinon critical value			Stationary/ non stationary
				1%	5%	10%	
ΔLR&D	2	Constant	-4.64	-3.6	-2.9	-2.6	I(1) stationary
$LR\&D_f$	1	Constant and trend	-5.61	-4.30	-3.57	-3.22	I(0) stationary
ΔLM	2	Constant and trend	-5.29	-4.33	-3.57	-3.22	I(1) stationary
Wea	2	Constant and trend	-6.94	-4.30	-3.57	3.22	I(0) stationary
$\Delta LTFP_{Ag}$	2	Constant	-4.54	-3.69	-2.98	-2.62	I(1) stationary

are stationary. Since differentiating leads to lose long-run relation among variables, so cointegration is a simple solution to solve the problem. In this part we use ARDL that is a new method in estimating dynamic models to retest the relation between agricultural R&D and productivity. Gained results of estimated ARDL model by Schwarts-Baysian model are in table 5.

Now after estimating model by using of dynamic coefficients, the existence of long-run relation between variables is tested. In order for testing we find t-statistics.

$$t = \frac{0/12-1}{0/038} = -23/04 \qquad (28)$$

The hypothesis of long-run relation existence among variables is true because the amount of calculated statistic is more than the amount of offered one by in the importance level of 95% (-4.76).

Dynamic ARDL results show all of estimated coefficients have positive effect on agricultural productivity. Coefficients are meaningful in the importance level of 5% and 10% and the model doesn`t have any problem of heteroscedasticity and correlation. Also according to $\varkappa^2$ test ($\varkappa^2$=1.2), residuals have been distributed normally. By using of correlation test for every two pairs of variables and comparing it with square root of coefficient of determination (0.9), we showed that all correlation coefficients are less than the square root of coefficient of determination and so model does not have the problem of collinearity.

In this model, agricultural productivity in addition to independent variables is in relation to the productivity of previous year. Productivity variable with one year lag has 0.12% positive effect on agricultural productivity. The estimation of long-run relation model is as equation 29.

LTFPagri=-29/04+ 0/17 LR&D + 0/89 LR&DF + 0/6 LIMPORA + 0/1WEA - 0/087T (-28/86) (2/06) (7/08) (6/15) (1/69) (-5/40) (29)

In equation 29 the numbers inside parenthesis are t-statistics of estimated model. All of coefficients except the coefficient of raining amount (that is meaningful in 90% level) are meaningful in 95% level. In this model the relationship among domestic and foreign R&D, import and weather condition with productivity are calculated in log-run. In log-run productivity is in relation to domestic R&D with 0.17 elasticity coefficient.

In above model the importance of foreign R&D on productivity is more than domestic

Table 5: Gained results of dynamic (1, 1, 0, 0,1) ARDL model

Variable name	Coefficient	Standard deviation	t-statistic
LTFP(-1)	0.12	0.038	3.09
Wea	0.05	0.031	1.6
Wea(-1)	0.045	0.029	1.5
LR&D	0.14	0.067	2.1
Limport	0.6	0.091	6.4
$LR\&D_f$	0.74	0.026	27.3
$LR\&D_f(-1)$	0.67	0.021	31.1
C	-25.55	0.74	-34.2
T	-0.07	0.011	-6.7
	R^2=0.99	DW=2.25	F(8,12)=561

Table 6: Correlation matrix for Collinearity test

	LTFP	LR&D	LR&Df	LIMPOR	HLR&D	WEA
LTFP	1	0.58	0.56	-0.24	0.67	0.76
LR&D	0.58	1	0.33	0.75	0.85	0.32
LR&Df	0.56	0.33	1	-0.73	0.41	0.09
LIMPOR	-0.26	0.75	-0.73	1	-0.61	-0.1
HLR&D	0.67	0.85	0.41	-0.61	1	0.29
WEA	0.07	0.32	0.09	-0.1	0.29	1

Table 7: Countries portion in agricultural import of Iran

Country	Agricultural import portion of Iran (2007)
European Union	36.35%
Australia	5.8%
Argentina	4.9%
Brazil	5.4%
Mexico	3.8%
China	11.12%
Thailand	6.4%
Pakistan	3.3%
South Korea	4.8%
Japan	4.3%
Turkey	4.3%
India	5.5%
Lebanon	2.1%

R&D. this matter is because of low investment of Iran in agricultural R&D. Short-run dynamic model showed foreign R&D with one year lag has 0.67 elasticity on agricultural productivity. As we expected raining has positive effect on agricultural productivity. This variable with one year lag has 0.04% effect on productivity. Agricultural import is another important variable that has meaningful effect on productivity, every 1% increase in agricultural import leads to 0.6% increase in agricultural TFP and if we have import in inputs and intermediate goods, this effect will increase more. Also this coefficient shows the importance of trade in economic equations.

Statistics show European countries have first position in exporting agricultural products to Iran but the portion of Asian and Latin American countries in exporting are second and third position. Diagram 1 shows the trend of agricultural import in Iran during studied period. As we see the most amount of import relates to first years after revolution. Imports after war changed from marginal goods to intermediate ones.

The biggest agricultural products and inputs producers and exporters in the world are United States of America, European Union, Australia, Brazil, China, Argentina, Thailand, Mexico, Canada, New Zealand and Pakistan. Table 7 shows import portion of Iran in agricultural products and inputs exporter countries.

Weather variable is another variable that because of the dependence of agricultural sector to water and rain has been entered to model. Weather variable shows the trend of raining for past 30 years in average form. According to Gutierrez (2002) one of main effective factors on agricultural productivity is weather conditions. Iran is located in a dry area and the average amount of rain during studied period is about 256 millimeter that in comparison with England and France is too low. The average of rain in 10 main countries of agricultural products producers is about 458 millimeter. In this study the effect of weather and raining conditions on agricultural productivity improvement is estimated 0.1. Definitely the effect of this variable in north and northwest of country because of more rain is more. Anyway every 1% change in raining amount leads to 0.1% change on agricultural productivity.

Time trend variable presents the negative ef-

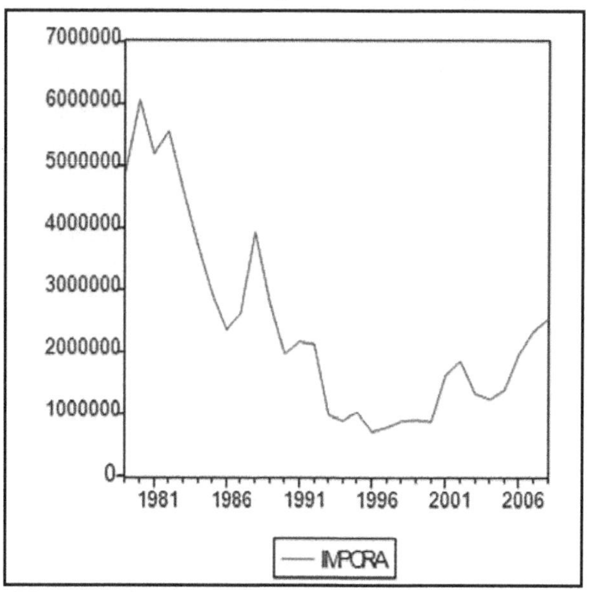

Diagram 1: Imports of Iran in agricultural sector (Thousand Dollar)

Table 8: Coefficients of short-run model and ECM model

Variable name	coefficient	Standard deviation	t-statistic
dWEA	0.049	0.031	1.6
dLR&D	0.14	0.067	2.1
dLIMPORA	0.59	0.091	6.4
dLR&Df	0.74	0.026	27.6
dC	-25.5	0.75	-34.2
dT	-0.07	0.011	-6.7
Ecm(-1)	-0.88	0.038	-22.6
	R^2=0.99	DW=2.2	F(6,14)=816.6

fect of effective factors on agricultural productivity that due to some econometric reasons have not been entered in model.

Because of convergence, we use error correcting model (ECM) in this section. In fact ECM relates short-run fluctuations of presented variables in model to long-run amounts of them. In this section to analyze the short-run relation between agricultural TFP and other studied variables, we used this model and the results of it are in table 8.

The results of ECM agricultural productivity show all studied variables are meaningful in 95% level (except weather variable that is meaningful in 90% level.) and are according to theoretical trend of matter. Coefficient of ECM (-1) in short-run model is -0.88 that is meaningful and based on expectation it`s negative. As we know this coefficient shows the unbalance of short-run to long-run. According to this coefficient about 88% of productivity unbalance in one period is adjusted in next period.

CONCLUSION AND RECOMMENDATIONS

A great deal of world researches costs have been allocated to modern and developed countries. Beside that done investments on agricultural R&D in Iran are low and in comparison with modern countries it`s nothing. High sensitivity of agricultural productivity toward international researches shows the importance of choosing trade partners with high agricultural R&D investments. So government should do the best to choose the best trade partners with high agricultural researches. On the other hand, due to the lack of relationship among domestic and foreign research centers during recent years, agricultural R&D does not have enough effect on production and agricultural productivity. Since in Iran there is no exact statistics of private sector investments in agricultural researches and the amount of private researches in comparison with government researches is low, in this study we use the information of government R&D investments. The average amount of agricultural R&D budget in Iran toward value added of this sector in studied years is 0.56 that has too distance with optimal amount (2). The average amount of agricultural R&D costs toward value added of it among the most important trade partners of Iran are 1.87 for Japan, 1.89 for Germany, 1.93 for England, 1.65 for South Korea, 1.2 for Brazil, 1.77 for Netherlands, 1 for Pakistan, 1.4 for India, 1.45 for China and 0.87 for Turkey.

As we can see in diagram 2, the trend of domestic R&D in agricultural sector during different years toward foreign trade partners R&D costs is too different. The variable of foreign R&D has ordered and stable trend with increasing movement during studied years. Drown time series for foreign researches has derived based on Coe and Helpman and Alston equation. Increasing trend of time series is because of high agricultural researches for European countries and other countries like Japan, China and Brazil. Domestic agricultural R&D costs during the first years after revolution and war were low and this amount was continued up to 1995. After 1995 government R&D costs had mild movement. Agricultural R&D costs in Iran in comparison with European trade partners are too low but when we compare these amounts with other developing trade partners as Pakistan, Turkey, Lebanon, Argentina and Mexico it`s not low.

One of other effective variables on agricultural productivity is import variable. We know import leads to production extent of marginal products and also economic growth of country. Import of

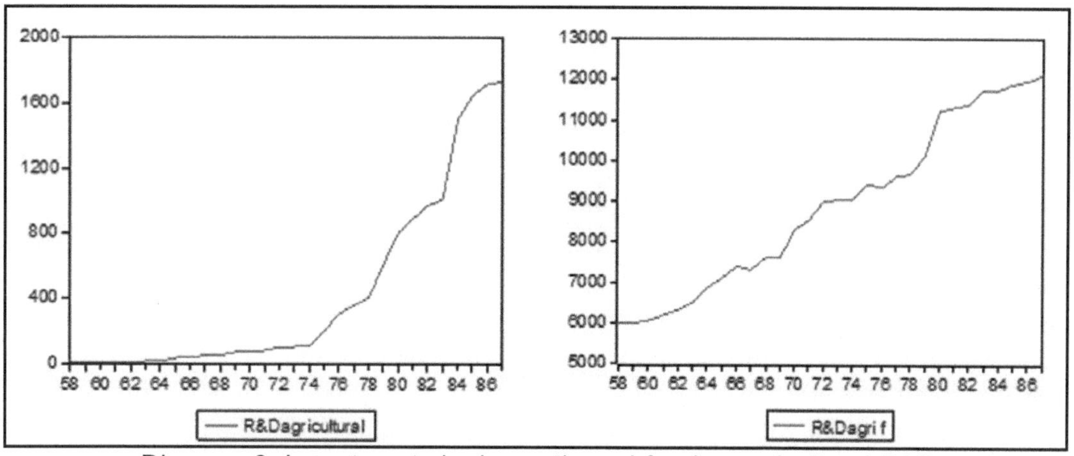

Diagram 2: Investments in domestic and foreign agricultural R&D (million dollar and Bilion Rial)

Table 9: Main combination of agricultural imports in Iran in studied period

Agricultural machinery and equipment	Agricultural goods and products
Chemical fertilizers	Grains and it`s products
Agricultural machinery	Fruit and vegetable
Poisons	Coffee, tea and cocoa
Veterinary drugs	Oil seeds
Regenerated varieties of plants and trees	sugar

intermediate and marginal agricultural products in Iran during 1979-2008 had average growth of 8.4% and during this period the portion of Iran in import was about 0.7%. The main imports of agricultural sector including basic goods are in table 9.

The average of agricultural import growth for oil seeds is 29.84% that had the most amount of growth during studied period. After oil seeds, grains with 28%, poisons with 26%, machinery with 23% and agricultural engineering equipment with 19% are in second to fifth positions.

Research results showed investment in domestic and foreign agricultural R&D has positive effect on agricultural TFP but the effect of foreign agricultural R&D is more than domestic R&D. of course import and openness degree of economy have positive and meaningful effect on agricultural productivity. Also we showed that suitable weather conditions beside new technologies are necessary for the growth of agricultural sector. According to above studies, some suggestions are presented to improve investment return in this sector.

1- Since the coefficient of government researches is relatively low, it`s necessary to allocate considerable portion of national product to R&D costs and increase research budget of agricultural sector to get to standard level.

2- Because of the importance of foreign agricultural researches in presented model, it`s necessary to pay attention to foreign trade partner selection because spill over of agricultural R&D in knowledgeable countries is more than developing countries.

3- Having more relationship among foreign and domestic scientific and research centers for getting newest technology and researches results is another important suggested policies of this paper.

REFERENCES

1- Aghion, P. (1995). A Model of Growth Through Creative Destruction, Econometrica, 12: 121-145.

2- Alston, J.(2003). Attribution and other Problems in Assessing the Returns to Agricultural R&D, Agri-Cultural Economics, 25: 21-45.

3- Amini, A. (2004). Measurement and Analysis of Effective Factors on Total Productivity of Economy in Iran, Peik Nour Publication, 2: 54-78.

4- Bagherzadeh, A. (2007). Theory of Economic Efficiency, Arad Publication, Tehran. Pp:189-204

5- Coe, D., & Helpman, E. (1997). International R&D spillovers, European Economic Review, 39: 12-30.

6- Evenson, R. (2000). Agricultural Research and Productivity Growth in India, IFPRI,Washington, 2: 65-80.

7- FAO. (2009). Report of R&D for Countries, Site: www.fao.org.
8- Griliches, Z. (1989). Productivity Puzzles and R&D, Economic Perspectives, 2: 31-50.
9- Gutierrez, L. (2005). Agricultural Labour Productivity in Some Countries, Agricultural Economics Review, 3: 12-24.
10- Gutierrez, L.(2005). Why is Agricultural Labor Productivity Higher in Some Countries than others? Agricultural Economics Review, 13: 20-31.
11- Huffman, W. (2006). Science for Agricultural; Long–Term Perspective, Iowa State University Press. Pp: 126-158
12- IFPRI. (2009). Agricultural Science and Technology Indicators (ASTI), Site www.asti.cgiar.org.
13- Jacobs, D. (2004). R&D Spill Overs to Agricultural Measurement and Application, Economic Policy, 11: 40-57.
14- Kandric, A., & Crimer, S. (1970). The Index of Productivity for Economics, Economic Policy Journals, 12: 44-62.
15- Khalilian, S. (2008). The Study of Effective Factors on Value Added in Industry and Agricultural Sectors, agricultural papers, Mashhad, 2: 28-43.
16- Komijani, A., & Shah Abadi, A. (2007). The Effect of Ddomestic and Foreign R&D Activities on TFP, Journal of Commercial Surveys, 1: 10-22.
17- Line, Y. (2006). Rate of Return to Public Agricultural Research in the Presence of Research Spillovers, American Agricultural Economics, 21: 88-98.
18- Noferesti, M., & Arab Mazar, A. (2005). The Sstudy of Macroeconomic Model in Iran, Economic Ministry, Tehran. Pp: 120-179.
19- Pesaran, M., & Shin, T. (1998). Working with Microfit 4.0.Oxford University Press. Pp: 120-148.
20- Schimmel Pfenning, H., & Leftly, A. (2006). Concepts of Productivity Measurement in Agriculture, Journal of Development and agricultural economics, 8: 55-78.
21- Shujat, A., & Iqbal, H. (2007). The Survey of TFP in Pakistan, Case Study of Agricultural Sector, Indian Agricultural Economics Journals, 16: 31-56.
22- Tokgoz, S. (2004). R&D spillovers in agricultural, Journal of Econometrics, 31: 12-41.

4

Investigation of East Azerbaijan Researchers and Extension Agents' View about the Role of Extension Services on Agricultural Development

Javad Mahmoodi Karamjavan

Abstract

The purpose of this study was investigation of East Azerbaijan researchers and extension agents' view about the role of extension services on agricultural development. The methodology was descriptive-correlation which is carried out by survey method. In this study, the sample population was 335 people including all researchers and extension agents working in the agricultural sector. To assess the reliability of the questionnaire, Cronbach's alpha coefficient was used and level over 0.75 for all sections of the questionnaire was measured. Its validity was established asking for the help from the technical experts. Results showed that the agricultural extension is an undeniable necessity in the agricultural development the effects of which has been positive and significant in the major variables of the agricultural development. The shortcomings are not due to the nature of extension rather they are because of improper execution, lack of support and similar reasons. The reduction of pest damage, improvement of the social status, contribution to boost handicrafts, acceptance of product insurance, recommendations for land leveling, creation of sale cooperation, awareness raising of the market, use of modern irrigation methods and several factors else was those extension role factors that the respondents agreed upon.

Keywords:
Azerbaijan, Agricultural Development, Extension agents, Researcher

Assistant Professor, Department of Agricultural Extension and Education, Tabriz Branch, Islamic Azad University, Tabriz, Iran.
** Corresponding author's email: Karamjavan@iaut.ac.ir*

INTRODUCTION

Inside the agricultural development system and its expanded subsectors, the establishment of the helpful and effective communication between these subsectors and the stakeholders is the responsibility of the agriculture extension. The main concern of the extension is a human being that manifests an individual, group and society. However, the main goal of the extension is the creation of the desirable changes in human as the key factor of development (Malekmohammadi, 1999). Despite the strong beliefs about the necessity and importance of the agricultural extension activities for the development of agricultural, agricultural extension systems in different parts of the world have been criticized for their performances. Undoubtedly, these criticisms have not been unsupportive (they have been backed up by evidence). However, it is comprehensible that some of the extension shortcomings are related to the internal factors whereas some are attributable to the inappropriate external attitudes which require configuration as well as agreement regarding their roles. Questioning the efficiency and effectiveness of extension activities, Centralized structure from top to bottom based on technology, Inappropriate service coverage, Carelessness to the vulnerable people, Poor access to facilities, Lack of access to the efficient and helpful knowledge based on the farmers needs and not being able to make use of it, Poor links with other agricultural institutions especially research centers, Insufficient attention to the sustainable development of agriculture, and etc are among the problems of extension worldwide (Qamar, 2002). Despite the obvious role of extension in the development, it seems that there is scarce approach towards the activities of agricultural extension in Iran. Furthermore, the activities performed by this educational institution, have been defined neither to the authors, nor to the researchers, practitioners, executives, farmers and etc (Zamanipour, 2001).

Thus the present study is an attempt to assess the views of researchers and extension agents about the role of extension services on agricultural development in East Azerbaijan province. It aims to provide the possible strategies for the betterment, regulation, modification and prediction of agriculture extension performance in this province. Establishing a connection between agriculture extension researches and agents is crucial to achieve the goals of agricultural development.

Daivic (2008) in his paper entitled 'Extension in African Saharan', Anderson (2008) in his report for World Bank 'Extensional Advisory Services' and Rivera (2001) in his book Global Perspective of Agricultural and Rural Extension, refers to the main reason for the failure of extension activities in the agricultural development process as the peoples problem in understanding the effects of the extension.

Thus the fundamental problem addressed in this study is that the role of extension in the agricultural development programs is unclear and untouchable. Moreover the effect of the extension activities in short term is undetectable (Malekmohammadi, 1999, Shahbazi, 2003) which in turn has led to the negligence of the significance and position of this institution in the agricultural developments. On the other hand, the unclearness of the results of the extension activities has made great inadvertent or deliberate efforts to eliminate it. Thus, the main questions of this research are: according to the respondents' ideas, does the agricultural extension have a role in the agricultural development of the region? And the researches and extension agents agree on what function(s) of the extension on the agricultural development. How is the agreement among respondents and they agree on what function and how they evaluate its effect on the agricultural development of the province.

MATERIALS AND METHODS

The research paradigm was quantitative. This research was an applied study. The methodology is descriptive-correlation which is carried out by survey method. In this study, the sample population were 335 people (the same as total population) including; all researchers and advocates working in the agricultural sector. To assess the reliability of the questionnaires SPSS software was used and Cronbach's alpha coefficient level over 0.75 for all sections of the questionnaire was measured. Its validity was established asking for the help from the technical experts. The dependent variable of the research was the ideas of the participants about the 'agricultural development' that is formed based on 60 indicators.

RESULTS

A. Descriptive results

Findings show that about 90 percent of respondents were male and 10 percent were women. The distribution of age frequency among agricultural experts was between 41 to 45 years old that makes 38.8% of all respondents. The income mean was 7,934,272 Iran Rials. There was a significant difference between the income of researchers and extension agents. The highest years of job experience among staff are 16-20 years which makes 36.7% out of total percent of the respondents. Regarding the farming experience of the respondents, about 70% of them has the history of agricultural activities. Regarding the relationship between the course of the study and the agricultural activities, 12 respondents had non-agriculture and non- related specialty in agricultural extension. 178 respondents (56%) had a bachelor's degree. 140 respondents (44%) had master or higher degree (PhD). 107 experts (34.7%) was members of social groups. The investigation of the communication between experts and farmers shows that more than 71% of the respondents have stated it as high and very high. 49 respondents (24.7%) had little communication with the farmers who are mainly the members of the researchers group. The obtained ordinal average was 3.97, which is equivalent to a high communication.

A.1. Current and future status of the extension functions in the agricultural development

According to table 1, the current priority gradation of extension activity in region includes; farmers' 'incomes improvement, the development of modern irrigation methods and the establishment of the local cooperation institutions for the farmers'

Regarding the future priorities, experts believe that the creation of the cooperatives for selling agricultural products, the establishment of the conversing and manufacturing industries and the production and improvement of the agricultural technologies are in priority respectively.

As illustrated in table 2, the priorities in the current roles of the extension are; contribution to the production of the needed technology in agriculture sector, matching and conformity of technology with farmers social situations and finally matching and adaptation of the technology with the farmers' economical conditions.

In addition, the future priorities of extension includes; transferring, expansion and developing needed technology, reflecting the effects of the technology distribution delivery to the authorities and contribution to the production of the needed technology in agriculture sector.

A.3. Educational- extension methods in agricultural development

According to table 3, among the priorities to present information for the agricultural improvement are; result demonstration method, method demonstration implementation and using radio and TV programs

A.4. Researchers and Extension agents' participation on research-extension plans

According to table 4, experts took participated in the farmer's education programs, at relevant scientific seminars and workshops, respectively.

A.5. The views of researchers and extension agents about agricultural extension and education

According to table 5, experts believe that extensional programs improve research activities. Furthermore, they contribute to solve farmers' problems. In other words, farmers can't use experts' recommendations without extension activities.

B. Analytical results

B.1. Factor analysis of role variables of agricultural extension and their effects on agricultural development components

In order to reduce the number of variables in classes, factor analysis was carried out. The results of test KMO= 0.851 and Bartlett's test = 11751.84 show that the data are suitable for factor analysis. Having separated the factor analysis and eliminating the heterogeneous variables, the variables in table 6 were classified as follows:

According to the conducted factor analysis, variables were reduced to 7 levels where the sustainability in agriculture and the development of human resource were considered as the first factors (with 15.36% of the variance expla-

Table1: Current and future status of extension function in agricultural development

Items	Priority		S. D.		Mean	
	future	Current	future	Current	future	Current
Rural civilization (water, electricity, phone, etc.)	52	33	1.3	1.07	2.9	1.9
Availability of water supply and drainage channels	25	50	1.16	1.32	3.59	2.32
Products transportation	13	44	1.01	.986	3.48	2.76
Post harvest storage facilities	20	55	1.15	0.81	3.72	3.05
Mechanization	31	7	1.17	1.16	3.44	2.81
Conversion and processing industries for agricultural products	2	28	1.05	1.12	4.08	2.56
Support producers in the form of insurance products	46	16	1.36	1.07	3.48	2.08
Restoration of lands	29	57	1.2	1.12	3.7	2.28
Integration of agricultural lands	11	5	1.12	0.86	3.96	2.92
Leveling of lands under cultivation	32	4	1.31	0.87	3.8	3.02
Increasing lands under cultivation	56	56	1.45	1.03	2.8	2.99
Development of electrical and mechanical wells	30	52	1.15	0.96	3.44	2.4
Availability of inputs required in production	28	54	1.15	1.06	3.56	2.05
provide Credits required by the Agricultural Bank	34	18	1.2	1.07	3.4	2.22
Development of modern irrigation methods	33	2	1.2	1.24	3.3	2.16
Expand the use of modern information tools	9	36	1.09	0.94	3.89	2.85
Production and improvement of agricultural technologies	3	39	1.03	0.99	3.8	2.6
Supply of agricultural technologies	7	49	1.09	1.06	3.9	2.97
Farmers access to new technologies	54	47	1.4	1.19	3.07	1.96
Farmers access to market information	49	9	1.3	1.03	3.09	1.69
Generation sale cooperatives for agricultural products	1	46	0.96	1.28	3.98	2.67
Guaranteed purchase of agricultural products	6	29	1.1	1.13	4.02	2.88
Control intermediaries and brokers in buying products	38	10	1.3	0.96	3.5	2.78
Prevent degradation of natural resources	12	17	1.6	0.96	3.7	2.69
Continuity and sustainability of benefits from agricultural production	55	19	1.31	0.87	2.8	2.41
The use of integrated pest management	47	51	1.3	0.92	3.2	1.69
Proper use of fertilizers	50	38	1.31	0.84	3.05	2.02
Proper use of chemical pesticides	41	37	1.2	0.97	3.2	2.3
soil fertility	53	11	1.3	0.86	2.4	2.17
Consideration to The groups of small farmers and women farmers	58	20	1.5	0.84	3.3	2.32
Development of agricultural production cooperatives	44	41	1.3	0.91	2.32	2.05
Improve professional training for farmers	19	34	1.15	0.99	3.7	2.46
Improve the education level of farmers	10	53	1.1	0.88	3.9	1.5
Develop technical skills of farmers in the production process	16	27	1.15	0.97	3.76	2.02
improve Management skills of farmers in the production process	24	25	1.21	0.79	3.75	2.12
Reduce waste at the time of planting	48	24	1.36	0.84	3.2	2.28
Reduce waste in product growing time	59	31	1.22	0.87	3.35	2.23
Reduce product loss during harvesting by using the Machines	36	23	1.2	0.89	3.3	2.42
Choosing the right time to harvest by farmers	5	40	1.01	0.89	3.7	2.09
increase Product prices by buyers due to improved product quality	15	26	1.07	0.86	3.59	2.29
Reduce contamination of products to various diseases	45	13	1.37	0.89	3.5	2.53
Reduce pest damage to products	43	8	1.37	0.89	3.56	2.6
Improved tillage operations	27	35	1.21	0.88	3.67	2.49
The use of animal manures	17	6	1.07	0.94	3.45	3.16
Using the modified crop and garden varieties	8	45	1.01	1.11	3.6	2.9
Appropriate time for planting	14	39	1.05	0.9	3.6	3.27
Correct use of water resources by farmers	18	60	1.16	0.91	3.7	2.6
Identify the type of fertilizer needed to production in the product during	22	8	1.13	0.88	3.6	2.67
Proper use of chemical pesticides	4	14	1.01	0.89	3.78	2.55
The use of micronutrient fertilizers in crop production	26	12	1.19	0.94	3.6	2.69
management and accounting of financial resources Related to farm	57	15	1.37	0.95	2.67	2.36
Provide appropriate incentives to motivate farmers to increase production	60	59	1.19	0.96	3.3	2.47
Creation of local institutions and farmers' cooperatives	40	3	1.27	0.99	3.36	2.34
Using initiative and creativity in production by farmers	23	42	1.13	1.7	3.5	2.47
Ensure investment in agriculture sector by farmers	42	22	1.26	0.88	3.21	2.38
Increase farmers' incomes	21	1	1.12	0.74	3.59	2.8
Increase the purchasing power of farmers	51	43	1.28	0.99	2.96	2.2
Increase the operation of production	39	60	1.25	1.05	3.4	2.09
Providing off- farm employment opportunities for farmers	35	30	0.92	0.95	4.3	2.57
Exports of agricultural products	37	32	0.84	0.97	3.34	2.27

Table2: Current and future role of extension in agricultural development

Items	Priority		S.D.		Mean	
	future	Current	future	Current	future	Current
Improve farmers' technical to cope with the changes in agricultural environment	9	16	0.85	1.2	4.2	3.3
Improve the managerial to cope with the changes in agricultural environment	7	13	0.84	1.14	4.17	3.3
Preparation attitude and willingness of farmers to cope with the changes in agricultural environment	6	14	0.84	1.19	4.2	3.3
The use of indigenous knowledge of farmers in agricultural development projects	11	19	0.83	1.26	4.2	3.3
Identification and assessment of educational needs of farmers	4	11	0.83	1.2	4.29	3.5
Identify problems and needs of farmers in production and feedback them to the related sectors	16	7	1.3	1.12	3.87	3.6
help to production Technologies needed to agricultural sector	3	1	0.81	1.04	4.2	3.9
Assimilating and adapting technologies with farmers' economic conditions	14	3	0.97	1.05	4.13	3.6
Assimilating and adapting technologies with farmers' social conditions	18	2	1.04	1.01	4.1	3.5
Assimilating and adapting technologies with farmers' technical knowledge	17	18	1.04	1.23	4.2	3.3
Assimilating and adapting technology to local customs and cultural level of farmers	13	8	0.95	1.07	4.2	3.4
Transfer , development of needed technologies	1	5	0.75	1.11	4.4	3.7
feedback of technology Supply effects to the concerned authorities	2	6	0.79	1.12	4.4	3.7
help To expedite the transfer and acceptance of findings and innovations	5	9	0.83	1.13	4.3	3.4
Preparation Appropriate context to consultation with 17the farmers for development programs	10	15	0.88	1.24	4.2	3.5
attract trust of Farmers for Implementation the developmental activities	19	10	1.2	1.4	3.9	3.9
help To attract farmers participation in the implementation of development activities	8	12	0.86	1.2	4.3	3.5
attract farmers participation in the evaluation of development activities	12	17	0.94	1.3	4.2	3.5
Increased cooperation between farmers , policy makers and experts in agricultural development activities	15	4	1	1.1	4.3	3.7

Table3: Extension methods in agricultural development from expertise' perspective

Method	Mean	S. D.
Result demonstration	4.81	1.09
Method demonstration	4.65	1.05
Radio- TV programs	4.71	1.11
On farm education	4.49	1.08
Educational films	4.48	1.33
Visiting sample farms	4.25	1.3
Extensional meetings	3.81	0.97
Select template and sample farmers	3.52	0.98
Educational workshops	3.46	1.29
Extensional publications	3.05	1.02

Note. Scale (Very low=1, low=2, moderate=3, high=4, very high=5)

nation) and infrastructural improvement (with 3.94% of the variance explanation) as the 7th factor.

B.2. Agreement study between researchers

The results of the 'agreement study' among experts about the role and functions of extension in the agricultural development have been summarized in table 7. The results show the agreement among researchers on the raised issues

B.3. Agreement study among extension agents

B.4. Agreement study among all respondents

Table 4: The rate of the researchers 'and extension agents' participation at plans

Effect of extension Methods in agricultural development	Mean	S. D.
participating in Farmers training	3.74	1.07
Participate in related academic seminars	3.61	1.13
Holding training workshops	3.59	1.17
Participation in research - extension projects	3.33	1.16
Participation in accelerate results project	3.05	1.29
Participation in research - adoptive projects	2.82	1.27
Participation in the "transfer of findings week" plan	2.25	1.07

Note. Scale (Very low=1, low=2, moderate=3, high=4, very high=5)

Table 5: The researchers' and extension agents' ideas about agricultural extension and education

Items	Mean	S. D.
Extension programs causes to improve the activities of research section	4.5	0.62
Extension programs help farmers to solve their problems	4.36	0.74
Without the extension activities, farmers do not use the opinions of experts	4.56	0.89
Without extension activities, do not establish good communication between executives and farmers	4.25	1.03
extension provides new and needed Information for farmers	4.22	0.79
Research sector, cannot teach new technologies and how to use them to farmers and extension is needed	4.19	0.84
If extension sector not exist, no part care the farmer	4.11	0.86
extensional programs are very useful and improve agriculture situation	4.09	0.88
Without the extension, cannot be informed of the effects of technology in agriculture sector	3.89	1.03
Training farmers has impact in agricultural development	3.68	1.04
Without the extension program, farmers cannot overcome new problems	3.36	1.05

Note. Scale (Strongly Disagree = 1, disagree = 2, Apathetic = 3, disagree = 4 strongly agree = 5)

In sum, among those extension role factors that the respondents agreed upon are; the reduction of pest damage, improvement of the social status, contribution to boost handicrafts, acceptance of product insurance, recommendations for land leveling, creation of sale cooperation, awareness raising of the market, use of modern irrigation methods, new information distribution, recommendations for optimal use of chemical fertilizers and pesticides, How to use suitable technology, insight preparation to synchronize the

Table 6: Factor analysis of extension role variables in agricultural development

Variable	Factor							
	F 1	F 2	F 3	F 4	F 5	F 6	F 7	
Prevent degradation of natural resources	.823							X1
Proper use of pesticides and fertilizers	.796							X2
Soil fertility ,attention to the livelihood farmers	.763							X3
Provide job counseling	.713							X4
Increasing technical and managerial skills and overcome to production problems	.684							X5
Market information		.756						X6
sale cooperatives provide		716						X7
helping to guarantee the purchase of production		.688						X8
interfaces control		.657						X9
help to the proper use of funds		.625						X10
using new Information tools			.633					X11
deliver technologies			.612					X12
Provide recommendations on the use of machinery				-.581				X13
help To increase income					.741			X14
reduce costs and increase purchasing power of input					.702			X15
increase exports					.689			X16
reduce Infectious to diseases and losses of pests						.958		X17
proper water consumption						.635		X18
reduce waste						.604		X19
Helping to boost handicrafts							.641	X20
packaging							.622	X21
industrial processing							.597	X22
storage silos							.569	X23

Note. F 1: Stability and development of human resources; F 2: Improved marketing; F 3: Providing technology; F4: Mechanization; F5: Economic development; F6: Waste reduction; F7: Infrastructure improvements

Table7: Kruskal- Wallis test results among extension agents about the extension role factors involved in agricultural development

variables	X^2	P-value	variable	X^2	P-value
Introduction of chemical fertilizers	7.204	0.206	help to provide job opportunities	3.99	0.59
Introduce corrective seeds	3.26	0.660	Help to improve export	4.52	0.47
Introduction of chemical pesticides	6.305	0.278	Diversification of production	7.98	0.15
Supply of appropriate technologies	2.824	0.727	Optimized use of fertilizer	1.32	0.93
Farmers access to technologies	6.514	0.259	Recommendations for soil fertility	6.44	0.26
Awareness of market information	2.566	0.767	Creation of production cooperatives	1.39	0.99
Create sale cooperatives	7.299	0.199	Improve job training	6.78	0.23
Degradation of natural resources	4.109	0.534	Increase the technical skills	6.66	0.24
Correct use of water resources	2.819	0.728	Improvement of management skills	5.44	0.36
optimize use of poison	4.183	0.523	Improving the social status	2	0.84
Encourage the construction of irrigation canals	9.558	0.089	power to overcome Problems	4.78	0.44
			Attention to small farmers	9.26	0.09
Use of Implant machinery	6.861	0.231	Improvement initiatives and creativity of farmers	2.35	0.79
usage The machines take care of the product	8.841	0.116	Assessment of educational needs	3.381	0.136
The use of harvest machinery	4.26	0.512	Identification of production problems	3.381	0.136
Help to improve packaging	4.765	0.445	Matching technology with the economic conditions of farmers	4.447	0.487
The insurance product	5.238	0.387	Provide feedback of technologies	8.368	0.137
Increase operation	9.161	0.103	Accelerate the adoption of findings and innovations	7.089	0.214
Land leveling	6.056	0.301	Substrate preparation in consultation with farmers	8.12	0.15
Integration of lands	4.377	0.496	Participation in the Implementation	2.571	0.766
help to the Contribute and restoration of lands	10.072	0.073	Identification of production problems	6.064	0.3
Preparation of farmers ' attitudes	10.1	0.072	attract Participation in design developmental activities	5.685	0.338
Loss reduction in the harvest stage	10.43	0.064	Enforcement partnerships between actors, researchers and farmers	3.865	0.569
Reduce pollution to disease	5.92	0.31			
Reduce pest damage	9.75	0.083	Preparing farmers ' attitudes for change	3.049	0.692
Reduce production costs	4.79	0.34	Assess the training needs of farmers	4.326	0.503
help to increase the purchasing power	9.01	0.1	Matching technology with the economic conditions of farmers	6.771	0.238
Improve income	5.69	0.33	Matching technology with the knowledge of farmers	6.718	0.242

P≤0/05

changes, reduction of diseases infection, reduction of the natural resources degradation, enhance technical skills, improvement of management skills, use of planting machines, use of harvesting machines, provide feedback for technology, contribution to improve ex-

Table 8: Kruskal- Wallis test results between researchers about the extension role factors involved in agricultural development

variables	X^2	P-value	variable	X^2	P-value
Introduction of chemical fertilizers	7.2	.2	Reduce production costs	4.7	.28
Introduce corrective seeds	3.2	.66	help to increase the purchasing power	9	.76
Introduction of chemical pesticides	6.3	.27	Improve income	5.6	.13
Supply of appropriate technologies	2.8	.72	help to provide job opportunities	3.9	.08
Farmers access to technologies	6.5	.25	Help to improve export	4.5	.09
Awareness of market information	2.5	.76	Diversification of production	7.9	.25
Create sale cooperatives	7.2	.19	Optimized use of fertilizer	1.3	.14
Degradation of natural resources	4.1	.53	Recommendations for soil fertility	6.4	.11
Correct use of water resources	2.8	.72	Creation of production cooperatives	1.3	.06
optimize use of poison	4.1	.52	Improve job training	6.7	.06
Encourage the construction of irrigation canals	9.5	.08	Increase the technical skills	6.6	.45
Use of Implant machinery	6.8	.23	Improvement of management skills	5.4	.36
usage The machines take care of the product	8.8	.11	Improving the social status	2	.25
The use of harvest machinery	4.2	.51	power to overcome Problems	4.7	.6
Help to improve packaging	4.6	.44	Attention to small farmers	9.2	.52
The insurance product	5.2	.38	Improvement initiatives and creativity of farmers	2.3	.36
Increase operation	9.1	.1	Assessment of educational needs	3.3	.32
Land leveling	6	.3	Identification of production problems	3.3	.58
Integration of lands	4.3	.49	Matching technology with the economic conditions of farmers	4.4	.25
help to the Contribute and restoration of lands	10	.07	Provide feedback of technologies	8.3	.08
Preparation of farmers ' attitudes	10.1	.07	Accelerate the adoption of findings and innovations	7	.12
Loss reduction in the harvest stage	10.4	.06	Substrate preparation in consultation with farmers	8.1	.25
Reduce pollution to disease	5.9	.3	Participation in the Implementation	2.5	.08
Reduce pest damage	9.7	.08	Participation in maintenance	6	.32

P≤0/05

Table 9: Kruskal- Wallis test results among all respondents about the extension role factors involved in agricultural development

variables	X^2	P-value	variable	X^2	P-value
Provide livestock recommendations	7.204	0.206	Attention to small farmers	9.161	0.103
Loss reduction in the harvested product	3.26	0.660	Improvement initiatives and creativity of farmers	6.056	0.301
Reduce pest damage	6.305	0.278	Correct use of water resources	4.377	0.496
Reduce production costs	2.824	0.727	optimize the use of poison	10.072	0.073
help to increase the purchasing power	6.514	0.259	Participation in the Implementation	2.939	0.401
Improve income	2.566	0.767	Participation in maintenance	1.395	0.707
help to provide job opportunities	7.299	0.199	Identification of production problems	5.624	0.131
Help to improve export	4.109	0.534	attract Participation in design developmental activities	5.618	0.132
Optimized use of fertilizer	2.819	0.728	Enforcement partnerships between actors, researchers and farmers	4.775	0.189
Recommendations for the use of micronutrient fertilizers	4.183	0.523	Preparing farmers ' attitudes for change	3.888	0.274
Creation of production cooperatives	9.558	0.089	Assess the training needs of farmers	7.594	0.055
Improve job training	6.861	0.231	Matching technology with the economic conditions of farmers	4.18	0.286
Increase the technical skills	8.841	0.116	Matching technology with the knowledge of farmers	5.114	0.164
Improvement of management skills	4.26	0.512	Provide feedback of technology	4.626	0.201
Improving the social status	4.765	0.445	Accelerate the adoption of findings and innovations	4.78	0.189
power to overcome Problems	5.238	0.387	Diversification of production	3.756	0.051

P≤0/05

Table10: Investigation of the relationship between the studied variables and agricultural development

Variables	*r*	P-value
Human Resources Development	0.793	***0.000
Supply technologies	0.494	***0.000
transfer Effects of technology supply to the relevant sections	0.328	***0.000
Increase participation of stakeholders	0.421	***0.000
Need assessment	0.591	**0.030
help to improve infrastructures	0.179	**0.020
Help to improve the economic status of farmers	0.317	***0.000
Reflect Problems to the relevant authorities	0.344	***0.000
Improved management of inputs and stability	0.238	***0.000
Networking between specialists, policymakers and farmers	0.326	**0.001
Reduce wastes	0.276	*0.003
Provide recommendations on the use of machinery	0.189	*0.041

* P≤0/05 ** p≤0/01 *** p≤0/001

ports, reducing product loss during harvest, reducing production costs, increasing income, invite for cooperation in execution, problem solving during production process and Participation in conservation, providing employment opportunities, technology transfer, partnerships among executives, researchers and farmer, educational need analysis of farmers and recommendations for the use of modern irrigation methods. According to table 10 there is high correlation among variables involved in the role of extension in the agricultural development.

According to table 11, the extension roles that respondents agreed upon contribute to 79.8% of changes in the agricultural development ($R^2 = 0.798$)

In the final step, Regression analysis was conducted on the all extension role variables. As table 12 illustrates the role of human resources development, technology transfer, the increase of the stakeholders participation, networking among experts, policy makers and farmers, waste reduction and contribution to the improvement of the farmers' economical conditions, are among the extension activities necessary for the agricultural development from the respondents perspective.

Table11: The role of extensional independent variables in the agricultural development

Model	R	R^2	R^2_{Ad}	S.E
Enter	0.893	0.798	75.3	0.195

Table12: The results of the role of independent variables on agricultural extension

	t	Beta	S.E	B	Sig
Constant	4.438	-	0.089	0.397	0.000
Human Resources Development	18.16	0.674	0.031	0.559	0.000
Supply technologies	9.963	0.456	0.043	0.208	0.001
transfer Effects of technology supply to the relevant sections	0.436	0.081	0.059	0.053	0.472
Increase participation of stakeholders	6.328	0.335	0.047	0.179	0.000
Need assessment	1.345	0.218	0.038	0.09	0.06
help to improve infrastructures	0.985	0.082	0.028	0.038	0.326
Help to improve the economic status of farmers	2.609	0.153	0.236	0.328	0.007
Reflect Problems to the relevant authorities	0.689	0.076	0.048	0.033	0.492
Improved management of inputs and stability	-1.925	-0.236	0.057	-0.089	0.08
Networking between specialists , policymakers and farmers	3.611	0.325	0.049	0.128	0.006
Reduce wastes	2.608	0.128	0.025	0.039	0.019
Help to improve marketing	-0.585	-0.049	0.057	-0.034	0.621
Provide recommendations on the use of machinery	1.107	0.059	0.032	0.035	0.27

P≤0/05

DISCUSSION

In sum, the agricultural extension is an undeniable necessity in the agricultural development the effects of which has been positive and significant in the major variables of the agricultural development. The shortcomings are not due to the nature of extension rather they are because of improper execution, lack of support and similar reasons.

The results of the study show the positive effect of the development of human resources on all aspects of agricultural development. It backs up the studies of Wielinga (2006), Anderson (2008), Malekmohammadi (1999) and Karbasioun (2007). According to the results of the present study, the main reason for the technological changes of the agriculture is the supplement and transfer of technology to the agriculture sector. Swanson (2003), Gebrekidan *et al.,* (2005) support the findings of the current study. The increase of partnership among agricultural executives due to the extension has a great role in agricultural development enhancement and contributes to the improvement of its process. The studies conducted by Leeuwis (2004), Anderson (2008), Norton *et al.,* (1997) point out to it as well. According to the results of this study, extension has an effect on the reduction of the agricultural waste via providing appropriate recommendations and counseling methods in the various stages of the production and harvest. The empowerment of farmers in how to use the various methods to combat pests and diseases and improve the quality of the production, contributes to the reduction and prevention. The role of extension is among the agreed roles that have been approved by studies of Hoang *et al.,* (2006). The findings show that the educational-extensional methods used to train the farmers has a significant role in enhancing the agriculture development which is supported by Asiabaka (2003), Karbasiyoun (2007).

ACKNOWLEDGEMENT

Hereby, I would like to appreciate the Islamic Azad University, Tabriz Branch, Iran, as the financial supporter of this study.

REFERENCES

1- Anderson, J.R. (2008). Agricultural Advisory Services. Background Paper for World Bank Development Report (2008). Agriculture for Development. Washington DC: The World Bank.

2- Asiabaka, C.C. (2003). Promoting Sustainable Extension Approaches: Farmer Field School (FFS) and its Role in Sustainable Agricultural Development in African. Lessons from Kenya Paper Presented at the Association of Third World Scientists, Njoro, Kenya, Egerton University.

3- Daivic, K. (2008). Extension in Sub- Saharan Africa: Overview and Assessment of Past and Current Models and Future Prospects. Proceeding of the 24 Annual Meeting. AIAEE.E.A.R.T.H. University of Costa Rica.

4- Gebrekidan, B., Yitayew A., Fekadu Y., Elias Z. & Habtemariam, K. (2005). Integrated Agricultural Development Strategies in the ANRS: Lessons from the AMAREW Project. Prepared for the "the Third

EAF-EARO International Symposium on Development Studies in Ethiopia", June 17-18, 2005, Addis Ababa, Ethiopia.

5- Hoang, L. Anh, Castella, Jean-Ch., & Novosad, P. (2006). Social Networks and Information Access: Implications for Agricultural Extension in a Rice Farming Community in Northern Vietnam. Agriculture and Human Values. 23: 513–527.

6- Karbasioun, M. (2007). Towards a Competency Profile for the Role of Instruction of Agricultural Extension Professionals Isfahan. Wageningen University.

7- Malekmohammadi, I. (1999). Agriculture and Natural Resources Extension. Iran, Tehran: UDC.

8- Norton, G., Taylor, W., Daniel, B., & McDowell, G. (1997). Investing in Agricultural Extension: The Case of Albania Lefter Daku Research Project Submitted to the Faculty of the Virginia Polytechnic Institute and State University in Partial Blacksburg, Virginia, Lefter Daku.

9- Qamar, M.K. (2002). Global Trends in Agricultural Extension: Challenges Facing Asia and the Pacific Region; FAO Regional Expert Consultation on Agricultural Extension (Bangkok, July 2002). FAO, Rome.

10- Rivera, W. M. (2001). Agricultural and Rural Extension Worldwide: Options for Institutional Reform in the Developing Countries. FAO.

11- Shahbazi, E. (2003). Rural Development and Extension (second edition). Iran, Tehran: University of Tehran.

12- Swanson, E.B. (2006). Extension Strategies for Poverty Alleviation: Lessons from China and India. Journal of Agricultural Education and Extension. 12(4): 285-299.

13- Wielinga, E. (2006). Rural Extension in Vital Network Changing Roles of Extension in Dutch Agriculture. Journal of International Agricultural and Extension Education. 7(1). http://www.aiaee.org/index.htm

14- Zamanipour, A. (2001). Agricultural Extension in Development Process (Third press). Iran, Birjand: University of Birjand.

The Effects of Non-Farm Business in Rural Sustainable Development: A Case Study in Sanandaj's Rural Areas, Kurdistan

Reza Movahedi[1]*, Hadi Fathi[2], Somaye Latifi[2]

Abstract

Rural economy has an important role in rural sustainable development in every region or country. However, the rural economy in Iran is too much depending on agriculture and development of other rural economy in particular non-farm activities, has been neglected by Iranian rural communities. Therefore, the aim of this study was to identify the role of non-farm business on rural sustainable development in Sanandaj Township. This study was a type of surveying research and employed both questionnaire and interview tools. A four-part questionnaire (with 48 questions) was developed in order to collect data from the respondents. Likert-type responses with five scales were used to assess different sections of the questionnaires. The questionnaires' reliability was tested by Cronbach's Alpha technique and it was 76% (α= 0.76). A number of 279 families, 181 families with and 98 without non-farm business were selected randomly in 2011. Results of the study showed that non-farm business can contribute to employment growth, income generating, and prevent seasonal and permanent migration from rural areas. Based on the results, non-farm business and non-farm industries need to be strongly invested by financial facilities through rural cooperatives and unions. Additionally, encouraging rural participation is essential in non-farms and industrial activities.

Keywords:
Non-Farm Business, Agriculture, Villagers, Rural Economy, Rural Sustainable Development, Sanandaj Township, Kurdistan

Abbreviations: Small and Medium Enterprises (SMEs), Non-Farm Business (NFB)

* *Assistant Prof. Agricultural Extension and Education Department., Bu-Ali- Sina University, Hamedan, Iran.*
[1] *M.Sc. Graduated on Rural Development, Bu-Ali Sina University, Hamedan, Iran.*
[2] *M.Sc. Graduated on rural development, Bu-Ali Sina University, Hamedan, Iran*
* *Corresponding author's email: r.movahedi@basu.ac.ir*

INTRODUCTION

Without any doubt, rural economy has an important role in rural sustainable development in every region or country. However, the rural economy in Iran is too much depending on agriculture so that development of other rural economy in particular non-farm activities, has been neglected by Iranian rural communities (Kalantari *et al.,* 2008).

Studies show that with the process of development, the share of non-farm income and employment in the total income and employment of the rural households has increased in the developing countries. There are also evidences to show that productivity and profitability in the non-farm sector is generally higher than in the farm sector (Fisher and Mahajan, 1998). In India, Jha (2006) found that in non-farm manufacturing activities undertaken in the organized and unorganized sectors, the employment rate, value-added, and capital has grown during the years 1984-1999. In China, developments in rural industries have affected many aspects of the regional economic landscape, i.e. Jiangsu's rural non-farm business has produced 70% of the province's total industrial output and 80% of its total industrial value added in 2000 (CREY, 2001, CSSB, 2001, Skinner, 2007). Athar *et al.,* (1994), showed comparison between the farm and non-farm income, that the non-farm income has transferred rural workers to non-farm sectors. According to other studies (Bhalla, 1990, Lanjouw, 2008, Miller *et al.*, 2009, Zhu and Jiang, 1993), the sharp increase in rural household incomes has mainly contributed to improvements in skills, knowledge, and capital from non-farm activities.

The context factor of "family" as a relevant factor of influence on competitive advantages and disadvantages, as well as the success of family businesses, has been widely ignored in research (Astrachan, 2010, Frank, *et al.*, 2010) – despite the reasonable assumption of its significant explanatory potential (Dyer, 2003).

Many researchers have shown that non-farm activities have an important impact on the distribution of income in rural areas. This impact depends on the specific types of non-farm activities involved and on the capacities of different types of households to access these activities (Barham and Boucher, 1998, Lanjouw, 2008, Escobal, 2001, Khan and Riskin, 2001, Leones and Feldman, 1998, Reardon and Taylor, 1996, Shand, 1987, Walkers, 2007). To improve rural income as a whole, participation in non-farm activities is highly selective and thus tends to increase income disparities, particularly in poorer areas. However, other researchers have shown that non-farm incomes can be inequality-reducing, particularly as the proportion of non-farm income in total income increases (Adams, 1994, 1999, Adams and He, 1995, Miller *et al.,* 2009, Pears, 2007, Skinner, 2007 and Stark *et al.,* 1986).

Allen *et al.,* (2005), revealed that rural non-farm incomes help reduce not only poverty but also inequality. This is because participation in non-farm activities differentially improves the income of the poorest households, while the best farmers remain in agriculture.

Testushi *et al.*, (2002), argued that the local marketplace, where enterprises can easily purchase materials and sell products to local traders, plays a critical role in easing the entry of new enterprises i.e., small and medium enterprises (SMEs).

According to Kalantari *et al.,* (2008), the factors that need to be planned in order to prepare appropriate conditions for rural sustainable development in Iran are: diversification of rural economy, supporting job creation activities, promoting rural sustainable tourism, business and industrial activities, and establishing a stable urban-rural network. Results of this research showed that the lack of diversification in rural economy, particularly in non-agricultural sector was the main challenge faced to sustainable rural development in Iran.

The main purpose of this study was to identify role of non-farm business on improving rural sustainable development. To achieve this, the following objectives will deal with:

1- Studying on non-farm business's effects on the rural families,

2- Determining non-farm business's capabilities especially on increasing rural non-farm income,

3- Recognizing effects of establishing non-farm businesses in terms of stuffs' consumption, gender, age, job satisfaction, and participation.

MATERIALS AND METHODS

The study has been conducted through a surveying research by using both questionnaire and interview tools. The validity of questionnaire was controlled by a group of rural development experts. Finally, a four-part questionnaire (with 48 questions) was developed in order to collect data from the respondents. Likert-type responses with five scales were used to assess different sections of the questionnaires. For these types of questions, the questionnaires' reliability was tested by Cronbach's Alpha technique within SPSS statistical software. The total mean reliability of scales for the questionnaire was 76% (α= 0.76).

Statistical population of the study consisted of villages in the central region in Sanandaj Township. This region contains one county (Arandan) and 64 villages. At first, for selecting samples' size, two types of villages with and without non-farm business were selected (five villages in each category). Next step was done for family samples. Based on Cochran formula a number of 279 families were selected randomly (N=857), 181 families had a non-farm business and others had no non-farm business (table 1). In determining effects of non-farm business, some of the effective socio-economic indicators on rural development were selected. These indicators were: family income, family costs, consumption stuff, immigration, job satisfaction, rural participation, using the media, insurance, literacy, and rural house quality and quantity. Data analyzing was done by using SPSS software and statistical tests used in the study were Mann-Whitney, Duncan's and t tests.

When comparing the differences between two unpaired groups in small sizes, whose samples take the same shape, the Mann Whitney non-parametric test is used (DIIA 2007). Therefore, to find the difference between the perceptions of two groups of either rural family with and without NFB in terms of job satisfaction and job stability the Mann Whitney U-test was used.

Duncan's test is used to determine whether three or more means differ significantly in an analysis of variance. It may be used regardless of whether the overall analysis of variance is significant. It assumes equal variance and is approximate for unequal group sizes (in this study family with and without non-farm business in terms of different family income and costs).

Table 1: Selected villages and samples

	Village	Household	sample
With NFB	Babariz	147	40
	Nawareh	132	34
	Arandan	157	46
	Gholian	128	33
	Sateleh	98	28
Without NFB	Alijan	34	18
	Farhad abad	41	21
	Mamokh olia	42	21
	Khakvarzi	24	15
	Gazardare	54	23
Total		857	279

RESULTS AND DISCUSSION

Income status: As seen in table 2, there is a clear difference in the income rate between villages that have non-farm business and villages that lack non-farm business. This difference is in such a way that the minimum and maximum income ranged 2 USD and 4 USD per day in the villages lacking non-farm business, whereas, the rate of income in those villages with non-farm business has been at most 6 USD and upper per day. Meanwhile, the income of villages lacking non-farm business depends strongly on agriculture. This leads the villagers migrate to neighboring cities in non-farming seasons.

Private and household utensils: The results indicate that the percent and rate of using private and household goods between villages with and without non-farm business are completely different. Given the results in table 4, although household and private equipment including color TV, washing machine, refrigerator, freezer, air conditioner and car, etc are common in both groups of villages, but their consumption percent is completely different; also, the use of goods such as cars (51.96%), computer (16.2%), furniture (19.2%), video recorder (10.23%), and vacuum cleaner (35.32%) is only common in villages with non-farm business.

Age groups of villages with and without NFB: According to the information of table 5, the average age in villages with non-farm business is much lower than villages without non-farm business. In other words, in villages with non-farm business, the families' heads are considered among the active population, and are in middle-aged group (21-30) and adult (age group of 31-40). But, in villages without non-farm business, the heads of the families are at the

Table 2: Percent of rural family income in terms of USD per day

	Family income	1-2	2.1-4	4.1-6	6 and Upper
Rural with NFB	Babariz	0.8	5.3	18.2	75.7
	Nawareh	1.2	10.4	33.9	54.5
	Arandan	2.6	13.5	31.8	52.1
	Gholian	0.4	4.4	17.3	77.9
	Sateleh	3.1	14.1	29.7	53.1
Rural without NFB	Alijan	52.7	47.3	0	0
	Farhad abad	54.6	45.4	0	0
	Mamokh		57.7	0	0
	olia	42.3			
	Khakvarzi	45.5	54.4	0	0
	Gazardare	40.2	59.8	0	0

Table 3: Comparing family income and costs in two groups of villages in terms of the mean

Variable	Groups	Test	z	Sig.
Family Income (in four levels)	With non-farm business	Duncan	8.120	0.000
	Without non-farm business		4.526	
Family costs (consumption/garment/travel)	With non-farm business	Duncan	3.814	0.010
	Without non-farm business		2.120	

middle-aged and old age group. Among the villages without non-farm business on average, 43.5% of population is at age group over 60, indicating the immigration of the adults from villages without non-farm business due to unemployment and economic problems.

The remaining at village and reversed immigration: To study the remaining at villages, the viewpoint of the villagers with job and non-farm business has been used in terms of creating new employment. The findings of the study reveal a meaningful difference between the type of employment in families with and without non-farm business. In such a way that villages such as Babariz, Gholian, Sateleh, Nawarreh, and Arandan which have been located near industrial regions, not only they have not lost their active population, but also have attracted some population based on the results of the study, almost 31.2% of population at these villages is the immigrants who have chosen these village as their residence place because of their work in industrial manufacturing units.

Job satisfaction: According to many economic theories, lack of productive employment at the villages could lead to lack of job security and ultimately to immigration from villages. In this vein, creating productive and permanent employment toward achieving job security and satisfaction is one of the most important objects of development in rural regions. According to information available, there is a meaningful relationship between non-farm business creation and job stability and satisfaction and based on studies conducted this relationship is at 99% level. Rural families of Sanandaj Township because of the possession of handicraft and semi-private and private non cultural industries are of

Table 4: Percent of consumption stuff in households with and without NFB

Consumption Stuff	Rural with NFB	Consumption Stuff	Rural without NFB
Television	98.24	Television	43.29
Refrigerator	97.65	Refrigerator	29.32
Automobile	51.96	Automobile	3.5
Laundry	39.35	Laundry	1.2
Air conditioner	36.21	Cooler	1.26
Vacuum	35.32	Vacuum	1.9
Furniture	19.2	Furniture	0.4
Computer	16.20	Computer	0.4
VCD	10.23	VCD	0.3

Table 5: Comparing the percent of active population in the villages with and without NFB

Age groups (year)		Under 20	21-30	31-40	41-50	51-60	61-70	70 and upper	Sum percentage
Rural with non-farm business	Babariz	16.2	21.5	20.1	18.3	13.2	9.7	0.9	100
	Nawareh	19.4	25.3	24.6	19.2	6.7	0.4	0.8	100
	Arandan	13.6	19.8	22.6	25.1	11.4	0.7	0.5	100
	Gholian	16.7	22.3	19.4	24.7	10.2	5.5	1.2	100
	Sateleh	21.4	18.3	15.2	26.9	12.6	4.9	0.9	100
Rural with	Alijan	8.6	7.3	11.7	15.2	23.1	19.2	14.9	100
	Farhad abad	4.6	8.7	2.1	25.3	27.2	15.9	16.2	100
	Mamokh olia	8.4	3.9	12.5	19.7	29.1	13.2	13.2	100
	Khakvarzi	7.9	5.6	6.8	23.6	32.1	14.2	9.8	100
	Gazardare	5.3	3.4	13.2	24.1	23.5	17.2	13.3	100

Table 6: Active population percent in the villages with and without NFB

Age groups (year)		Under 30 year	31-40	41-50	51-60	61-70	Upper of 70
Rural with non-farm business	Babariz	23.2	24.6	33.1	11.16	7.05	0.89
	Nawareh	28.9	28	32.1	6.8	3.4	0.8
	Arandan	18.3	39.7	26.4	12	3	0.6
	gholian	25.3	30.1	27.3	11.3	3.9	2.1
	Sateleh	16.2	32.5	30.4	12.6	7.7	0.6
Rural without non-farm business	Alijan	5.6	4.2	11.7	13	24.1	41.4
	Farhad abad	2.1	8.7	4.2	16.8	30.1	38.1
	Mamokh olia	1.8	12.2	14.5	12.8	25.1	33.6
	Khakvarzi	2.4	11.7	13.1	10.6	29.3	32.9
	Gazardare	2.3	10.4	7.2	12.3	24.9	42.9

more job satisfaction compared to villages lacking non-agricultural non-farm business. So, their motivation for permanency in rural regions is much higher.

The household's participation: Encouraging the motivation of the rural participation which is usually in the financial, non-financial, and advisory forms, was another outcome of establishing non-farm business in this study. The most important effect of household's participation is related to land sharing, providing water, electricity and other basic infrastructures. According to statistical analyzing of table 8, it can be noted that with increasing the income of villagers, the percent of the participation increases. According to the results the percent of financial participation of those villages with non-farm business is much higher in Sanandaj. Furthermore, there was a significant difference between villages with and without non cultural industries in regard to financial and non-financial participation. And due to the increase in income, 71.2% of rural families with non-farm business have had financial participation; while this figure in village without non-farm business is only 19.6%.

The use of mass media: TV, radio, newspaper, magazine and books are considered the most important mass media, that have an important role in improving information and awareness of rural people, especially farmers, and experts believe that these media are vary effect on their awareness and insight. Based on t-test in this study, there is a significant difference in the rate of using these mass media between villages with and without non agricultural business. According to table 9, the average use of mass media at village without and with non agricultural industries is respectively: radio (0.75 versus 3.56), newspaper (0.26 versus 2.03), and book (0.24 versus 1.25) that reveals clear differences.

Table 7: Comparing the villages with and without NFB in terms of job satisfaction

Grouping variable	Test variable	Test	U	Sig.
Rural with/without NFB	Job satisfaction	Mann-Whitney	621.23	0.000
Rural with/without NFB	Job stability	Mann-Whitney	212.41	0.01

Table 8: Percent of household's participation

Participation Type	Rural with non-farm business	Rural without Non-farm business
Financial	71.2%	19.6%
Non-financial	8.6%	61.5%
Advisory	20.2%	18.9%

Insurance services (treatment, retirement, etc): Following the law of construction estates and non-farm industries, the workers of these units are being covered by hygiene, treatment and social security insurances. According to the results, 68.54% of rural families with the non-farm business in the sample community were under insurance coverage whilst, 12.1% of rural families without non-farm business were under insurance coverage.

Quality and quantity of rural housing (dwell): Results about quality and quantity of rural housing revealed that 78.3% of buildings in villages with non-farm business has been constructed using high quality materials and engineering methods, while this figure in villages without non-farm business reaches 36.4%. This fact is completely true about villages in sample community, and in addition to the difference in the housing quality, this difference is meaningful regarding the area too. The results also showed 71.7% of rural housing with non-farm business was more than 100 m2, and this figure in villages without non-farm business was 32.4%.

CONCLUSIONS

Results revealed the villages with non-farm business have been in general younger than the other families without non-farm business. This shows a better situation of villages with non-farm business in terms of age that leads the younger doesn't immigrate to cities. In this vein, a few cases of villages with non-farm business had even inverse immigration from cities to the villages. Moreover, because of gaining more income by villages with non-farm business; they have been consuming better stuffs than others. This result can be more or less seen in the studies conducted by Barham and Boucher, 1998, Lanjouw, 2008, Escobal, 2001, Khan and Riskin, 2001, Leones and Feldman, 1998, Reardon and Taylor, 1996, Shand, 1987, Skinner, 2007, and Walkers, 2007.

Table 9: Comparing the use of mass media in the villages with and without NFB

Variable	groups	t	d.f	Sig.
Radio	A	2.69	122	0.004
	B	2.84	101.9	0.003
Television	A	2.14	122	0.041
	B	1.56	103.1	0.048
Journal and newspaper	A	5.14	122	0.000
	B	5.19	83.36	0.000
Book	A	3.87	122	0.002
	B	5.36	75.23	0.003

A= the villages without non-farm business, B= the villages with non-farm business

About literate level, results showed a higher level of the villages with non-farm business than those without any business in terms of the mean. A similar result was found about the number of insurance holders in comparing the two groups of villages. Results revealed the most houses of the villages with non-farm business had an area more than 100 quadrate meters and they were in good quality, while more than 50% of houses in villages without non-farm business were less than 100 quadrate meters and in old and demolished situation.

Based on the results and in order to improve non-farm business in rural areas it can be recommended that new local factories and cooperatives should be established by mobilizing small capitals and savings of rural people in order to improve rural non-farm economy.

Regard to this point that non-farm businesses have had a positive effect in improving the rural employment, income, and quality of living; it is strongly recommended that needed conditions, seed beds, and facilities are provided for entering rural people to non-farm activities by both governmental and community-based organizations.

Encourage and develop non-farm business infrastructure is another important thing that is necessary to improve socio-economic status of villages. Non-farm business and non-farm industries need to be basically invested by financial facilities through rural cooperatives and unions. Additionally, encouraging rural participation whether financially or non-financially is essential to develop non-farms and industrial activities. The rural participation in establishment and development rural networks and small enterprises can be a proper solution to improve non-farm business in the Sanandaj's villages.

REFERENCES

1-Adams, R. H. (1994). Non-Farm Income and Inequality in Rural Pakistan: A Decomposition Analysis, The Journal of Development Studies, 31(1): 110-133.

2- Adams, R. H. (1999). Non-farm Income, Inequality, and Land in Rural Egypt. Policy Research Working Paper No.2178.UN: The World Bank.

3- Adams, R. H., & He, J.J. (1995). Sources of Income, Inequality, and Poverty in Rural Pakistan, Washington D.C.; IFPRI Research.

4- Allen, F., Qian, J., & Qian, M. (2005). Law, Finance, and economic growth in China. Journal of Financial Economics, 77(1): 57–116.

5- Astrachan, J. H. (2010). Strategy in Family Business: Toward a Multidimensional Research Agenda. Journal of Family Business Strategy, 1(1): 6–14.

6- Athar, H., Lanjouw, P., & Stern, N. (1994). Income Inequalities in China: Evidence from Household Survey Data, World Development, 22(12): 1947-1957.

7- Barham, B., & Boucher, S. (1998). Migration, Remittances, and Inequality: Estimating the Effects of Migration Income Distribution, Journal of Development Economics, 55: 307-331.

8- Bhalla, A. S. (1990). Rural-Urban Disparities in India and China, World Development, 18(8): 1097-1110.

9- CREY (China Rural Enterprises Yearbook), (2001). Beijing: China Agriculture Press.

10- CSSB (China State Statistical Bureau), (2001). Fifty Years of City Development in New China. Beijing: Xinhua Publishing House.

11- DIIA (Division of Instructional Innovation and Assessment) 2007: Retrieved from: http://www.utexas.edu/academic/diia/assessment/iar

12- Dyer, W.G. (2003). The Family: The Missing Variable in Organizational Research. Entrepreneurship Theory and Practice, 27: 401–416.

13- Escobal, J. (2001). The Determinants of Non-Farm Income Diversification in Rural Peru, World Development, 29(3): 497-508.

14- Fisher, T., & Mahajan, V. (1998). The Forgotten Sector: Non-Farm Employment and Enterprises in Rural India. London: Intermediate Technology Publications.

15- Frank, H., Lueger, M., Nose, L., & Suchy, D. (2010). The concept of "Familiness" Literature reView and Systems Theory-Based Reflections. Journal of Family Business Strategy, 1(3):119-130.

16- Jha, B. (2006). Rural Non-Farm Employment in India, A coordinated and Consolidated Unpublished Report, Submitted to Ministry of Agriculture, GOI, New Delhi.

17- Kalantari, K., Shabanali Fami, H., Asadi, A., Qasemi, I., & Chubchian, S. (2008). Major Challenges of Iranian Rural Communities for Achieving Sustainable Development. American Journal of Agricultural and Biological Sciences, 3(4): 724-728.

18- Khan, A.R., & Riskin,C. (2001). Inequality and Poverty in China in the Age of Globalization. New York: Oxford University Press.

19- Lanjouw, P. (2008). Rural Non-Agricultural Employment and Poverty in Latin America. Washington DC: World Bank.

20- Leones, J. P., & Feldman, S. (1998). Non-Farm Activity and Rural Household Income: Evidence from Philippine Microdata. Economic Development and Cultural Change, 46(4): 789-806.

21- Miller, D., Lee, J., Chang, S., & Le Breton-Miller, I. (2009). Filling the Institutional Void: The Social Behavior and Performance of Family vs. Non-Family Technology Firms in Emerging Markets. Journal of International Business Studies, 40(5): 802-817.

22- Pears, R. (2007). Location, Development, Industry. Oxfords: Blackwell

23- Reardon, T., & Taylor, J. E. (1996). Agro-Climatic Shock, Income Inequality, and Poverty: Evidence from Burkina Faso, World Development, 24(5): 901-914.

24- Shand, R. T. (1987). Income Distribution in a Dynamic Rural Sector: Some Evidence from Malaysia. Economic Development and Cultural Change, 36: 35-50.

25- Skinner, N. M. (2007). Social and Environmental Regulation in Rural China. Geoforum. 34(2): 267-281

26- Stark, O., Taylor, J. E., & Yitzhaki, S. (1986).

Remittances and Inequality. Economic Journal, 96(383): 722-740.
27- Tetsushi, S., Hu, D., & Keijiro, O. (2002). Process of Cluster Formation in China: A Case Study of a Garment Town. The Journal of Development Studies, 39(1): 118–139
28- Walkers, V. (2007). Policy Innovation for Rural Sustainable Development. New York: USA Rural Policy Research Institute.
29- Zhu, L., & Jiang, Z. (1993). From Brigade to Vilplage Community: The Land Tenure System and Rural Development in China, Cambridge Journal of Economics, 17: 441-461.

6

Trade Liberalization, Exchange Rate Changes, and the Competitiveness of Carbohydrate Staple Markets in Nigeria

Maduabuchi Mkpado[1*] and C.J. Arene[2]

Abstract

This study examined the effects of trade liberalization and exchange rate changes on carbohydrate staples in Nigeria. Secondary data published from 1974 to 2006 on cassava exports and rice imports as well as non-tradable carbohydrate staples were used. It was hypothesized that exchange rate changes and trade liberalization via price relatives, trade intensity and nominal protection coefficient have affected prices of carbohydrate staples in Nigeria. Results show that the effects of trade liberalization on prices of non tradable carbohydrate staples were mixed. Trade liberalization accounted for most changes in the price of non tradable rice than other crops. World prices positively affected the prices of maize and non tradable rice. Also increasing exchange rate will lead to increase in price of non tradable rice due to price competition with imported rice. The intensification of liberalization exercise from the removal of quantitative restriction to use of tariff, among other recommendations, should be encouraged because it can be a remedy to the negative impact of increase in the trade intensity and erosion of nominal protection coefficient on prices of the non tradable crops.

Keywords:
Trade Liberalization, Exchange Rate, Carbohydrate Staple Markets, Nigeria

[1] Department of Agricultural Economics and Extension Federal University of Oye-Ekiti, Nigeria
[2] Department of Agricultural Economics University of Nigeria, Nsukka
** Corresponding author's email: manotejah341@yahoo.com*

INTRODUCTION

Carbohydrate foodstuffs, derived from cereals and tubers, are major parts of food consumed in Nigeria (Ogundele, and Okoruwa, 2006). Food insecurity in the country is often worsened by the ban on importation of rice and wheat. This is because Nigeria's population is growing rapidly at the rate of 3 percent per annum while the growth rate of food production is at one percent per annum (Achike, 2004). The liberalization of trade during the Structural Adjustment Programme (SAP) era, subsequent removal of some import prohibition and input subsidy and implementation of programmes to improve domestic food production for possible export have exposed small scale farmers, who are major crop producers, to more economic pressure in competing with large scale farmers elsewhere. Arene and Mkpado (2004) opined that this economic pressure is more severe when the people prefer imported goods to domestic ones. Changes in trade policy may have offered some form of protection to these local farmers. For instance, prior to the adoption of ECOWAS common trade tariffs, rice, which is the major imported carbohydrate staple, was heavily taxed such that about ₦5 billion from rice tax could be invested into agriculture during annual budgets (FGN, 2004). Trade liberalization has resulted in exportation of cassava - a carbohydrate crop from Nigeria since 1990 (Eurostat, 2000). Currently, Nigeria is the world largest producer of this crop for more than 7 years now, producing 37,504,100 tons in 2010 (FAOSTAT, 2005, FAOSTAT 2010).

In order to achieve the millennium development goals, Nigerian government has instituted the National Economic Empowerment Development Strategy (NEEDS), and more recently, a seven point agenda policy instrument. The objectives of the policy instruments with respect to agriculture include achieving food security, poverty eradication and creation of a responsive modern and globally competitive sector that will ensure higher income for farmers and rural dwellers. To achieve this laudable objective, the federal government of Nigeria has put in place a number of programmes and projects for increasing agricultural production such as the presidential initiative on cassava, fadama development projects, national special programme on food security, and National Export Promotion Council which encourages Nigerian exports and so on. There is need to assess the sustainability of such government interventions within the liberalization era by determining the effects of exchange rate and trade policy variables that indicate liberalization namely varying tariff rates, nominal protection coefficient and trade intensity on domestic pricing system as it affects target clients of the programmes and projects for national development and poverty reduction. The ability of small-scale carbohydrate food crop farmers to attain profit maximization goals or satisfy their subsistence food security needs is influenced by the prevailing policy environment especially in this era of globalizing economy.

Increasing agricultural output and exports are among the rationale for the introduction of SAP, liberalization policy and devaluation of Nigeria's currency. The effects of SAP and policies to cushion it have continued to affect macroeconomic variables especially the pricing system in Nigeria in relation to international trade. The prices of agricultural products influence their production and revenue of farmers due to cost of production, depreciation and abolition of agricultural marketing boards. The domestic price of agricultural commodities is essentially a function of exchange rate policy, trade liberalization and substitution of one crop for another especially with respect to crops that serve similar purposes such as those of the class of cereals and tubers, which can serve as sources of carbohydrates. Fluctuations in prices of agricultural output increase risks and uncertainties associated with the industry and possible investment.

Agricultural trade policies with respect to SAP can be divided into 3 phases namely pre SAP, during SAP and post SAP trade policies. The 1960s-1985 was the pre SAP era characterized with highly regulated exchange rate and quantitative restrictions. Import and export duties as well as controlled exchange rate were the major trade policy instruments prior to and many years after independence. Due to the suspension of dollar-gold convertible on 15 August 1971, Nigeria adopted a new system of exchange rates with effect from 23rd August 1971. This was classified into two namely contracts dominated by US dollars and contracts dominated by pound sterling. The Central Bank of Nigeria (CBN) initially maintained a fixed buying and selling rate for the naira which increased/improved with the oil boom). The system of trying to fix exchange rate dominated the pre SAP era; this policy tried to reduce the magnitude of changes of the exchange rate. However since the introduction of SAP, exchange rate changes have been more pronounced. This could have been affected by the devaluation of the naira. However, there has been gradual increase in exchange rate since the introduction of SAP from ₦2.02 at the end of 1986 to ₦127 dollar by the end of 2006, while during the pre SAP era it was less than one naira to the dollar.

The main explicit instruments of pre-SAP era policies were export duties, taxes and centralized marketing (Marketing Board). Export duties ranged from 5 and 60 percent from 1960-1970. But by 1973 export duties were abolished in order to revive agricultural export, which was affected by the "Dutch disease" as a result of oil boom. Import duties on food commodities such as maize, rice, wheat and sorghum were raised to between 50 and 100 percent from 1978 to 1982. Government subsidies on fertilizers and other agro-chemicals, improved seeds and capital equipment particularly tractors were about 50 percent for tractors and 85 percent for others (Oyejide, 1986). Quantitative restriction in the form of import and export bans placed on certain agricultural commodities as well as reinforced centralized marketing to improve government revenue characterized 1981 to early 1986. Table 1 describes trade experiences.

The SAP era (1986-94) and afterwards featured the devaluation of the naira (Table 1), abolition of import and export licensing requirement except for fertilizer and few other commodities between 1986 and 1988. Foreign exchange control system by C.B.N was abolished but replaced with Bureaux de change as authorized dealers. The marketing board was scraped in 1986. This was followed by further reduction in export duties and the removal of export prohibitions for many agricultural crops except food grains (cereals). The effect of customs and excise tariff consolidation decree of 1988 is extended list of banned imports by 1991 to about 20 percent of industrial and 30 percent of agricultural products (Soludo, 1995). Import duties were designed to discourage importation of non-essential raw materials as well as inputs and commodities that have local substitutes, especially in the agricultural sector. Import prohibition was thus shortened from 76 to 16 items (Falusi, 2005). This list favoured rice importation and some products of rice, maize and wheat (Ogunkola, 2003). Between 1989 and 1991 tariff rate rose from 100 to 300 percent on food stuffs, foot wears, transport equipment and chemicals (Soludo, 1995). Outright prohibition was on many food staples excluding rice (Falusi, 2005).

Nigerian government had de-emphasized the use of import prohibitions since 1995 (during post SAP policies) by replacing it with a new seven-year tariff reform with frequent adjustment and changes in the tariff structure. The high import duties in 1995 were reduced after 1999 (Falusi, 2005). But by 2004 the tariff rate averaged about 25 percent with some exceeding 100 percent. The introduction of ECOWAS common tariff has further reduced tariff structure from 100-25 percent (FGN, 2004). Nigeria maintained a 150 percent ceiling rate binding on all agricultural products. In general, recourse to

Table 1: Trends in exchange rate trade liberalization variables and tradable carbohydrate trade performance in Nigeria (1974 – 2006)

Year	Exchange Rate	Price of Cassava Chips In US$	Price of Rice in US $	Quantity of Rice Imported (In '000 Metric Tones)	Cassava Export (In Metric '000 Tones)	Nominal Protection Coefficient for Cassava	Nominal Protection Coefficient for Rice
1974-77	0.628	177.0	357.8	117.4	0.00	0.57	0.95
1978-81	0.589	145.0	404.8	559.6	0.00	1.11	1.32
1982-85	0.763	155.8	202.0	450.0	0.00	4.67	2.32
1986-89	4.491	151.5	265.5	305.0	0.00	7.60	2.83
1990-93	19.925	166.3	289.3	316.5	10.40	1.10	8.0
1994-97	81.100	145.3	307.8	423.6	200.50	0.90	2.02
1998-2001	95.783	97.3	231.8	990.7	18.00	1.50	1.51
2002-2006	128.27	82.6	227.0	1438.1	1338.30	1.32	1.19

Source: Calculated from ESCB/FAO (2000), The Guardian (2007), NBS Trade Summary

quantitative restriction on imports is on the decline though the ban on importation of maize, sorghum, millet, wheat flour, vegetables, plastic articles and all types of meat exist (Ogunkola and Bankole, 2005). On the other hand, government has been trying to expand export with establishment of NEPC. As mentioned earlier one of the food crops that benefited from this arrangement since 1990 is cassava.

Effects of trade policy and its liberalization on agriculture have attracted the attention of researchers. But the experience of carbohydrate food crops especially in Nigeria has been neglected. For instance Osuntogun, *et al.,* (1993) examined some strategic issues for promoting Nigeria's non-oil export while Nwosu (1992) made a general assessment of SAP on Nigerian agriculture. Ihimodu (1993) examined SAP in all context of enhancing agricultural development. Adubi and Okunmadewa (1999) examined the relationship between price and exchange rate volatility on export and import prices in the context of dynamic adjustment involving currency devaluation in Nigeria. Studies that have tried to fulfill this specific need include those by Chidebelu *et al.,* (1998), Philips (1996), Ayichi (1997), Okoli and Okoye (2005), Arene and Okafor (2000), Arene and Odusolu (1998), and Arene (2000), These studies have concentrated on effects of price and exchange rate on export of one or a few of the following crops: cocoa beans, groundnuts, cotton, palm produce and cassava. Only the study by Okoli and Okoye (2005) involved cassava. It appears that the experiences with food crops are neglected. Also studies on effects of exchange rate and its volatility or risk have concentrated on its impact on tradable crops. But since exchange rate changes could pass through to domestic prices, there is need to investigate its effect on the prices of non tradable crops.

Opinions of researchers still vary on the effects of trade policy on agriculture. To illustrate, Mwase (1998) noted that liberalization and privatization of agricultural marketing was a major 'U'-turn in Tanzania's cashew economy, which resulted in easier access to foreign exchange and inputs, privatization of input purchases, processing and export of cashews which enhanced competition, increased producer prices and prompt payment to farmers. Kidane (1999) noted that in Ethiopia farmers responded positively to devaluation via increase in Real Exchange Rate (RER) by diverting both human and material resources at their disposal to the production of coffee, which had continued to be the major source of foreign exchange. While in Ghana, the reverse was the case because Assuming-Brempong (1994) noted that exchange rate changes and trade policies, which tend to protect importable crops, have adversely affected the production of exportable crops by reducing incentives for production of exportable crops relative to non tradables.

This work is specifically aimed at providing empirical evidence of the effects of trade liberalization on prices of carbohydrate staples in Nigeria. The questions this study addressed that are relevant to government objective of increasing food production and ensuring food security in Nigeria are: Will Nigerian carbohydrate crop farmers still have price advantage in the presence of competition with imported foodstuffs serving similar purposes? Will there be a drastic shift and abandonment of the carbohydrate food crops enterprises with complete reliance on rice importation? What is the prospect of cassava export in relation to rice importation in improving the prices of non tradable carbohydrate staples in Nigerian? Specifically, what are the effects of trade liberalization variables on prices of carbohydrate staples in Nigeria? The extent to which trade liberalization policies have affected prices of carbohydrate staples alongside changes in per capita income (PCY) is still a grey area to economists and policy makers. The ability of small-scale carbohydrate food crop farmers to attain profit maximization goals or satisfy their subsistence food security needs is influenced by the prevailing policy environment especially in this era of globalizing economy. The need, therefore, to empirically examine the effects of trade liberalization and exchange rate changes on prices of crops and hence ascertain their implications on the livelihood of farmers in Nigeria cannot be over emphasized.

The broad objective of the study was to examine the effects of trade liberalization and exchange rate changes on prices of carbohydrate staples in Nigeria. It aims at identifying economic variables that are associated with trade liberalization and exchange rate changes that affect prices of non tradable carbohydrate staples. Specifically the study aims at:

1- examining the effects of trade intensity and exchange rate changes on prices of carbohydrate staples in Nigeria.

2- determining the impact of price of importable rice relative to price of exportable cassava on prices of carbohydrate staples in Nigeria.

Theoretical/Analytical framework, the Model and methodology

It is true that both farm gate prices and domestic prices are distorted by agricultural and trade policies. Trade policies under liberalization exercise affect farm gate prices more than agricultural policies because government policies will remove subsidies to farmers which will expose farmers to market force which trade policies influence (Tshibaka, 1986; Oyejide, 1986). With respect to Nominal Protection Coefficient, trade policy can be classified as protective or otherwise. Nominal Protection Coefficient measures the deviation of domestic price from boarder price. It is not necessarily tariff but a composite designation of government's interventions that can influence prices such as legislative prohibition, tariffs, exchange rate, and so on. When the value of Nominal Protection Coefficient (NPC) is one it implies liberalization and efficient use of resources in the subsector. The NPC can indicate liberalization or otherwise as a result of government intervention (see equation 19 for methods of its calculation).

Sectorial and sub sectorial agricultural trade policy have been characterized by a trend from quantitative restriction to use of tariff during the period under study. An index can be constructed with respect to liberalization exercise in the country. Its interpretation will be possible and more encompassing and informative in measuring episodes than use of dummy variable. The sub sector also witnessed the use of exchange rate and prices as trade policy variables that have played parts in the sub sector besides some level of trade protection.

With a small open economy producing three types of goods in the carbohydrate crop family namely exportable crops, importable crops and non tradable crops; foreign prices, nominal exchange rates, export subsidies/taxes and import duties determine the domestic nominal prices of exportable (P_x) and importable (P_y); while supply and demand factors (which are a function of trade and exchange rate policies) determine the domestic nominal prices of non tradable goods, P_z. The economy is characterized with low capital/human ratio. This paper adopted the use of price relative approach as employed by Oyejide (1986), Asuming-Brempong, (1994), Reinika (1994) and advocated by Collier *et al.,* (1999). One of the advantages of the use of price relatives in the context of Low of One Price (LOP) is the ability of the law to reflect the effects of exchange rate on prices if the world prices are expressed in domestic currency. But to reflect the effect of exchange rate, the study has carefully expressed the price of the tradable crops in their foreign currency. The basic regression model is presented as follows (the statistical suitability of inclusion of these variables namely NPC, OP, Ex and ID in the model is seen from the fact that the estimated results have tolerable Durbin Watson statistics and the coefficient of determinations is not very high):

In the context of our study, Cassel's law can be mathematically represented as follows:

$$P_x = P^*_x E_o (1 - t_x) \quad (1)$$

$$P_y = P^*_y E_o (1 + t_y) \quad (2)$$

Where: P_x = Price paid to the exportable crop producer in domestic currency (domestic currency influenced by exchange rate liberalization).

P^*_x = World price of exportable crop in foreign currency

E_o = Nominal/official exchange rate

t_x = Export tax (export tariff rate)

P_y = Price paid to importable crop producer in Naira (domestic currency).

P^*_y =World price of importable crop in Dollar (foreign currency)

t_y = Import tariff rate.

With a small open economy producing three types of goods in the carbohydrate crop family namely exportable crops, importable crops and non tradable crops; foreign prices, nominal exchange rates, export subsidies/taxes and import duties determine the domestic nominal prices of exportable (P_x) and importable (P_y); while supply and demand factors (which are a function of trade and exchange rate policies) determine the domestic nominal prices of non tradable goods, P_z. Establish a relative price structure from (1) and (2) by introducing farm gate prices of non tradable commodities (P_z):

$$P_x/P_z = \left(P^*_x/P_z\right) E_O (1 - t_x) \quad (3)$$

$$P_y/P_z = \left(P^*_y/P_z\right) E_O (1 + t_y) \quad (4)$$

Equations (3) and (4) indicate that the real exchange rate and tariff provide a measure of the relative price of importable and exportable to home goods in the Nigerian economy. Demonstrate that domestic prices of importable crops relative to exportable crops depend on world prices, trade regimes and tariff rate or other policy measures by dividing (4) with (3).

$$Py_Z/P_X = \left(P_Y^* / P_x^*\right) \frac{1+t_y}{1-t_x} \quad \cdots\cdots\cdots(5)$$

Equation (5) illustrates that price relatives of importable to exportable in domestic currency is a function of world prices and trade policy-tariffs. If the ratio $[P_y^*/P_x^*]$ is a constant and the exchange rate is fairly stable each year, which can lead to internal equilibrium, then the following conditions will hold

$Q_t = Q_x[P_x/P_z] + Q_z[P_y/P_z]$..(6)

Where $Q_x^! > 0$ and $Q_z^! < 0$

Q_T = Total output

Q_X = output of exportable

Q_z = output of importable

$C_z[P_y/P_z] = Q_z[P_x/P_z]$..(7)

Where C_Z = consumption of non tradable

In order to improve internal equilibrium, government may manipulate trade policy instrument (tariff) by introducing distortions/subsidy such that difference exist between t_x and t_y. If import duties are higher than export duties, the equilibrium price P_Z of non tradable will rise by an amount that is less than t_y but greater than t_x ($t_y > \acute{P}_Z > t_x$). The difference between t_y and t_x can be decomposed into $(t_y\text{-}d)$ and $(d\text{-}t_x)$ as:

$t_y - t_x = (t_y - d) + (d - t_x)$ .... (8)

It implies that producers in the import competing sector receive an implicit subsidy given by $(t_y\text{-}d)$ rather than dealing with nominal import tariff rate t_y, where as $(d\text{-}t_x)$ represents implicit tax (subsidy) associated with export; d, represents increase in the price of non tradable. The import and export duties will determine relative prices which will induce substitution to give $\acute{P}_Z$. Since P_Z existed at t_x, and change in $\acute{P}_Z$ is as a result of increase in t_y, then d is represented as

$d = t_x + w\,(t_y - t_x) = wt_y + t_x\,(1-w)\,t_x$..................(9)

Where w is a parameter measuring substitution with respect to prices, which will be discussed later.

Recall that from (7), demand/consumption of non tradable equals its output. Q_Z and C_Z can also be defined as:

$Q_z = q_z\,[P_y/P_z, P_x/P_z, K, L, T]$..................(10)

$C_z = C_z\,[P_y/P_z, P_x/P_z, Y]$..................................(11)

Where: K = Capital; L = Labour; T = Technology; Y = income or GDP

K, L, and T represent the productive capacity of the economy, while Y pays for what is produced. They can be held together as constant to examine the comparatively static properties of the model where the primary interest is the movement of relative price. Thus, after an initial displacement, the system achieves a new equilibrium where:

$$\hat{Q}_Z = \hat{C}_Z = (\eta_Y - E_Y)\left(\hat{P}_Y / \hat{P}_Z\right) + (\eta_X - E_X)\left(\hat{P}_X / \hat{P}_Z\right) = 0 \quad (12)$$

Where: η_y and η_x = demand elasticity for non tradable with respect to the prices of importable and exportable E_y, and E_x the corresponding supply elasticity (supply elasticity of non tradable) with respect to the price of importable and exportable, respectively.

^ = Proportion change

Equation (12) can be simplified as:

$\psi_y(\acute{P}_Z - \acute{P}_y) + \psi_X(\acute{P}_Z - \acute{P}_X) = 0$..................(13)

Where $\psi_y = \eta_y - E_y$; $\psi_x = \eta_x - E_x$

But change in domestic consumption includes: $\psi_y(\acute{P}_y - \acute{P}_X)$, because total consumption C_T in the economy is given as:

$C_T = C_Y[P_Y/P_Z] + C_Z[P_Y/P_Z] + C_X[P_X/P_Z]$.......(14)

Where C_Y = Consumption of importable

C_Z = Consumption of non tradable

C_X = Consumption of exportable

The C_T is affected by demand and supply elasticities as a result of substitution among the carbohydrate staples, such that:

So that by rearranging and simplifying (15), it gives

$$\psi_y\left(\hat{P}_z - \hat{P}_Y\right) + \psi_X\left(\hat{P}_Z - \hat{P}_X\right) + \psi_Y\left(\hat{P}_Y - \hat{P}_X\right) = 0 \quad (15)$$

So that by rearranging and simplifying (15), it gives (16)

$$(P_Z - P_X) = w\,(P_Y - P_X) \quad (16)$$

Where $W = \Psi_Y/\Psi_Y + \Psi_X$ (with $0 \leq w \leq 1$) is the parameter representing substitution coefficient referred to in equation (9). Equation (16) has captured the effects of possibility of substitution on price competition. Equation (16) can be rewritten as:

$$D\,Ln\,[P_Z/P_X] = W\,D\,Ln\,[P_Y/P_X] \quad (17)$$

Where: D represent the derivative of the natural logarithm of the variables in bracket. Integrating equation (17) and assuming that w is a constant gives.

$$Ln[P_Z/P_X] = A_o + W ln\,[P_Y/P_X] + e \quad (18)$$

Where: A_o = constant; e= error term

Equation (18) captured the substitution effect as a result of price competition between importable commodity and non-tradable in the presence of exportable crop. Per capital income (PCY) has to be introduced to capture some income effects on prices. Other variable that affect prices are exchange rate, trade intensity, index of the liberalization exercise and the Nominal Protection Coefficient (NPC). The Nominal Protection Coefficient (NPC) may denote the interaction between tariffs and exchange rate as it measures the deviation of domestic wholesale price from world market price. NPC is presented as:

$$NPC_j = P^d_j/P^b_j \quad (19)$$

where: P^d_j = The domestic price commodity *j*
P^b_j= The border price of commodity *j* expressed in domestic currency.

When these variables are taken into account the estimable equation becomes:

$$Ln\,P_z/P_X = a_0 + b_0 \ln P^*_Y/P^*_X + C_0 \ln \frac{NPC_y}{NPC_x} + D_0 \ln OP + E_o \ln Ex + F_o \ln ID + G_o \ln PCY + e \quad (20)$$

Where: W= b_0 and C_0
P^*_X= World price of exportable crop in foreign currency
P^*_Y= World price of importable crop in foreign currency
PCY = Per capita income measured in dollar
NPC_y = Nominal protection coefficient of local rice
NPC_x = Nominal protection coefficient of exportable cassava
P_z = Farm gate prices of non-tradable expressed in dollar.
P^d_j = The domestic price of commodity, *j*
P^b_j = The border price of commodity, *j*, expressed in domestic currency.
Op = trade intensity (rice import + cassava export)/ GDP of agricultural crop sub sector
Ex = exchange rate
ID[1] = sum of liberalization index with respect to sub sector (effective ban without documentation of illegal trade = 0, quantitative restriction or ban with documentation of illegal trade =1, use of tariffs = 2. This variable is important given that the use of tariff has been at ad valeom)

The estimable equation (20) was applied to 5 different non tradable crops namely maize, millet, sorghum, local rice and yam, respectively; their prices were presented as P_{Z1}, P_{Z2}, P_{Z3}, P_{Z4}, and P_{Z5}, respectively. The variables are all logarithmic transformation. PC-GIVE statistical software was used in the analysis. The analysis began with examination of time series properties of the variables used in the study. The time series properties were investigated and their order of integration determined using the Augmented Dickey Fuller (ADF) unit root test. The ADF regression takes the form:

$$DX_T = \beta_0 + \beta_1 X_{T-1} + \beta_i \sum DX_{T-i} + \ell \quad ...(21)$$

D = difference operator/ difference term
X= individual variable at a time T
β = coefficient; = error term

[1] *Specifically with respect to rice importation Ogundele and Okoruwa 2006 identified the policy environment in the sub sector. They outlined pre ban, ban and post ban periods with respect to rice. The pre ban period was from 1971-1985, while the ban period was from 1986-1995. However, they maintained that illegal trade made the ban to be quantitative restriction instrument. Available data and literature support cross border trade during that period (FAO 2000). But from 1995-2007, Nigerian government has used tariff as main liberalization instrument. On the other hand cassava moved from limited tradable to tradable crop. From 1974 -1989, it was a limited tradable crop while from 1990 till date it is a tradable crop. One can easily identify a trend from quantitative restriction to use of tariff during the period under study. This is used as index in this study.*

The null hypothesis is that $\beta 1 = 0$. Rejection of the null hypothesis suggests that the series is non-stationary; and has to be differenced at least once in order to make it stationary. The ADF test was performed up to 2-lag length. Next was the ADF and Johansen test for co integration relation (Johansen, 1992). Where co integration exists, error correction model (ECM) is estimated. If not, the analysis continues without the ECM mechanism

The values of the regression coefficients have to indicate nature of the effects of trade liberalization variables (measured as price relatives of importable to exportable, and nominal protection coefficients) and per capita income on prices of non tradable relative to price of cassava, which were the aim of the study.

3.1 Coverage: The study covered a period of 32 years from 1974 to 2006. Data collected for the period included price of imported rice, price of exported cassava, and the prices of the following non tradable crops: maize, millet, sorghum, local rice and yam.

3.2 Data Collection: Secondary data were obtained from International Center for Trade and Statistics Data Base (COMTRADE), UNCTAD's Trade Analysis and Information System (TRAINS),. This was used to guide the authors with respect to HS-Code. The FAO Statistics Data Base, Publications of Central Bank of Nigeria, Nigerian Ports Authority, National Bureau of Statistic, Nigerian Export Promotion Council, and The Guardian were used to trace information on trade and marketing of carbohydrate staples. These helped to provide information on HS-6 tariff line. Data such as domestic and border prices of the following crops: maize, millet, sorghum, local rice, foreign\imported rice, cassava, yam and cocoyam were collected. Data on trade regime policies such as tariff structures, cost of insurance and freight, and in cases where free on board were applied, were also sourced.

Data analysis

Normality test

Normality test was carried out to determine the reliability of the emerging result. The normality test shows that all the variables had normal distribution. This is because the Chi-square values are significant at 5% probability level. The result was presented in table 2.

Time series property

The unit root tests are presented in Tables 3. They showed the order of stationary of the variables for different periods in years that the variables were used for the analysis. Since the variables are of the order of integration, it requires text for establishment of proper criteria for inclusion of error correction mechanism in the model (Table 3).

Determination of appropriate condition for inclusion of error correction mechanism

Table 4 showed the variables whose Augmented Engle- Granger Test supported co-integration.

All the residual variables for the period 1974-2006 were stationery and their ADF supported co integration. Thus, appropriate conditions for co integration relationship between dependent variables and corresponding independent variables has been ascertained for estimation of error correction model where necessary (Johansen, 1992)

Descriptive statistics result

An examination of the price changes in line the major economic policy changes which can be grouped under pre SAP, during SAP and post SAP is presented graphically in figure i.

The pre SAP era witnessed the introduction of the marketing boards. From 1974 to 1985 (before the introduction of SAP in 1986), there were minimal price changes in all carbohydrates staples in Nigeria. From 1986 to 1994 –during the SAP era, the prices increased gradually, with the price of local rice above others. The immediate post SAP era (from 1995- 1999) witnessed greatest increase in prices of the commodities, with the price of millet exceeding others. This period is followed by a gradual decrease in the prices of millet, sorghum and yam while the prices of maize and cassava increased slightly.

Estimated results

Effects of trade liberalization and exchange rate changes on price of non-tradable maize

The regression equation for estimation of the effects of trade liberalization and exchange rate changes on prices of non-tradable maize in Nigeria is presented in table 5. The F-ratio, R^2 and Durbin Watson are 3.066, 0.48 and 1.85 respectively. These show that the result is acceptable because these statistics are within the acceptable limits. The significant explanatory variables are price of imported rice relative to price of exportable cassava and nominal protection coefficient of non-tradable rice relative to that of exportable cassava.

Table 2: Results of normality test of the data

Variables	Chi-square Value	Level of Significant
P_Y^*/P_X^*	8.0569	0.0001**
NPC_y/NPC_x	6.0981	0.0476*
Open	39.734	0.000**
P_{Z1}/P_{XT}	50.343	0.0000**
P_{Z2}/P_{XT}	29.820	0.0000**
P_{Z3}/P_{XT}	21.481	0.0000**
P_{Z4T}/P_{XT}	6.2544	0.0438*
P_{Z5T}/P_{XT}	18.013	0.0001**
Exchange rate	21.413	0.000**
Index	43.465	0.000**
PCY$	23.813	0.0000**

Source: Computed From Field Data; *, ** = Sig. at 5% and 1 % levels respectively.

Table 3: ADF unit root test

Variables	At data level	At 1st difference
DL P_Y^*/P_X^*	-6.3897**	-4.7878**
NPC_y/NPC_x	-5.7630**	-4.6715**
Open	-3.4946**	-2.8515**
P_{Z1}/P_{XT}	-6.7561**	-3.7733**
P_{Z2}/P_{XT}	-6.0706**	-3.7719**
P_{Z3}/P_{XT}	-7.2310**	-5.5748**
P_{Z4T}/P_{XT}	-5.8814**	-4.6154**
P_{Z5T}/P_{XT}	-6.6245**	-4.8998**
DL Open	-6.4930**	-5.4823**
DD Exchange Rate	-3.6056**	-2.8868**
DL Index	-2.8827**	-2.8581**
PCY$		

Critical values: 5% = -1.954; 1% = -2.649; ** P<0.01
Source: calculated from data

Price of imported rice relative to price of exportable cassava (P_Y^*/P_X). Has a positive effect on the price of non-tradable maize. A similar result was obtained in Ghana by Asuming-Brempong (1994). He reported that the domestic price of sorghum relative to maize will increase by 52% as a result of 1% rise in domestic price of rice relative to maize (both of which are tradable cereals) (Table 5).

The nominal protection coefficient of non tradable rice relative to that of tradable rice (NPCY/NPCX) had a negative impact on the price of non tradable maize. Increase in the nominal protection of non tradable rice will tend to reflect on increase in its price but such price deferential will be at the detriment of price of non tradable maize. It implies that people would tend to prefer the consumption of non tradable rice to maize given price competition. This could be because local rice is a closer substitute to imported rice than maize.

The significant error correction mechanism shows that the speed of price adjustment in the long run is 62%. It implies that maize has a fairly high speed of adjustment to trade policies.

Effects of trade liberalization and exchange rate changes on prices of non-tradable millet

The regression equation for estimation of the effects of trade liberalization and exchange rate changes on price of non tradable millet in Nigeria is presented in table 6. The result is acceptable because the F-ratio, R^2 and Durbin Watson are respectively, 2.4598, 0.43 and 1.61 which are within the acceptable limits. The significant explanatory variables are degree of openness and trade liberalization index (Table 6).

The degree of openness is negatively related to the price of non tradable millet. It indicates that as trade in the sector increases, substitution of imported millet with imported rice adversely affects the price of millet. It implies that peoples' consumption preference shifts from millet, thus, a reduction in the price of millet.

Trade liberalization index which is used to designate a shift from quantitative restriction to use of tariffs had a positive impact on price of non tradable millet. As tax (tariff) on a commodity tends to push its price upwards, millet can benefit from such price increase.

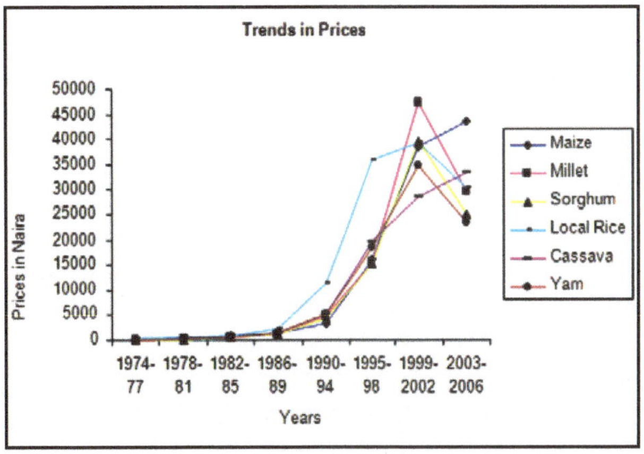

Figure 1: Trends in prices of carbohydrates staples in Nigeria; Source: Calculated from data

The significant error correction mechanism shows that the speed of price ad*justment in the long run is 42%. This means that the adjustment of price of millet to trade polices is relatively slow.

Table 4: ADF unit root test for determination of variables whose ADF values support cointegration test (Augmented Engle- Granger Test for Co integration)

Variables	At data level	At 1st difference
$\text{Ln } P_{Z1}/P_{XT} \text{ Ln } P_Y^*/P_X^* \text{ Ln } \frac{NPC_y}{NPC_x}$; ln Open; ln Index ln *PCY*	-3.4946**	-2.8515**
$\text{Ln } P_{Z2}/P_{XT} \text{ Ln } P_Y^*/P_X^* \text{ Ln } \frac{NPC_y}{NPC_x}$; ln Open; ln Index ln *PCY*	-2.5231*	-2.5589*
$\text{Ln } P_{Z3}/P_{XT} \text{ Ln } P_Y^*/P_X^* \text{ Ln } \frac{NPC_y}{NPC_x}$; ln Open; ln Index ln *PCY*	-2.7194**	-2.5462*
$\text{Ln } P_{Z4T}/P_{XT} \text{ Ln } P_Y^*/P_X^* \text{ Ln } \frac{NPC_y}{NPC_x}$; ln Open; ln Index ln *PCY*	-4.4439**	-3.4111**
$\text{Ln } P_{Z5T}/P_{XT} \text{ Ln } P_Y^*/P_X^* \text{ Ln } \frac{NPC_y}{NPC_x}$; ln Open; ln Index ln *PCY*	-3.0174**	-2.3947*

Critical Values: 5% = -1.954; 1% = -2. 649; *P<0.05, **P<0.01 Source: Calculated from data

Effects of trade liberalization and exchange rate changes on price of non-tradable sorghum

The regression equation for estimation of the effects of trade liberalization and exchange rate changes on price of non-tradable sorghum in Nigeria is presented in table 7.

The result is acceptable because the F-ratio, R^2 and Durbin Watson are 3.473, 0.51 and 1.77 respectively which are within the acceptable limits. The significant explanatory variables are degree of openness and trade liberalization index. This result is similar to what was obtained in the case of millet; hence similar explanation may hold for sorghum. The significant error correction mechanism shows that the speed of price adjustment in the long run is 36%. This means that the adjustment of price of sorghum to trade policies is relatively slow. It may be informative to note that the cultivation and consumption of millet and sorghum in Nigeria do not cut across all cultures

Effects of trade liberalization and exchange rate changes on prices of non tradable rice

The regression equation for estimation of the effects of trade liberalization and exchange rate changes on price of non tradable rice in Nigeria is presented in table 8. The result in table 8 is acceptable because the F-ratio, R^2 and Durbin Watson are respectively, 5.24, 0.62 and 1.65 which are within the acceptable limits (Table 8).

The significant explanatory variables are price of imported rice relative to price of exportable cassava (P_Y^*/P_X), and nominal protection coefficient of non tradable rice relative to that of exportable cassava (NPCY/NPCX), exchange rate (EX), and trade liberalization index (Index).

Price of imported rice relative to price of exportable cassava (P_Y^*/P_X), has a direct effect on the price of non tradable rice. Asuming-Brempong (1994) also had a similar report. He noted that the domestic price of millet relative to maize will increase by 0.80% as a result of 1% rise in the domestic price of imported rice relative to maize.

The nominal protection coefficient of non-tradable rice relative to that of tradable rice (NPCY/NPCX) also had a positive effect on the price of non tradable rice. Increase in the nominal protection of non-tradable rice will tend to reflect on increase in its price and decrease in the relative price of exportable cassava. This is in line with Oyejide's (1986) report that much

Table 5: Effects of trade liberalization and exchange rate changes on price of non-tradable maize

Independent variable MAIZE	Coefficients/ Marginal effects and t- values of the independent variables			
	Dln P_{Z1T}/P_{XT}	Stand. Error	t-value	t-probability
D Ln P_Y^*/P_X^*	0.51657	0.26o76	1.921	0.0465
D Ln $\frac{NPC_y}{NPC_x}$	-0.49851	0.28661	1.739	0.0953
DLn Open	-0.10713	0.13497	0.794	0.4355
DDInEX	-0.0693	0.47744	0.020	0.9846
Dln Index	0.16156	0.48107	0.336	0.7400
D Ln *PCY*	0.020086	0.11131	0.180	0.8584
Ecm-1	-0.61877	0.18587	3.290	0.0029
Constant	0.071012	0.11101	0.640	0.5287
F-ratio	3.066			0.0197
R^2	0.48			
Durbin Watson	`1.85			
A0	0.5447			
RSS	6.825			

Calculated from data: 1974-2006; D= Deference Operator; Ln =Log

of the protection for import competing activities has been at the expense of the exportable sector.

Exchange rate (EX) has a positive relationship with the price of non-tradable rice. This crop (non tradable rice) is the closest substitute to tradable rice. Thus it is expected that exchange rate pass through will be fully expressed in this case. This could be through prices of imported intermediate goods, which are reflected by the share of imports and through prices of domesti-

Table 6: Effects of trade liberalization and exchange rate changes on price of non-tradable millet

Independent variable MILLET	Coefficients/ Marginal effects and t- values of the independent variables			
	Dln P_{Z2T}/P_{XT}	Stand. Error	t-value	t-probability
D Ln P_Y^*/P_X^*	0.16562	0.44958	0.368	0.7160
D Ln $\frac{NPC_y}{NPC_x}$	-0.11464	0.22738	0.5042	0.6189
DLn Open	-0.045178	0.02848	1.816	0.0809
DDInEX	-0.52756	0.38863	0.136	0.8932
Dln Index	0.63130	0.38516	1.639	0.1148
D Ln *PCY*	0.072657	0.08845	0.821	0.4198
Ecm-1	-0.42151	0.17443	2.496	0.0240
Constant	0.06310	0.088379	06.4	0.4948
F-ratio	2.4598			0.0487
R^2	0.43			
Durbin Watson	1.61			
A0	0.4103			
RSS	3.7039			

Calculated from data: 1974-2006 D= Deference Operator; Ln =Log NS=Not Significant.

Table 7: Effects of trade liberalization and exchange rate changes on prices of non-tradable sorghum

Independent variable sorghum	Coefficients/ Marginal effects and t- values of the independent variables			
	Dln P_{Z3T}/P_{XT}	Stand. Error	t-value	t-probability
D Ln P_Y^*/P_X^*	0.161725	0.41881	0.380	0.7029
D Ln $\frac{NPC_y}{NPC_x}$	-0.05879	0.2171	0.270	0.7827
DLn Open	-0.02066	0.01064	2.021	0.0019
DDlnEX	-0.38998	0.36945	1.056	0.3021
Dln Index	0.69699	0.35581	1.959	0.0624
D Ln *PCY*	0.09367	0.082124	1.173	0.2526
Ecm-1	-0.35509	0.17234	3.546	0.0508
Constant	0.06603	0.081838	0.046	0.4280
F-ratio	3.473			0.0110
R^2	0.51			
Durbin Watson	1.77			
A0	0.40131			
RSS	3.7039			

Calculated from data: 1974-2006; D= Deference Operator; Ln =Log, NS = Not significant

cally produced goods whose resources for their production competed with the exportable. Kwanashie *et al.,* (1998); Adubi and Okunmadewa, (1999) and Kidane, (1999) as well as Okoli and Okoye, (2005) have noted that exchange rate was a major determinant of exports. This study upholds the concept of exchange rate pass through.

Trade liberalization index (Index) which is used to designate a shift from quantitative restriction to use of tariffs, had a positive effect on price of non tradable rice. As tax (tariff) on a commodity tends to push its price upwards, so non tradable rice can benefit from such price increase.

The significant error correction mechanism shows that the speed of price adjustment in the long run is 82%. This means that the adjustment of price of non tradable rice to trade policies is very fast, hence it is the closest substitute to tradable rice

Effects of trade liberalization and exchange rate changes on price of non-tradable yam

The result is acceptable because the F-ratio, R^2 and Durbin Watson are respectively 3.5, 0.52 and 1.57 which are within the acceptable limits. The significant explanatory variables are degree of openness and trade liberalization index. This result is similar to what was obtained in the case of millet and sorghum; hence similar explanation may hold for yam as shown in table 9.

Increase in the nominal protection of imported rice over exportable cassava negatively affected the price relative of non tradable yam It implies that increase in the nominal protection coefficient of imported rice leads to a decrease in the price incentive for production of non tradable yam; and an increase in the price incentive for production of exportable cassava. The result illustrates that the nominal protection coefficient, is eroded by trade liberalization exercise except in non tradable rice and cassava, which has taken advantage of world price. This has indicated a shift in resource allocation to favour cassava production and export as well as non tradable rice. The result is in consonance with the result of the study by Asuming-Brempong (1994), which shows that the effects of trade liberalization in Ghana were mixed because some crops were favoured while some were not. Increase in the protection of importable crop tends to increase the relative price of non tradable, which would tend to decrease the price incentive for production of exportable. Increase in the trade intensity has negative impact on the prices of non–tradable yam. This indicates that as the volume of imported substitute (rice) increases people will prefer its consumption to that of the mentioned non-tradable yam (see table 9).

Table 8: Effects of trade liberalization and exchange rate changes on price of non-tradable rice

Independent variable LOCAL RICE	**Coefficients/ Marginal effects and t- values of the independent variables**			
	Dln P_{Z4T}/P_{XT}	**Stand. Error**	**t-value**	**t-probability**
D Ln P_Y^*/P_X^*	0.62435	0.2743	2.277	0.0324
D Ln $\frac{NPC_y}{NPC_x}$	0.17693	0.10352	1.733	0.0901
DLn Open	0.018859	0.65510	0.288	0.7760
DDLnEX	0.14450	0.06210	1.851	0.0867
DLn Index	0.23774	0.11851	1.885	0.0490
D Ln *PCY*	-0.00228	0.053589	0.048	0.9664
Ecm-1	-0.81855	0.23081	3.546	0.0017
Constant	0.00245	0.05381	0.046	0.9640
F-ratio	5.2401			0.0011
R^2	0.62			
Durbin Watson	1.65			
A0	0.2623			
RSS	1.5826			

Calculated from data: 1974-2006; D= Deference Operator; Ln = Log

The results do not reveal much on the consumption preference of the populace. It could be that income of the majority has not been able to allow them make substantial substitution given the poverty level of the country; besides these food crops are staples and it is expected that their demand will be income inelastic as it is with many agricultural crops.

The significant error correction mechanism shows that the speed of price adjustment in the long run is 62%. This means that the adjustment of price of non-tradable rice to trade polices is fast, and hence it is affected by trade policy.

Summary and conclusion

The results illustrated the Nigerian experience of trade liberalization in the carbohydrate sub sector. The findings are in line with existing theory that effects of trade liberalization on prices of non tradable is ambiguous in the short run. Nigeria's Structural Adjustment Programme (SAP) has focused on liberalization of commodity markets in all sectors of the economy. Particularly in the liberalization and growth phase of adjustment process which started in 1986, flexible exchange rates for the naira were affected, and administrative prices for major tradable crops were abolished. The macroeconomic environment generated under the SAP and its effect on small-holder farmers who produce the bulk of Nigeria's agricultural output have direct implications for resource allocation and aggregate agricultural output. This study has provided empirical evidence of the policy and price linkages in the carbohydrate sub-sector of Nigeria's economy. More specifically, the effects of trade liberalization and exchange rate changes on prices of carbohydrate staples in Nigeria have been examined.

Effects of trade intensity and exchange rate changes on prices of carbohydrate staples and impact of price and other of economic variables associated with trade liberalization on prices of non tradable carbohydrate staples were identified through the examination of the effects of trade intensity and exchange rate changes on prices of carbohydrate staples, and determination of the impact of price of importable rice relative to the price of exportable cassava on prices of carbohydrate staples. Using secondary data, the study shows that trade liberalization and exchange rate changes, together with other trade policies e.g. nominal protection co-efficient, tariffs, and price relative have had substantial impact on prices and by implication on resource allocation among carbohydrate staples in Nigeria.

There was absolute increase in output from 1986-1989 to 2002-2006. This is indicative of the impact of trade liberalization policy which characterize the periods from 1986 to 2006. In terms of pricing, there was a gradual increase in absolute prices and the per capital income which are similar to increases in exchange rate changes during the period. This is indicative of relative decreases in prices and per capital income due to devaluation of the naira, increased production cost, and abolition of marketing boards which encouraged export drive (increased output), but discouraged absolute increases in output and income as a result of high cost of marketed (or purchased) inputs e.g. fertilizers.

In terms of price relatives the effects of trade liberalization and exchange rate changes on prices of carbohydrate staples in Nigeria have been mixed. Trade liberalization accounted for most of the changes in the prices of non tradable maize and local rice but not a determinant of price of non tradable yam, increase in the nominal protection co-efficient for rice over exportable cassava negatively affected the price relative for non tradable maize and yam. It is also a positive determinant of the price of non tradable rice. Increase in trade intensity has a negative impact on prices of non tradable millet, sorghum and yam. There may not have been much revelation on the consumption preference of the populace as indicated by the insignificant effect of consumer price index.

It is, therefore, concluded that maize and local rice farmers can increase prices of their products with increase in price of imported rice while yam farmers may not take such an advantage. The implication of increase in nominal protection coefficient for imported rice is a decrease in the price incentive for the production of non tradable maize and yam and an increase in price incentive for the production of exportable. Protection is eroded by trade liberalization exercise except in non tradable rice and cassava which have taken the advantage of world prices. A panacea for falling prices of non tradable millet, sorghum and yam as a result of increased trade intensity can be through moving from quantitative restriction to use of tariffs indicated by the trade liberalization index which had a positive effect on prices of the stated non tradable crops. Low per capita income of consumers has not permitted substantial demand; besides, these food crops are staples and it is expected that their demand is income inelastic as it is with many food crops.

Based on the findings from the study, it is recommended that:

1- Government initiatives on cassava and rice production should be extended to the studied non-tradable carbohydrates staples in order to increase their output and processing for possible

Table 9: Effects of trade liberalization and exchange rate changes on prices of non-tradable yam

Independent variable YAMS	Coefficients/ Marginal effects and t- values of the independent variables			
	Dln P_{Z5T}/P_{XT}	Stand. Error	t-value	t-probability
D Ln P_Y^*/P_X^*	0.080035	0.36831	0.217	0.8299
D Ln $\frac{NPC_y}{NPC_x}$	-0.3835	0.18544	2.068	0.0500
DLn Open	-0.01753	0.008470	1.980	0.0846
DDLnEX	-0.13934	0.33151	0.420	0.6782
DLn Index	0.24061	0.30874	0.779	0.4437
D Ln *PCY*	0.04711	0.71729	0.568	0.5758
Ecm-1	-0.61622	0.21258	2.899	0.0081
Constant	0.3194	0.072196	0.442	0.6623
F-ratio	3.55			0.0098
R^2	0.52			
Durbin Watson	1.57			
A0	.35125			
RSS	2.8379			

Calculated from data: 1974-2006; D= Deference Operator; Ln =Log, NS = Not significant.

exportation so that such crops can take advantage of international prices.

2- Liberalization exercise should be intensified through the relaxation of quantitative restriction to use of tariff in order to correct the negative impact of increased trade intensity and erosion of nominal protection on prices of the non tradable crops.

ACKNOWLEDGEMENTS

The authors are very much obliged to African Economics Research Consortium (AERC) for financial, material and technical assistance. They also acknowledge with thanks the useful suggestions and comments from Group D resource persons; and are grateful to the researchers in the same group for their contributions to the improvement of the study that led to development of this paper.

REFERENCES

1- Achike, A.I. (2004). Macro-economic Picture of Nigerian Agriculture: Strategic Analysis of Development Constraints and Priorities in Southern Nigeria. Updates on Agriculture and conflict July, USAID Mission Abuja Pp. 3-14

2- Achike, A.I., Agu, C., & Oduh, M. (2005). Impact of Common External Tariff and Economic Partnership Agreements on Agriculture in Nigeria. Technical Report submitted to OXFAM GB.

3- Adubi, A.A., & Okunmadewa, F. (1999). Price, Exchange Rate Volatility and Nigeria's Agricultural Trade Flows. A Dynamic Analysis. *AERC Research* Paper 87, Nairobi, Kenya.

4- Arene, C.J. (2000). Exchange Rate Policies and the Production of Major Cash Crops in Nigeria. Modeling, Measurement and Control, 22 (3): 25-36.

5- Arene, C.J., & Okafor, A.O. (2000). Exchange Rate Liberalization as a Policy Instrument for Stimulating Non-Traditional Agricultural Exports in Nigeria Economic Affairs. 46(4): 234-247.

6- Arene, C.J., & Odusolu, G.O. (1998). Exchange Rate Reform as a Policy Instrument for Stimulating Smallholder Agricultural Investment Demand and Output Supply in Nigeria Economic Affairs. 43(4): 232-240

7- Arene, C.J., & Mkpado, M. (2004). Counter Urbanization and Agricultural Production in Nigeria Journal of Rural Development. 23(1): 73-81.

8- Asuming-Brempong, S. (1994). Effects of Exchange Rate Liberalization and Input Subsidy Removal on the Competitiveness of Cereals in Ghana Issues in African Rural Development. 2, Pp. 43-59.

9- Ayichi, D. (1997). An Econometric Study of Export Supply Responses of Major Nigerian Agricultural Commodities to Price and Exchange Rate Regimes Ph.D. Thesis, Department of Agricultural Economics, University of Nigeria, Nsukka.

10- Chidebelu, S.A.N.D., Ayichi, D., & Okoye, C.U. (1998). The Impact of Price and Exchange Rate Fluctuations on Volume of Agricultural Export Commodities. The Case of Natural Rubber, 1970-1990. Ibadan. NISER/SSCN National Research in Nigeria.

11- Collier, P., Greenaway, D., & Gunning, J.W. (1999). Trade liberalization: A Methodological Framework in Oyejide A. *et al.,* (eds) Regional integration and trade liberalization in Sub Saharan Africa Vol. 1 Macmillan Press Ltd London

12-ESCB (2000). International Trade in Cassava Products An African Perspective Basic Food Stuffs Service (ESCB) of the F.A.O. Commodity Trade Division Report September

13- Falusi, A.O. (2005). Macro-Economic Reforms and Agricultural Development in Nigeria: A Preliminary Assessment. Keynote Paper of 19th Annual Farm Management Association of Nigeria Conference Proceedings Pp. 2-16.

14- FAO (2004). Food and Agriculture Organization Year Book.

15- FAO (2005). Production Year Book FAO Rome

16- FAO (2006). Roots. Tubers, Plantains and Bananas in human Nutrition 2005-2006. FAO\corporate Document Repository Retrived from http://www.ajol.info/index.php/ajb/article/viewFile/60344/48581..

17- FAO (2006). Commodity Market Review 2005- 2006. FAO\Corporate Document Repository Retrived from http://www.fao.org/docrep/008/a0334e/a0334e00.htm

18- FAOSTAT (2005). Output of cassava. Retrived from http://www.faostat.org/stats

19- FAOSTAT (2010) Output of cassava. Retrieved from http://www.faostat.org/stats

20- FGN (2004). Federal Republic of Nigeria "2005 Appropriation Bill Speech Vanguard. Pp, 14-16 www. Vanguard Media Ltd.

21- FOS (2004). Federal Office of Statistics Annual Abstract of Statistics, Abuja, Nigeria.

22- FOS 1990-2006. Nigerian Foreign Trade Summary Federal Office of Statistics /National Bureau of Statistics Abuja

23- Ihimodu, I.I. (1993). The Structural Adjustment Programme and Nigeria's Agricultural Development. Monograph Series No 2 National Center for Economic Management and Administration, Ibadan, Nigeria.

24- Johansen, S. (1992). Testing Weak Exogeneity and Order of Cointegration in UK Money Demand Journal of Policy Modeling. 14, 313 – 334.

25- Johansen, S., & Juselius, K. (1992). Testing

structural analysis of co integration vectors. Journal of Economic Dynamics and Control, 12, 231 – 54.
26- Kidane, A. (1999). Real Exchange Rate Price and Agricultural Supply in Ethiopia: The Case of Perennial Crops. AERC Research Paper 99.
27- Kwanashie, M., Ajilima, I., & Garub, A. (1998). The Nigerian Economy: Response of Agriculture to Adjustment Policies. AERC Research Paper 78.
28- Kwanashie, M., Garbu, A., & Ajilima, I. (1997). Policy Modeling in Agriculture: Testing the Response of Agriculture to Adjustment Policies in Nigeria. AERC Research Paper 57.
29- Mwase, N. (1998). Economic Liberalization and Privatization of Agricultural Marketing and Input Supply in Tanzania: A Case Study of Cashew Nuts. AERC Research Paper 86
30- NBS 2000-2004. National Bureau of Statistics, Annual Report Various Issues, Abuja, Nigeria.
31- NEPZA (2006). Nigerian Export Processing Zone Authority: Free Trade News, 2nd Edition, Vol. 2, P. 4
32- Nigerian Customs, Excise Consolidation Decrees Various Issues Abuja Nigeria.
33- Nwosu, A.C. (1992). Structural Adjustment and Nigerian Agriculture: An initial Assessment Staff Report No AGES 9224. Economic Research service USDA, Washington, D.C.
34- Ogundele, O.O., & Okoruwa, O.V. (2006). Technical Efficiency Differentials in Rice Production Technologies in Nigeria. AERC Research paper 154.
35- Ogunkola E.O., Bankole, A.S., & Adewuyi, A.O. (2005). An Evaluation of the Impact of Trade and Investment Policy Reform in Nigeria" A Final Report submitted to AERC Nairobi Kenya
36- Okoli, E.N., & Okoye, C.U. (2005). Analysis of Factors Influencing Cassava Exports in Nigeria Proceeding, of 19th Annual Farm Management Association of Nigerians Conference Pp, 68-72
37- Olusegun, A.J. (2006). Dynamic acreage response of selected individual crops to price risk in Nigeria *African Journal of Economic Policy* 13 (2): 81-104
38- Osuntogun, A.C., Edordu, C., & Oramah, B.O. (1993). Promoting Nigerian Non-oil Export: Analysis of some Strategic Issues Final Report. AERC, Nairobi, Kenya.
39- Oyejide, T. A. (1986). Effects of Trade and Exchange Rate Policies on Agriculture in Nigeria Research Report No 55. Washington D.C., International Food Policy Research Institute.
40- Phillip, D.O.A (1996). Responsiveness of selected Agricultural Export Commodities to Exchange Rate Devaluation in Nigeria: An Econometric Analysis Economic and Financial Review, 34, No. 2.
41- Reinika, R. (1994). How to Identify Trade Liberalization Episodes. An Empirical Study on Kenya CSAE Working Paper
42- Soludo, C.C. (1995). An Evaluation of Structural Adjustment Programmes in Africa: Issues, Methods and Consequences for Nigeria in Contemporary Issues in Social Sciences edited by J.I. Onuoha and J. O. C. Ozioko, Acena Publishers AP, Pages 59 – 103.
43- The Guardian (2007). Nigeria's Agricultural Export Jumps to $111.2 Million in the First Quarter. The Guardian News Paper Tuesday July 9 P. 21
44- Tshibaka, T. B. (1986). The Effects of Trade and Exchange Rate Policies on Agriculture in Zaire; Research Report 56, Washington D.C., International Food Policy Research Institute.
45- Tunji, A. (2001). An Overview of the Nigerian Rice Economy The Nigerian Institute of Social and Economic Research (NISER), P.M.B. 5, University Post Office. Ibadan– Nigeria Retrived from http://www.unep.ch/etu/etp/events/Agriculture/nigeria.pdf

7

The Effect of Socio-Personal Factors on Economic and Managerial Variables in Shahroud Commercial Dairy Farms

Hamed Kashfi [1*], Ahmad Reza Yazdani[2], Farhad Shirani Bidabadi[2], Mehrdad Latifi [3]

Abstract

The purpose of this research was to study the effects of four personal and social variables including the age, literacy, job experience and size of herd on different variables such as implementing of managerial strategies in transition period, relevant costs of these strategies, average production per head and gross income of milk sale per head. For this purpose all required information were collected through questionnaire distribution among the owners of commercial dairy farms of Shahroud district. This questionnaire was included three parts. Relevant information about personal and social indices was inserted in first part. Second part included of relevant information about performance and non-performance of considered managerial strategies in transition period and third part was about relevant information about production and economic records. Finally all information about 50 herds was obtained and compared as well. Data analysis was through applying Multiple Linear Regression (MLR) method. Also it was possible to specify any relation among independent and dependent variables through calculation of Pearson Correlation Coefficient. Finally it was revealed that social and personal factors have a significant effect on the implementation of management strategies and other independent factors.

Keywords:
Personal and Social indices, Managerial strategies, transition period, Managerial and Economic indices

[1] M.Sc. Animal Husbandry Management, Gorgan University of Agricultural Sciences and Natural Resources.
[2] Members of the Scientific Board of Gorgan University of Agricultural Sciences and Natural Resources.
[3] Shahroud's Central Veterinary Medicine Labratoary.
** Corresponding author's email: Hami2006_hk@yahoo.com*

INTRODUCTION

Transition period, from 3 weeks before upto 3 weeks after parturition of dairy cows, is critically important to health, production, and profitability of dairy herds. Most health disorders occur during this time. There are lots of serious changes for dairy cows in transition period from the end of pregnancy up to the beginning of lactation period. This period was really important for many years and for a lot of researchers due to the effects on the health, production and profitability by focusing on nutrition and management of dairy cows. (Drackley, 1999) In this research current conditions of animal husbandry centers and any effects of the mentioned variables have been studied on different managerial and economic indices. There are different studies that have pointed out to the effect of some personal and social factors which are effective on the economic and managerial indices at commercial dairy farms. It is in a way that mentioned factors have an effective role in applying of modern technologies. For instance, Carley and Fletcher, (1986) stated that older cattlemen have more resistance against new technologies. Regarding the academic records again it was mentioned that bearing a high school or university degree by managers of animal husbandry centers have a positive effect in selection of managerial methods. In a research made by Alqunaibet *et al.,* (1995) on dairy herds at Saudi Arabia, it was revealed that any increase in the size of herd makes better economic situation for benefiting from technology.

Also in another research by Hanson *et al.,* (1998) on dairy herds at North America under intensive grazing and open grazing, it was revealed that cattlemen with higher profitability had little job experiences but higher academic records. In another research made by Vandermersh and Mathijs (2002) at husbandry animal centers of Belgium, it was revealed that the age of managers and other managerial factors had no more effects on risks and intends of them in progress of different groups. Regarding the effect of job experience Blelik and Rajcaniova (2004) stated that higher work experiences do not have a significant effect on technical efficiency on the farm. Latruffe *et al.,* (2004) considered academic records and education as two important factors in upgrading technical efficiency level of animal husbandry units. In another study in Slovakia's farms by Blelik and Rajcaniova (2004), it was specified that literacy and education level have a positive but not significant effect on efficiency of the farmers. They concluded in the same research that generally any increase in size of the production units may firstly decrease margin costs and increases the profit return. But any oversize of the agency may cause a reduction in return profit and increase of margin costs. Latruffe *et al.,* (2004) stated that in case of any increase in size of herd, there will be an increase in technical efficiency.

Demircan *et al.,* (2006) has also concluded that the size of herd and number of herd's dairy cows have a positive effect on managerial methods of the herd. In case of any increase in the size of the herd, there will be an increase in gross profit but it is not significant. Furthermore, regarding the effect of owner's age on profitability, Gloy *et al.,* (2006) stated that older managers at animal husbandry centers have little efficiency. Ramezani (2009), Ortega *et al.,* (2007), Dagistan *et al.,* (2009) have also stated that higher job experience may increase the efficiency and productivity of managers. Ceyhan and Hazenci (2010) found out that experiences and education levels has no relations with economic efficiency. The objectives of this research are to evaluate the effects of above social and personal factors on important managerial and economical factors in commercial dairy farms in transition priod.

MATERIALS AD METHODS

The present study has been done from January 2011 up to April 2011 in commercial dairy herds of Shahroud district. Data was collected through direct interview with dairy farmers. More than 65 dairy farms located at Shahroud district all of them were selected for interview, about 15 questioners were considered as invalid due to different mistakes and thus data obtained from 50 questionnairs were processed. The questionnaire included different questions about personal and social factors, age of farmers, education level, work experiences and size of the dairy herd. Upon collection of these information it was possible to change it into quantitative information through grading system for further analysis (Table 1). The next part of this questionnaire included relevant information about managerial strategies of transition period and effective managerial policies for prevention of important metabolic disorders in that period. The mentioned managerial strategies include 14 items separated in accordance with the effect of different disorders.

According to the type of disorders, these strategies include five managerial strategies effective on Ketosis, three managerial strategies effective on Acidosis prevention, 3 managerial strategies effective on milk fever and two managerial guidelines effective on displaced abomasums and finally one managerial strategy effective on retained placenta incidence rate (Table 2).

In addition all information about production rate and gross income per head was gathered through the production records of farmers.

Data analysis was made by using SAS software with application of Multiple Linear Regression model (MLR). the outlines of model are explained below:

$$Y1=\beta01+\beta11X1+\ldots+\beta q1Xq+\varepsilon1$$
$$Y2=\beta02+\beta12X1+\ldots+\beta q2Xq+\varepsilon2$$
.
.
.
$$Yp=\beta0p+\beta1pX1+\ldots+\beta qpXq+\varepsilon p$$

$$\rho_{x,y}=\frac{\operatorname{cov}(x,y)}{\sqrt{(\operatorname{var}x)(\operatorname{var}y)}}=\frac{\sigma_{xy}}{\sigma_x\sigma_y}$$

Also the correlation coefficient among independent and dependent variables have been calculated by using Pearson correlation coefficient as below:

RESULTS AND DISCUSSION

Age

Age is a factor with double effects as studied in present research. This means that higher age may cause a reduction in productivity due to a reduction in body abilities. On the other hand higher age may cause higher experiences therefore there is an increase in productivity as well. The results of variance analysis (Table 3) showed that age has a significant effect on application rate of managerial strategies in transition period and application costs of

Table 1: Managerial strategies in transition period

Type of disorders	Managerial strategies in transition period
Ketosis	1-Body Condition Score management
	2-Encouraging the animal to use more dry mater (Consuming feeding management):
	3-Benefiting from Propylene Glycol supplement
	4-Benefiting from Niacin supplement
	5-Benefiting from the high quality feeding in transition periods (the importance of feeding quality):
Acidosis	6-Pay attention to the importance of ration's fiber and its rate with concentration:
	7-Increasing the number of feeding
	8-Using Sodium bicarbonate
Milk fever	9-Using anion salts for reducing of ration's DCAD
	10-Limiting the calcium rate in close-up rations
	11-Injection of vitamin D metabolites (AD3E)
Abomasums Displaced	12-Using forage with suitable length and lack of sudden usage of high concentration (Pay attention to the fiber and physical form of the rations)
	13-Increasing the consuming feed after calving
Retained Placenta	14-Injection of E-Selenium two weeks prior to calving

Table 2: Method of grading the personal and social indexes

Grade / Criteria	1	2	3	4	5	6	7
Academic level of the manager	Illiterate	Primary school	3rd Grade of middle school	High School diploma	Associate Degree	Bachelor of Science and higher	
Age (Year)	60 and higher	50-60	40-50	30-40	20-30		
Experience level (Year)	0-5	6-10	11-15	16-20	21-25	26-30	30 and higher
Size of herd (Head)	20-50	51-150	151-300	300 and higher			

implementing strategy (p<0.01). This means that by age increase, the rate of application of managerial strategies in transition period and cost of implementation of these strategies increased significantly. These results were in compliance with the findings out of a research made by Emami and Meybodi (2005). But it may reject the findings out of the researches made by Carley and Fletcher (1986) who may believe that older cattlemen have more resistance against new technologies. The result showed that age has a positive and direct effect on average milk production and net income resulted from it (P<0.01), which can be attributed to increase in work experience as well as increase in applying of managerial strategy. The obtained results were reject the findings of Gloy *et al.*, (2002) and Demircan *et al.*, (2006) and also Tauer and Mishra (2006) who have stated that age increase has a negative effect on milk production and gross income.

Furthermore, the results of variance analysis among age groups pointed out that there is no significant statistical difference in milk production , gross income and application costs of managerial strategies in transition period (p>0.05) between different age groups (Table 3).

By calculating the correlation coefficients (Table 4) between the variables of age and other indices, it was specified that there is a positive relation between the age and average rate of applying any managerial strategies in transition period (r = 0.53). It is in a way that any increase of age may cause an increase in applying of these strategies. There was also positive relation between age and average production (r = 0.62) along with gross income of milk sale per head (r=0.67). These findings may approved the results of researches published by Emami and Meybodi (2005), Carley and Fletcher (1986) and Kimberley and Johanson (1989).

Work experience

Analysis of variance, regression interpretation and average comparison showed that work experience had a significant effect on application of managerial strategies, average costs of applying the strategies, average rate of milk production and gross income of milk sale per head (P>0.05). The results were not in agreement with the findings of Blelik and Rajcaniova (2004) who stated that higher work experience

Table 3: The obtained results out of Regression and comparisons of the average for personal and social variants

Economic and Manageri variants	Effective Personal and Social variables on economic and managerial indexes							
	Age		Academic records		Work experience		Size of herd	
	F-Test	T-Test	F-Test	T-Test	F-Test	T-Test	F-Test	T-Test
Implementation rate of managerial strategies	44.07**	6.64*	111.08**	10.54*	0.38^{ns}	-0.61^{ns}	7.88**	2.81^{ns}
Implementing costs of managerial strategies	19.60**	4.43^{ns}	41.02**	6.40*	0.01^{ns}	0.11^{ns}	13.20**	3.63^{ns}
Production average	30.95**	5.56^{ns}	85.28**	9.23*	0.00^{ns}	0.05^{ns}	14.94**	3.87^{ns}
Gross income	29.61**	5.44^{ns}	83.38**	9.13*	0.00^{ns}	0.02^{ns}	14.88**	3.86^{ns}

** p<0.01 * p<0.05 ^{ns} p>0.05

Table 4: Correlation coefficients calculated between personal / social and managerial-economic variables

	Age	Experience	Academic records	Size of herd	Average implementation	Implementation costs	Average Production	Gross income
Age	1.00							
Experience	-0.47	1.00						
Academic records	0.58	-0.23	1.00					
Size of herd	0.16	0.24	0.38	1.00				
average implementation	0.69	-0.08	0.83	0.37	1.00			
implementation costs	0.53	0.01	0.67	0.46	0.82	1.00		
average Production	0.62	0.006	0.79	0.48	0.86	0.90	1.00	
Gross income	0.61	0.002	0.79	0.48	0.85	0.89	0.99	1.00

** p<0.01 * p<0.05 ^{ns} p>0.05

has not a significant effect on technical efficiency of farms. But the mentioned results were in tune with the findings of Ramezani (2009), Ortega *et al.,* (2007) who stated that higher work experience may increase the efficiency and productivity as well. Also the findings of Ceyhan and Hazenci (2010) who have stated that experience and educational level have no relation with economic efficiency are not in compliance with the results of the present research. The results of average comparison (Table 3.) between different job experience groups showed that there is not a significant difference from view point of applying the managerial strategies, application costs, average production and gross income out of milk sale per head (P>0.05).

Upon calculation of the correlation coefficients (Table 4), it was obvious that there was a negative relation between work experience and age (r = -0.47) and also negative relation between academic records and work experience (r = -0.23). Correlation coefficients of work experience and Implementation costs, average production and gross income from milk sale were respectively 0.0022, 0.006 and 0.01 which are very small from statistical view point.

Academic records

Regarding the academic records of farmers and after variance analysis and further average comparisons, it was revealed that academic records of managers has a significant effect on managerial and economical parameters of production units, application of managerial strategies, performance costs, average milk production and gross income out of milk sale per head (P<0.01). These obtained results were in compliance with the findings of researches made by Carley and Fletcher (1986) and Johanson *et al.,* (1989) but it may reject the findings of Vandermersh and Mathjis (2002), Blelik and Rajcaniova (2004). The results of an average comparisons between different groups of the academic records (Table 3) showed that there is a significant difference between application of managerial strategies, implementation costs, average milk production and gross income from statistical viewpoint (P<0.05).

Any consideration of correlation coefficients (Table 4) among above mentioned variables and other items showed that there is a relation between academic records and application of managerial strategies of transition period r=0.83 which is a sign of further effects of academic records in acceptance and applying of modern managerial methods in Dairy industry.

In addition, there was a positive correlation coefficient between academic records and implementation costs of the managerial strategies in transition period (r = 0.67).

The results of correlation analysis indicated that correlation coefficients between literacy level and production average was (r = 0.79) which is an indicator of the fact that academic level has a positive effect on implementation of managerial strategies and relevant implementation costs of the startegies which may increase milk production per head. Correlation coefficient for academic level and gross income of milk sale was r = 0.79 which showed that level of education has a positive effect on average milk production which may significantly increase the gross income from milk sale per head.

Size of herd

The results of variance analysis and correlation coefficients showed that herd size has a significant effect on productinal and economical parameters in this research ($P<0.05$). These findings were in compliance with the obtained results by Alqunaibet *et al.,* (1995) who explained that any increase in herd size may cause better economic outcome of technology. Also the mentioned findings may confirm the results that made by Blelik and Rajcaniova (2004) and Latruffe *et al.,* (2004) who stated that any increase in the size of agency may cause an increase in marginal costs, profit and technical efficiency. Average comparisons among different size of the herd showed that there is no significant difference between the studied groups for considered variables accordingly ($P>0.05$).

Calculation of the correlation coefficients (Table 4) showed that there was a positive relation between the size of the herds and average implementation of managerial strategies ($r = 0.37$).

The correlation coefficient of the herd size and application costs of managerial strategies in transition period was $r = 0.46$, correlation coefficient between herd size and average milk production was $r = 0.48$ and it was $r = 0.48$ between herd size and gross income out of milk sale per head.

The results of correlation coefficient analysis showed that the academic records have the highest effects on dependent variables as far as personal and social factors is concerned.

REFERENCES

1- Alqunaibet, M.H., Limam, M.M.T., & Sofian, E.E. (1995). Testing for Returns to Scale in Dairy Farms in Saudi Arabia. J. King saud univ. Journal of Agricultural Science. 7 (1): 3-11.

2- Blelik, P., & Rajcaniova, M. (2004). Scale Efficiency of Agricultural Enterprises in Slovakia. Journal of Agricultural Economics. 50 (8): 331-335.

3- Carley, D.H., & Flechter, S.M. (1986). An Evaluation of Management Practices Used by Southern Dairy Farms. Journal of Dairy science. 69: 2458-2464.

4- Ceyhan, V., & Hazneci, K. (2010). Economic Efficiency of Cattle-Fattening farms in Amasya province, Turkey. Journal of Animal and veterinary Advances. 9(1):60-69.

5- Dagistan, E., Koc, B., Gul, M., Parlakay, O., & Goksel, A. (2009). Identifying Technical Efficiency of Dairy Cattle Management in Rural Areas Through a Non-Parametric Method: A case study for the east Mediterranean in turkey. Journal of Animal & Veterinary Advances. 8 (5): 863-867.

6- Drackley, J.K. (1999). Biology of Dairy Cows During the Transition Period: the final frontier. Journal of Dairy Science. 82: 2259–2273.

7- Demircan, D., Binici, T., Koknaroglu, H., & Aktas, A.R. (2006). Economic Analysis of Different Dairy Farm Sizes in Burdur Province in Turkey. Journal of Animal Science. 51(1): 8-17.

8- Emami, A. (2005). Measuring Principles of Efficiency and Productivity. Commercial Researches - and Studies Institute. P 275

9- Gloy, B.A., Hyde, J., & Ladue, L. (2002). Dairy Farm Management and Long- Term Farm Financial Performance. Agricultural Resource Economics. 31 (2): 233-247.

10- Hanson, G.D., Cunningham, L.C., Morehart, M.J., & Parsons, R.T. (1998). Profitability of Moderate Intensive Grazing of Dairy Cows in the Northeast. Journal of Dairy Science.81: 821–829

11- Latruffe, L., Balcombe, K., davidova, S., & Zawalinska, K. (2004). Determinants of Technical Efficiency of Crop and Livestock Farms in Poland. Journal of Applied Economics. 36: 1255- 1263.

12- Ortega, L.E., Ward, R.W., & Andrew, C.O. (2007). Technical Efficiency of the Dual-Purpose Cattle System in Venezuela. Journal of Agricultural and Applied Economics, 39 (3): 719-733

13- Ramezani, A. (2009). Considering and Analysis of Technical Efficiency of Broiler Farms- North Khorasan Province with Relevant Data Analysis. Master of Science thesis- Gorgan Agriculture & Natural Resources University.

14- Tauer, L.W., & Mishra, A.K. (2006). Dairy Farm Cost Efficiency. Journal of Dairy Science. 89: 4937-4943.

15- Vandermersh, M., & Mathijs, E. (2002). Do Management Profiles Matter? An Analysis of Belgian Dairy Farmers. 10th EAAE Congress.

Effect of under Irrigation Management on Potato Performance Components

Fakhrodin Ghasemi Sahebi[1], Masoud Hekmat[2], Eman Pourkhiz[3]*

Abstract

To evaluate the effect of water, tape drip and furrow irrigation methods on the performance components and features of Agria cultivar potato, an experiment was carried out in the agriculture faculty of Kermanshah. Repeated three times, this study was carried out in the form of full random blocks with the main factor of different amounts of irrigation water in three levels (50, 75, and100 percent of cumulative evaporation from an A- class evaporation pan) and the secondary factor of irrigation method (including drip and furrow irrigations). At 3 repetitions, results showed that the minimum performance (19.168 tons per hectare) was related to the drip irrigation method with 50 percent evaporation from the A-class pan and the maximum one (34.455 tons per hectare) was related to the drip irrigation method with 100 percent evaporation from the A-class pan. The rate of irrigation and effectiveness of method on the number of the main stem per square meter and number of the main stem in the bush were not significant. The minimum percent (number of tubers) of tubers smaller than 35mm (26.47percent) and the maximum tuber production (37.17 percent) in the size of food and market-friendly (tuber bigger than 55mm) were attained by drip irrigation method with 100 percent evaporation from the A-class pan. The impact of Irrigation method on and the specific gravity of tubers and starch (dry matter) was and was not significant, respectively.

Keywords:
Tape drip irrigation, Furrow irrigation, Potato, Dry matter

[1] *M.Sc., Irrigation and Drainage, Razi University, Kermanshah, Iran.*
[2] *Department of Agricultural Management, Shoushtar Branch, Islamic Azad University, Shoushtar, Iran.*
[3] *Young Researchers Club, Ghaemshahr Branch, Islamic Azad University, Iran.*
* *Corresponding author's email: Masoud.hekmat1@gmail.com*

INTRODUCTION

Potato is one of the glandular products having an important role in human and animal nutrition. It is cultured in different climates and is the fourth product after wheat, rice, and popcorn; in the subject of producing countries number, it is the second product after popcorn (Dokhani and Rabiee Motmaen, 2001).

Studies showed that potato is very sensitive to the aridity stress .moisture stress has a bad effect on the potato`s growth and performance. Moisture stress, at the chlorophyll growth stage and before the tubers formation stage, reduces the leaf area, number of root system`s secondary branches, bush height, and the overall growth of the plant`s green cover. The second stage of potato growth is the tuber formation stage in which aridity reduces one of the performance components which is the number tuber per each bush; the tubers` mean size and specific gravity is highly reduced, too. The third stage of the potato growth is the tubergrowth stage. At this stage, the product`s quality and performance is highly affected by the aridity stress. The last stage is the ripening stage. At this stage, the green cover of the plant senesces and the tuber skin becomes thick, the water need is reduced, and even the moisture stress, after harvesting, increases the tuber resistance to the pests (Mosavi Fazl and Faeznia, 2003).

Produced tubers` size and weight distribution depends on each gland`s growth and the competitive relation between the tubers. All tubers of a bush do not, simultaneously, start the dry matter aggregation. This differences are resulted from the tuber formation time and the location of the tuber on the bush causing the distribution of the gland`s size and weight (Elizabeth, 1992). It seems that despite the reduction of a bush`s mean weight of each tuber caused by the increase of the produced tubers number, a larger number of tubers gratifies the weight reduction and the weight of all tubers is more in such bushes; but the market-friendly of tubers with a lesser mean weight would be reduced (Wurr and Allen, 1994).

Yuan *et al.,* evaluated the effects of different irrigation regimes on the potato growth at the drip irrigation. Irrigation water quantity was considered as 125, 100, 75, 50, and 25 percent of evaporation from the water level in a ceramic evaporation pan (0.2 m diameter). Plant height, biomass amount, total fresh glandular products, and market-friendly tubers (more than 85 g) were increased by increasing the irrigation water. Plant height and the total glandular product at the treatment of 125 percent evaporation from the evaporation pan were close to the 100 percent one. Increasing the irrigation water not only have the number of tubers increased, but increased the mean weight of the tubers, too. Irrigation water increased the quantity of the tubers but reduced their quality (Yuan *et al.,* 2003).

Nadler and Heour (1995) believed that the specific gravity of the tubers is reduced and the reducing sugars rates are increased by reducing the irrigation water. This reduction in the gland`s specific gravity is caused by a reaction to the under irrigation during the long term of growth season.

This study is aimed at evaluating the effect of different irrigation water quantities and tape drip and furrow irrigation methods on the performance, performance components, and features of potato (dry matter and specific gravity).

MATERIALS AND METHODES

The experiment, to compare the performance of potato (agria cultivar) under tape drip and furrow irrigation methods, was carried out in the agriculture faculty`s research farm of Kermanshah in the spring of 1386. Farm`s soil texture was salty clay. Repeated three times, this study was carried out in the form of full random blocks plan.

The main factor was different amounts of irrigation water in three levels of 50, 75, and100 percent of cumulative evaporation from an A-class evaporation pan and the secondary factor was two irrigation methods including drip and furrow irrigations. Each check was 13 m length and 3m width in which 4 lines were cultivated. The distance of culture lines was 75cm and the bushes distance on the row was 30cm. volume contours were used to measure and control the amount of influent irrigation water of each plot in each treatment. In measuring the gross irrigation requirement performance efficiency for the furrow method was 40 percent because the water was running off from the end of check.

In the tape drip irrigation method volume contours 30 m length T tapes were used and regulating valves were used to measure and control the amount of irrigation water in each treatment. Application efficiency for the tape drip method

Table 1: Total quantity of irrigated water during the plant growth period in each method

treatment	100 percent of water requirement	75 percent of water requirement	50 percent of water requirement	Furrow irrigation
Total quantity of the irrigated water (m^3) in each hectare	9749.62	7312.215	4874.81	21357.207

was 90 percent. Data of evaporation from the A-class evaporation pan located in the weather station near the agriculture faculty was used to calculate the plant water requirement. The potential rate of evaporation and transpiration was calculated by the following equation:

$$ET = K_p . E_{pan} \quad (1)$$

In which

ET: evaporation and transpiration (in millimeter) between two irrigations

K_p: evaporation pan coefficient

E_{pan}: rate of evaporation (in millimeter) from the pan between two continuous irrigations

The amount of water for each treatment was then attained by multape driplying the coefficients related to the water level of treatment (50, 75, and 100 percent) by the evaporation and transpiration calculated by equation1.

Agricultural operations such as weeding, fertilizing, and spraying were equally carried out based on the customs and requirements of the farms in each experimental unit. At the end of the growth season, two centric 6 m adjacent lines were selected from each check and their potato tubers were harvested. Based on the size, tubers were divided in three groups of smaller than 35, 35 to 55, and bigger than 55 mm. samples of each treatment were prepared and sent to the laboratory to carry out the qualitative tests of product such as measuring the specific gravity and the percent of dry matter.

Measurement of the specific gravity

Specific gravity was attained from division of samples - weight and volume. Tubers were randomly selected, peeled, chopped to very little pieces, and mixed to determine the dry matter. About 5 g of this mixture was poured in the ceramic bushes which have reached the fixed weight at 105° C in the oven; after 6 hours in the oven with air flow, it reached the fixed weight. Then, the ceramic bushes were cooled down in the desiccators, weighted again, and the percentage of the dry matter was calculated (Dokhani and Rabiee Motmaen, 2001). Consumptive water volume in different irrigation water treatments is illustrated in table 1. SAS software was used for statistical analysis and was compared to Duncan. Measure and control the amount of irrigation water input to each plot in each treatment was done by volumetric meters.

RESULTS AND DISCUSSION

Tubers` performance

Table 2 below shows the effect of the irrigation water rate on the tuber performance was significant at a 1 percent level. The minimum performance related to the 50 percent treatment was 19.168 tons per hectare and the maximum one related to the 100 percent treatment was 34.455 tons/ha as demonstrated on table 3. The 75 percent treatment and the furrow irrigation

Table 2: Analysis of variance of potato tuber`s performance

Source changes	degree of freedom	Bush performance(kg)	performance ton per hectare
repetition	2	0.0156	6.1351
treatment	3	0.4251**	354.0292**
error	6	0.0368	27.6018
sum	11	0.4777	378.7662
CV(%)		8.85	8.16

ns, no significant; ** p≤0/01, * p≤0/05

Table 3: Mean Comparison of potato tuber`s performance

experimental treatment	Bush performance (kg)	performance ton per hectare
50 percent cumulative evaporation from the pan	0.62167 c	19.168 c
75percent cumulative evaporation from the pan	0.95967 b	25.938 b
100 percent cumulative evaporation from the pan	1.13667 a	34.455 a
Furrow irrigation	0.82600 b	25.582 b

Means followed by same letter (s) in each column are not significantly different ($p < 0.05$)

treatment, based on Duncan groups, classified in one group. This experiment showed that the more water is provided for the plant, the more performance of tubers is resulted.

Size of the tubers

Percent of tubers smaller than 35m

The effect of irrigation water on the percent of the smaller than 35mm tubers was not significant as illustrated on table 4. Minimum losses (small tubers, 26.47 percent) were resulted from drip 100 percent treatment and the maximum ones (37.39 percent) were resulted from the furrow irrigation treatment.

Percent of 35-55 mm tubers

The effect of irrigation water on the percent of 35-55mm tubers was not significant as showed on table 4. The size of 35-55 mm is considered as seed size in classifying the tubers size. If the objective is to produce the potato seed size, the maximum seed tuber (47.87 percent) can be produced by the 50 percent drip method.

Percent of bigger than 55 mm tubers

Table 4 shows the effect of irrigation water on the percent of bigger than 55 mm tubers was not significant. More than 55 mm tubers are considered as the food size and market-friendly ones.

Table 4. Analysis of variance of tuber percent in the groups of smaller than 35mm, 35-55mm, and bigger than 55mm (Mean Squares)

Number of the main stem

Table 4 illustrates the effect of irrigation water on the number of the main stem per meter square was not significant. The number main stem depends on the size, physiologic age of the seed gland, and the genetic characteristics of the cultivar - number of eyes in the seed tuber (Allen, 1992). Maximum number of the main stem per square meter in the 100 percent treatment was 25.08 as shown on table 6.

The number of tubers

Number of tubers per meter square

The effect of irrigation water on the number of tubers was not significant (table 6). The maximum tuber number per meter square in the 50 percent water treatment was 35086 and the minimum one in the 75 percent treatment was 34.06. Shock *et al.,* (1993) showed that there a great correlation between stolen and the number of tubers.

Percent of the dry matter

The effect of irrigation water on the dry matter was not significant shown on table 9 but the percent of the dry matter reduced by increasing the

Table 4: Analysis of variance of tuber percent in the groups of smaller than 35mm, 35-55mm, and bigger than 55mm (Mean Squares)

Source changes	degree of freedom	smaller than 35mm	35-55mm	bigger than 55mm
Repetition	2	356.94	7.875	326.44
treatment	3	270.86ns	211.13ns	370.14*
error	6	388.61	505.60	274.83
sum	11	1016.42	724.60	971.42
CV(%)		26.26	22.31	23.98

ns, no significant; ** p≤0/01, * p≤0/05 .ger than 55mm (Mean Squares)

Table 5: Mean comparison of in the groups of smaller than 35mm, 35-55mm, and bigger than 55mm

Source changes	smaller than 35mm	35-55 mm	bigger than 55mm
50 percent cumulative evaporation from the pan	26.427 a	47.877 a	25.700 ab
75percent cumulative evaporation from the pan	31.677 a	40.457 a	27.867 ab
100 percent cumulative evaporation from the pan	26.470 a	36.357 a	37.170 a
Furrow irrigation	37.933 a	39.857 a	22.153 b

Means followed by same letter (s) in each column are not significantly different ($p < 0.05$)

Table 6: Analysis of the variance of the number of main stem and tubers (Mean Squares)

Source of variation	degree of freedom	Number of stem in the bush	Number of stem per meter square	Number of tuber in the bush	Number of tuber per meter square
Repetition	2	2.50	54.62	2.2386	45.93
Treatment	3	2.56*	63.04^{ns}	1.86^{ns}	44.93^{ns}
Error	6	1.63	51.50	10.32	260.35
Sum	11	6.707	169.17	14.42	351.22
CV(%)		12.25	13.73	19.29	19.29

ns, no significant; ** p≤0/01, * p≤0/05

Table 7: Mean comparison of the main stem and tubers numbers

Experimental treatment	Number of stem in the bush	Number of stem per meter square	Number of tuber in the bush	Number of tuber per meter square
50 percent cumulative evaporation from the pan	4.0300 ab	19.753 a	7.097 a	35.863 a
75percent cumulative evaporation from the pan	4.1933 ab	21.297 a	6.193 a	31.060 a
100 percent cumulative evaporation from the pan	5.0267 a	25.080 a	6.713 a	33.903 a
Furrow irrigation	3.8033 b	19.217 a	7.197 a	35.710 a

Means followed by same letter (s) in each column are not significantly different ($p < 0.05$).

irrigation water quantity. It is believed that by increasing the quantity of irrigation, percent of the dry matter is increased (Mohammadi and Faeznia, 2001). If the rate of dry matter is very low oil absorption would be high and soft chips can be produced (Dokhani and Rabieae Motmaen, 2001).

Gland`s specific gravity

The effect of irrigation water on the Gland's specific gravity was significant at a 1 percent probability level. The maximum specific gravity (1.11 g/cm^3) was related to the tape drip 100 percent treatment. Increasing the irrigation water, the specific gravity of the tubers was also increased, which is ion accordance with the findings of Yuan *et al.,* (Yuan *et al.,* 2003).

RECOMMENDATION

1- Reciprocal effect of both chemical and organic fertilizers on the potato in different weathers of the province and at different soils should be evaluated.

2- Water stress should be evaluated at the three stages of germination, growth, and gland formation.

3- Effects of placing the tape at different depths on the water use efficiency and the potato product performance should be evaluated.

4- Drip tape drip irrigation for different varieties of potato should be analyzed.

5- Effect of different irrigation methods and regimes on the qualitative characteristics of potato should be evaluated.

6- Regimes (1.25 and 0.25) of 1 time cumula-

Table 8: Analysis of variance of specific gravity and arid matter (Mean Squares)

Source of the variations	degree of freedom	Specific gravity (g/cm³)	dry matter (percent)
Repetition	2	0.00002	8.05
Treatment	3	0.00112*	13.78^{ns}
Error	6	0.00006	12.58
Sum	11	0.00130	34.43
CV(%)		0.29	6.974

ns, no significant; ** p≤0/01, * p≤0/05

Table 9: Mean comparison of specific gravity and dry matter

Experimental treatment	Specific gravity (g/cm³)	Dry matter (per-cent)
50 percent cumulative evaporation from the pan	1.085433 d	19.168 c
75percent cumulative evaporation from the pan	1.102067 b	25.938 b
100 percent cumulative evaporation from the pan	1.1134 a	34.455 a
Furrow irrigation	1.093333 c	25.582 b

Means followed by same letter (s) in each column are not significantly different ($p < 0.05$)

tive evaporation from the evaporation pan should be evaluated.

REFERENCES

1- Allen, E. J. (1992). The Effect of Row width on the Yield of Three Potato Varieties, J. Agric. Sci. camb. 79: 315-321.

2- Almekinfers, C.J.M. (1991). Flowering and tTrue Seed Production in Potato (*Solanum tuberosum* L.): Effect of Stem Density and Pruning at Lateral Stems, J. Potato Res. 34: 379-388.

3- Dokhani, Sh., & Rabeae Motmaen, L. (2001). Evaluation of the Variation Rate of Glucose and Organic Acids of Potato Varieties (Moren, Mafona, and Agria) in Esfahan Province Dduring Storage with High Efficiency Chromatographic Method, journal of agriculture and natural resources sciences and technologies, 5th edition, 1:161-171.

4- Elizabet, G.C. (1992). Structure and Development of Potato Plant, PP. 65-146. In: Harris, P.M. (Ed), the potato crop: Scientific basis for improvement, Chapman and Hall, London.

5- Haverkort, A.J., DeWart, M.V., & Bodlaender, K.B.A. (1990). Inter-relationship of the Number of Initial Sprouts, Stems, Stolons and Tubers per Potato Plant, Potato Res. 33: 269-274.

6- Mohammadi, A., & Faeznia, F. (2001). Effect of Moisture Stress on the Growth and Performance of Two Potato Varieties, Experimental Report of Semnan Agriculture Research Center (Shahroud)

7- Mosavi Fazl, H., & Faeznia, F. (2003). Effect of Water and Azot Fertilizer Different Quantituies on the Qualitative and Quantitative Characteristics of Potato, Papers of the 11th Conger of the Iran National Committee of the Irrigation and Drainage.

8- Nadler, A., & Heuer, B. (1995). Effect of Saline Irrigation and Water Deficit on Tuber Quality, Potato Res. 38: 119-123.

9- Shock, C.C., Holmes, Z.A., Strieber, T.D., Eldredge, E.P., & Zhang, P. (1993). The Effect of Timed Water Stress on Quality, Total Solids and Reducing Sugar Content of Potatoes, Am. Potato J. 70(3): 227-241.

11- Van Heemst, H.D.J. (1996). The Distribution of Dry Matter During Growth of a Potato Crop, J. Potato Res. 28(1): 55-66.

12- Wurr, D.C.E. (1994). Effect of Seed Size and Spacing on the Yield and Grading of Two Main Crop Potato Varieties, J. Agric. Sci., Camb. 82: 37-52.

13- Wurr, D.C.E. & Allen, E.J. (1974). Some Effect of Planting Density and Variety on the Relationship Between Tuber Size and Tuber Dry-Matter Percentage in Potatoes, J. Agric. Sci., Camb. 82: 227-252.

14- Yuan, B.Z., Nishiyama, S., & Kang, Y. (2003). Effect of Different Irrigation Regimes on the Growth and Yield of Drip-Irrigated Potato, Agric Water Manage. 63: 153-167.

9

Analysis of Rice Import Trend and it`s Economic Factors: Case of Iran

Zabiholah Gholami

Abstract

Rice comes second after wheat in Iran`s food consumption economy. However Iran is one of the greatest rice importer countries all over the world because of it`s rising population and recent growth in GDP. This paper presents an analysis of cointegration test between rice import and it`s economic factors over the period 1990-2011, employing Engle-Granger model. At first, Dickey-Fuller test shows that all variables are non-stationary at data level, so their first difference (that are stationary series) are used. Secondly, Engle-Granger testing presents existence of a long-run relationship between rice import and it`s economic factors including per capita GDP, foreign exchange rate and domestic price. At last, rice import model was estimated using OLS method which proves that all independent variables are significant at high level and the sign of coefficients are consistent with theoretical expectations So that import of rice positively correlates with Per capita GDP, domestic price and negatively correlates with foreigh exchange rate. With respect to increasing demand for rice, government should keep domestic policies for the rise of rice production so that the need for import falls in the long-run.

Keywords:
Iran, Rice import, Economic factors, Cointegration

[1] *Department of Economics, Islamic Azad University, Qaemshahr Branch.*

* *Corresponding author's email: gholamirudi@gmail.com*

INTRODUCTION

Rice has been an important cereal for human being over the history. Farmers first started growing rice in East and South Asia as long as 15,000 years ago, when people began to settle in river deltas and domesticated wild rice. Remains of early cultivated rice have been found in the Yangtze valley dating to about 8500 BC. Today rice is grown practically everywhere, except Antarctica. Rice is grown on flooded land and on dry land, in tropical rain forests of Africa and in arid deserts of the Middle East, on coastal plains and on the Himalayan mountains. In the year 2011, the world produced about 854 million tones of paddy rice. Most of that - about 771 million tones- was grown in Asia. It has been estimated that half the world's population subsists wholly or partially on rice. Ninety percent of the world crop is grown and consumed in Asia. Rice provides 27% of people's energy intake and 20% of their dietary protein. Its by-products are used for making straw and rope, paper, wine, crackers, beer, cosmetics, packing material, and even toothpaste.

In Iran, however, Rice is the only major cereal crop that is primarily consumed by humans directly as harvested, and only wheat and corn are produced in comparable quantity. Rice production is very difficult work and required a large number of field as well as several experts. Iranian people eat an average of 40 kilogram of rice every year. Iran ranks eleventh producer of rice in the world with an annual production 2000000 tones (2011), but the third importer of rice in the world.

Studying the trend of rice production and import during the period is very fruitful. Table 1 shows rice economy in Iran during 1986-2011. As table 1 shows, rice domestic production has an increasing trend as a result of increase in area under rice and use of high yield varieties. Despite this fact, domestic production has been always left behind consumption and this gap has been usually filled by import. High population and national income growth – resulting from sharp increase in postwar oil income – may be

Table 1: Rice economy, 1986-2011.

Year	Production (ton)	Import (ton)	Export (ton)	Price index($)
1986	981000	586556	-	-
1987	1605000	431988	-	0.135
1988	1216000	621709	-	0.137
1989	1484000	587383	-	0.152
1990	1776000	538713	-	0.155
1991	1784000	492679	-	0.247
1992	1803000	807442	-	0.246
1993	1419000	338287	-	0.239
1994	1854000	881503	-	0.351
1995	1981000	793657	-	0.355
1996	2357000	559708	20	0.398
1997	2364000	943837	2116	0.474
1998	2281000	1157508	3437	0.452
1999	2259000	481551	197	0.658
2000	2301000	1146698	1630	1.57
2001	2685000	915229	26	1.064
2002	2350000	637498	231	1.09
2003	2771000	631293	12	2.17
2004	2348000	1021836	425	3.11
2005	1971000	1167217	488	1.94
2006	1990000	698925	184	2.38
2007	2931300	1047499	448	3.02
2008	2888000	8750180	327	3.14
2009	1300000	6712638	-	4.11
2010	2931400	1044659	-	3.51
2011	2000000	1216192	-	-

Source: Iranian Statistical Headquarter

the main cause of such rise in rice consumption as the common food among Iranian families. Rice import is also considered as the best policy to control the domestic market as price goes up sharply. (i.e., see years 2007 and 2009). That is why government has maintained increasing rice import in order to equilibrate the domestic market. For instance, largest import belongs to year 2008 as a result of noticeable fall in domestic production and rise in domestic price.

The most important issue here is to determine which factors can significantly explain such great increase in rice import among variables such as domestic price, foreign exchange rate, domestic production and so on. This research tries to analyze Iranian rice import trend and it`s economic factors.

MATERIALS AND METHODS

Time-series data is needed to be integrated in the long period of time. The study of short-run dynamics is often done by first eliminating trends in the variables, usually by differencing. This procedure, however, throws away potential valuable information about long-run relationships about which economic theories have a lot to say.

Unit root test

Most of economic series have a unit root in the long-run .This fact has developed the theory of non-stationary time series analysis. A time series is stationary if the series mean, variance and autocovariances are independent of time. Variance and covariance of non-stationary series are changing over time; hence standard t tests in regression are no longer valid. Dicky and Fuller proved that if a series has a unit root , then t statistic for parameter estimates in a regression is no longer distributed as a student t. In order to avoid problem of non-stationary we should use the first or higher differentiated data. The Augmented Dickey-Fuller (ADF) test developed by Dickey and Fuller (1979) and Dickey and Fuller (1981) and Philips (1987). This test is based on the following regression:

$$\Delta y_t = \alpha_0 + \alpha_1 y_{t-1} + \alpha_2 t + \sum_{i=1}^{n} \beta \Delta y_{t-i} + u_i$$

$$u_t \sim \text{iid}\ (o, \sigma_\varepsilon^2) \qquad (1)$$

where Δ is the difference operator, y is the natural logarithm of the series, t is a trend variable, α and β are the parameters to be estimated and u_i is the error term.

In equation above series is non-stationary when α_1 is zero:

H_0: $\alpha_1 = 0$

H_1: $\alpha_1 \neq 0$

In ADF unit root test the null hypothesis shows that the series is non-stationary and has a unit root. It is shown as: $y_t \sim I(1)$. H_1 hypothesis shows that the series is stationary and shown as: $y_t \sim I(0)$.

If the t-statistic on the α_1 coefficient exceeds the critical value, the series are stationary. Critical values for this t statistic are given in Mackinnon (2010).

Cointegration test

Although economic series might be non-stationary in the long-run, there may exist a linear combination of the individual series that are stationary. Cointegration is a statistical concept to test for long-run relationship among various non-stationary series. The purpose of the cointegration test is to determine whether a group of non-stationary series is cointegrated or not. The theory of cointegration developed in Granger (1983) and elaborated in Engle and Granger and Weiss (1983) and Moenning (2001) addresses this issue of integrating short-run dynamics with long-run equilibrium. The stationary linear combination is called the cointegrating equation and may be interpreted as a long-run equilibrium relationship among the variables. Granger (1987) pointed out that a linear combination of two or more non-stationary series may be stationary.

Suppose *xt* and *yt* are two random walks and hence are not stationary.That is, $yt \sim I(1)$ and $xt \sim I(1)$. Then *yt* and *xt* are said to be cointegrated if there exist a leaner vector of *yt* and *xt* that is stationary and we can find out that there is a long-run relationship between the trends in *yt* and *xt*. In general, it is expect that a linear combination of *xt* and *yt* would also be a random walk Yet, the two series may have the property

Table 2: ADF test for data (level and difference).

	Variables	t-value	Mackinnon value	p-value
Level	Rice import	-3.03	-3.24	0.000
	Per capita GDP	3.03	-3.24	0.000
	Domestic price index	-3.05	-3.25	0.000
	Foreign exchange rate	-1.93	-2.64	0.000
Difference	Rice import	-3.28	-2.99	0.000
	Per capita GDP	-4.79	-4.44	0.000
	Domestic price index	-5.91	-4.46	0.000
	Foreign exchange rate	-3.25	-3.01	0.000

Source: Research results

that a particular linear combination of them is stationary.

Several tests are proposed for examining the hypothesis that two time series are not cointegrated. The simplest one is based on testing the stationary of the error terms.

Using ADF test for the the error terms of equati

$$\Delta u_t = \alpha_0 + \alpha_1 u_{t-1} + \alpha_2 t + \sum_{i=1}^{n} \beta \Delta u_{t-i} + \varepsilon_i \quad (2)$$

where α and β are the estimated parameters and ε is the error term. The number of lags (n) chosen in equation (2) should be sufficient to ensure that the error term ε, is white noise. The choice of n is based on the modified Lagrange Multiplier (LM) statistic. The test for cointegration involves the significance of the estimated coefficient. Again the null hypothesis is that the error terms are non-stationary and acceptance of this hypothesis indicates that the series are not integrated.

VECM Correction

A vector error correction (VEC) model is a restricted VAR that has cointegration restrictions built into the specification, so that it is designed for use with non-stationary series that are known to be cointegrated. The VEC specification restricts the long-run behavior of the endogenous variables to converge to their cointegrating relationships while allowing a wide range of short-run dynamics. As the VEC specification only applies to cointegrated series, you should run the Johansen cointegration test prior to VEC specification. The cointegration term is known as the error correction term since the deviation from long-run equilibrium is corrected gradually through a series of partial short-run adjustments. Pesaran & Smith (2001).

RESULTS

It is not valid to perform OLS regression on non-stationary variables Because of the potential for spurious regression. For example, if the variables are all trending upwards then any correlation between them may simply reflect this fact rather than a genuine correlation between the variables. But we often want to test if there is a relationship between variables that are non-stationary. For this reason, we should use cointegration test to see if the variables have a common trend. If all the variables are trending upwards in a similar fashion, then they are genuinely correlated in some way. In this part of research, we examine the stationary of series involving in the rice import function including: rice import (as dependent variable), foreign exchange rate, Per capita GDP and domestic price index (as independent variables). Table 2 shows the result of ADF test for series above in their level and first differences. The results of the unit root tests demonstrate that all series are non-stationary in levels, but stationary in first differ-

Table 3: ADF test for residuals.

	Variable	t-value	Mackinnon value	p-value
Level	Error term	0.000	0.000	0.00

Source: Research results

Table 4: Result of estimation of rice import equation: 1986-2011.

Variables	Parameters	Standard Error
Per capita GDP	0.021 ***	0.038
Domestic price index	0.53 ***	0.22
Foreign exchange rate	-1.94 **	0.073

R^2=98 Method: OLS Number of Observation = 26 D.W=1.99

ences (they have a unit root). The t-statistics for all series are greater (less negative) than the critical values in levels leading to the acceptance of the null hypothesis of non-stationarity. Whereas the first difference results show that the null hypothesis can be easily rejected at the 5% significance level and hence acceptance that all the series are stationary when first differenced. The results show that all the series tested are not stationary in (log) levels but are stationary after being differenced once, fulfilling a necessary condition for cointegration. All the series are therefore assumed to be integrated of order one.

Testing for cointegration involves testing the residuals from an OLS regression for stationarity. A cointegrating relationship can only exist between I(1) variables, i.e. variables whose first difference is stationary. So unit root tests must be performed in order to establish the order of integration of the variables. Given a set of I(1) variables, testing for cointegration involves two steps: first, estimate an OLS regression on rice import equation; second, examine a unit root test on the residuals of this regression to study whether they are stationary or not. If the error term is tending to increase over time, the variables are not related - they are not cointegrated. As table 2 obviously shows the residuals of rice import equation are stationary using ADF test, because the t-statistics for residuals are less than the critical values in level. So, the null hypothesis of a unit root is rejected at the 5% significance level, the residuals are stationary, and thus a long-run, cointegrating relationship exists between the variables.

Estimation of rice import function

In the final step, rice import function is estimated using OLS method while confident about the correction of results. Although all series in rice import function are non-stationary, regression would not be spurious, because Engle-Granger test firmly demonstrated that all series are cointegrated. Estimation results of rice import function over the period 1986-2011 are presented in table 4 below. The overall fit of the equation is very good ($R^2 = 0.97$) and variables tend to have the expected signs at the highest level of significance.

As expected, rice import is positively correlated per capita income GDP and domestic price index a negatively correlated with foreign exchange rate. 1, 2 and 3 stars above parameters show significance at 10, 5 and 1 percent level respectively. Per capita GDP growth in a country with high income elasticity of demand for food- such as Iran- raise the demand for food considerably. On the production side, inelastic supply due to technological restrictions leads to a huge gap between demand and supply of food. Consequently the price of food goes up and import rise is inevitable in order to equilibrate the domestic market. On the other hand, money depreciation leads to reduction in foreign exchange rate, lowered the price of imported rice and raised the import.

CONCLUSION

Cointegration test between some variables is a test for existence of a long - term relationship between them. This research presented an analysis of cointegration test between rice import and it`s economic factors over 1986-2011, using Engle-Granger test. In the first step of results, a Dickey-Fuller test showed that all variables are non-stationary in levels, but stationary in their first differences. In the second step of results , Engle-Granger test confirmed the stationary of residuals of rice import function leading to the acceptance of cointegration between rice import and it`s economic factors including per capita GDP, foreign exchange rate and domestic price index. Finally rice import function was estimated using OLS method which demonstrated that all independent variables were significant at high level and the sign of coefficients were all consistent with theoretical expectations. As we discussed the negative effect of foreign exchange rate on rice import, authorities must protect trade barriers as an economic stimulus for farmers and reduce inflation rate which rises foreign exchange rate and consequently the cost of rice import.

REFERENCES

1- Dickey, D.A. & Fuller, W.A. (1981). Likelihood ratio statistics for autoregressive time series with a unit root, Econometrica, 49: 1057-72.

2- Dickey, D.A. & Fuller, W.A. (1979). Distribution of the estimators for autoregressive time series with a unit root, Journal of the American Statistical Association, 74: 427-431.

3- Engle, R.F., Granger, C.W.J. (1987). Cointegration and error correction: representation, estimation and testing. Econometrica, 55: 251- 276.

4- Granger, C.M.J. (1983). Forecasting white noise, in applied time series analysis of economic data, Bureau of the Census, Washington, DC.

5- Granger, C.W.J., & Weiss, A.A. (1983). Time series Analysis of Error-Corrective Models, in Multivariate Statistics (Academic Press, New York).

6- MacKinnon, J.G (2010), Critical values for cointegration, queen"s economics department working paper no 1227.

7- Moenning, C. (2001), Testing for the existance of a long- run relationship between UK consumption, income and inflation using Pesaran et al. Bounds test Procedure: A comparative Study.

8- Pesaran, H., Shin, Y., & Smith, R.J. (2001), Bounds testing approaches to the analysis of level relationships. Journal of applied Econometrics, 16: 289- 326.

9- Phillips, P.C.B. (1987). Time series regression with unit root, Econometrica, 55(2), 227-302.

Economic Analysis of Water Demand in Greenhouses of Khash Township

Javad Shahraki [1] ***and Neda AliAhmadi*** [2*]

Keywords:
Irrigation water demand, Price elasticity of demand, Succession elasticity of demand, Translog cost function

Abstract

The main goal of this study was to investigate the way of affecting of water in production of cucumber. From aspect of econometrics, mutual relationships of production function and expenses have been analysed in which Translog cost function has been used. This function has been estimated using the conditional input demand functions, the shefard theorem, chemical fertilizer, animal fertilizer, labour force, seed, and pesticide in the framework of a system of equations by taking advantage of the irrelevant regression method. Applied inputs is for 151 cucumber Beneficiaries in Khash in farming year of 2011-12 that insider and crossing tendency of input demand for this product has been investigated through collecting questionnaires and using obtained coefficients. The results show that the translog cost model is a good fit compared to the data of research. According to the reviews, demand for water has a minor Insider-succession tendency more than one which shows possible stretch of demand function rather than the price of the inputs. Hence, appropriate pricing policies can be used to take a positive step in preventing the extreme usage of this input and directing beneficiaries for optimum use of these inputs.

[1] *Assistant Professor of Agricultural Economics, University of Sistan and Baluchestan.*
[2] *Master of Agricultural Economics in MA great of the University of Sistan and Baluchestan.*
* *Corresponding author's email: ahmadi_15877@yahoo.com*

INTRODUCTION

Water is the Nowadays water is one of the main scarce input of natural resource and life on Earth. More than 70% of the Earth surface is covered with water (nearly 360 million out of 510 million cube meter). Despite of this enormous volume of water, only 2% of the Earth's water is fresh and drinkable and the remaining in unusable due to a variety of soluble salts, especially uneatable ones. From this two per cent of the fresh water, more than 90 per cent is in the form of frozen in two poles and is located away from the human access (Iran hydrology). Management and optimal decision making about water is one of the most important human concerns today due to increasing population and limited water resources.

Undoubtedly, providing Sufficient and guaranteed irrigation water for plants is the main axis of human challenge because one of the main consumers of world water resources is agricultural sector. (Kouhpayee, 2004). In water management we did not create balance just control, motivate and limit partners at both supply and demand sides. One of the management practices of increasing water efficiency in demand section, that specify the water resource value and has non-organic structures, is establishing actual price of water in both supply and demand sides because determination of actual price of water will considerably help optimal allocation of water among different products and also to logical and proper consumption of water which will finally increase application and efficiency of water. Continuing process of unreal and non-economical determining of price of water will escalate extreme consumption and indiscriminate waste. In other words, determine the actual water price will cause water usage considering allowed capacity from underground and surface water of aquifers and quotas. (Ghaffari Shirvan, 1998). On the other hand Managing water demand involves better and more effective harnessing of water which is possible through legislation of laws, planning rules and regulations, using economic tools, programming and supervising as well as contributing of beneficiaries. So the main objective of water demand management is controlling the amount of demand and efficient application of water meaning development in depth. Considering the limitation of to limitation of water supply in the Iran, it is clearly seen that emphasizing on demand management is a necessary issue to direct water sources to sustainable usage. Creating new perspectives consistent with sustainable development objectives of resources among beneficiaries and controlling the amount of water taken through existing tools can be highlights of water -demand management.

Including national and international studies have been done include:

Torkamani and Kalani (2001) have used multi-product Translog cost function to estimate cost functions and demand of inputs at the same time in agriculture of Fars province. Results show that Translog cost model is a good fit of the investigated data and chemical fertilizer is a complementary for consumed seeds and workforce, machineries, and chemical fertilizer have replacement relationships. Also, all prices stretching of demand for inputs is less than one indicating no stretching of demand for them.

Hajir Keyani and HajiAhmad (2002) have investigated the estimation of demand functions for production inputs and supply of water and dry-farming wheat in Iran agriculture. In this research Translog flexible-profit function as well as profit equations, supply and demand of production inputs using equation system seemingly duplicated unrelatedly were used. And results indicate that all stretching obtained from estimation of coefficients for both water and DEIM wheat have signs corresponding theoretical expectations. And all self-stretching for input demand indicate negative inverse relationship between price and amount of production inputs.

Abedi (2005) used multi-product Translog cost function to estimate cost functions and demand of inputs at the same time in agriculture of Zayande Roud and results show that Translog cost model is a good fit of the investigated data and according to investigations from the pattern, minor self-replacement and crossing stretching is less than one however it was near to one for water. Additionally, self-price and crossing stretching of demand for all inputs is less than one.

Bostani and Mohammadi (2006) investigated productivity and demand function of water in production of sugar cane in Gelid district. Results show that Cobb Douglas production function of consumed water inputs, poison, machineries, planted area and funds have meaningful effects on production in which consumed water and cost of machine operations is positive. Average of positive productivity and final water uses are 0.89 and 0.304, respectively. According to R^2 estimated production function is able to determine 85 per cent of changes in production. Also price stretching of demand is more than one which illustrates stretchable demand function rather than input price.

Shajari *et al.* (2008) investigated management of water demand using policy of water -price determination in Jahrom date farms and case study of Shahani date. Results show that there is a positive and meaningful relationship between number of productive date trees in the garden, number of used workforce and frequency of irrigation and amount of date produced. Final productivity and final production value of water in drip and GHARGHABI irrigation is 0.194 kg and 204.06 kg, respectively and was calculated 67.23 Rials. In final price of water, price stretching for water for drip and water irrigation was obtained -3.035 and -2.093 respectively which shows that demand for water in both methods is stretchable.

Al.Qunaibet and Johnston (1985) estimated demand function of water using time- series data in Kuwait. According to results, price stretching of water comparing to the price was calculated -0.9 .So they mentioned that increasing the price of water can lead reduction of water consumption. Diagnosis coefficient was also more than 80 per cent in this estimation. Glas *et al.* (1989) investigated agricultural structure of Northern Ireland using multi-product Translog cost function. This model contains two product of livestock and its derivations as well as plants and four inputs including fund, workforce, chemical fertilizer and related index of sum of livestock food and imported livestock to the region. They, using the mentioned model, estimated replacement stretching among inputs, self and crossing stretching of demand of inputs, parameters taken from scale and rate of Hiks technology change for each year. In this study, they concluded that agricultural structure of Northern Ireland is not homothetic and inputs of fund is stretchable while workforce is not stretchable and input includes sum of livestock food, seeds and livestock import is also not stretchable .

Ali and Parikh (1992) used estimation of Translog cost function as well as extraction of functions of cost shares using Shephards lemma in order to investigate the relationships among inputs .Equations of cost share along with total cost function were estimated using limitations of symmetry and unity by SUR method and the authenticity of these limitations were investigated by Coefficient of langraz.

Following goals are taken in to consideration:

The main purposes of this research are:

1-Estimating Translog cost function of greenhouse cucumber of Khash

2-Estimation of Cucumber water demand function in greenhouses of Khash

3-Determining insider minor elasticity, Allen crossing and insider price stretching and crossing of demands of water inputs through Translog cost function

MATERIALS AND METHODS

To estimate water – demand function and determine price elasticity of water demand for greenhouse cucumber ,first ,production function of cucumber is estimated like Translog production function.

$$Y = \alpha_0 \prod_{i=1}^{n} X_i^{\alpha_i} e^{\frac{1}{2}\sum_{i=1}^{n}\sum_{j=1}^{m} \gamma_{ij} \text{Ln}X_i \text{Ln}X_j} \quad (1)$$

Where Y is Output, a_0 is efficiency, X_i and X_j are i, j, a_i and γ_{ij} of unknown parameters. Logarithmic form of this function is as follows.

$$LnY = Ln\alpha_0 + \sum_{i=1}^{n}(\alpha_i Lnx_i) + \frac{1}{2}\sum_{i=1}^{n}\sum_{j=1}^{n}(b_{ij} Lnx_i Lnx_j) \quad (2)$$

Each production function has one cost function. So, form of Translog cost function is as follows:

$$\ln c = \ln a_0 + \Sigma a_i \ln p_i + a_Q \ln Q + 1/2\Sigma\Sigma\gamma_{ij} \ln p_i \ln p_j + \Sigma\gamma_{iQ} \ln Q \ln P_i + 1/2\, \gamma (\ln Q)^2 \quad (3)$$

Where I,j=1,……,N

In which in above equation, Pi and Pj are price of inputs and C is the total cost.

Based on the characteristics of neoclassic production theory which says that cost share is addable, the following limitations are considered for demand function.

a) unity assumption $\Sigma a_i=1, \Sigma\gamma_{iQ}=\Sigma\gamma_{ij}=\Sigma\gamma_{iQ}=0$

b) symmetric assumption $\Sigma\gamma_{ij}=\Sigma\gamma_{ji}$

On the other hand , because the total of cost shares is one ($\Sigma s_i=1$) ,using these conditions on cost function, we need n-1 share and the prices in main function will change in to relative. To avoid becoming zero for koarians Variance matrix, one of equations of cost share is omitted and it is normally omitted in experimental tasks which has the minimum contribution to production costs (Mcgeehan,1993).

To estimate the cost function, ISUR has been applied. The reason to use this technique is existence of unity between impaired sentences in equations of cost share. Since total of cost share is one, if the left side of demand equations are summed ,this sum should be one. Thus, the sum of impaired sentences of equations is zero. On the other hand, independent variables (price of inputs and amount of production) are the same in all demand equations. As a result, all variables which have not been entered will appear in impaired sentences and will unity among impaired sentences. The common way to estimate functions is that one of equations of cost share is taken ways from the simultaneous –function category, and parameters of other functions are estimated and then relevant parameters of taken-away equation are calculated based on other parameters of equations. Since sum of equations of cost share is one, omitting any equation is optional (we deleted poison equation here). Regressions seemingly duplicated unrelatedly tend to maximum method of being true and are independent from omitted equation.

Elasticity of substitution and price elasticity of demand

One common index to measure level of reaction to price changes of production input is Allen AES –Evzaza V_{ij} between to inputs of I and j which are obtained from first and second derivation of cost function to price of I and j inputs.

$V_{ij} = c\ c_{ij}/c_i c_j$

For logarithm cost function, Allen AES can be written as follows:

$V_{ij} = b_i + s^2_i - s_i / s^2_i$

And self-replacement stretching is as follows:

$V_{ij} = b_{ij} + s_i s_j / s_i s_j$

There is a positive replacement between production inputs meaning the replacement of these inputs. While, negative replacement stretching determines that inputs are completing each other. One main gaol of this study is estimating replacement stretching among production factors to investigate the level of replacement of complementary of each pair of production inputs in agricultural sector. Investigation of this issue has interesting applications in policy making in investment.

One factor that can illustrate the technology in one production unit in summary is price stretching of demand for production inputs (Eij) which is defined as follows:

$E_{ij} = \partial \log X_i / \partial \log P_i$

Where production amount (Xi) and the price of all production inputs (Pi) are fixed.

Price self-stretching of demand (Eij) of a production input shows relative changes of input demanded of that input resulting from relative changes of that input. Price crossing stretching of demand (Eij), i ≠j measures relative changes in amount of an I input being demanded because of relative changes of price of j input.

We can show that Allen AES is related with price stretching of demand for production factors. Price crossing stretching of demand for production inputs can be written as follows:

$E_{ij} = S_j V_{ij}$

And price self-stretching of demand is written as follows:

$E_{ij} = S_i V_{ii}$

Where I, j=L, H, SH, W, S, B (Sherafat,1996)

In this research price of inputs and products is determined according to market price. Share of costs has been calculated through dividing paid costs for that factor to total of varying costs.

RESULTS AND DISCUSSION

Estimation of function of water-input demand for cucumber using Translog cost function Estimation results of Translog cost function coefficient using regression seemingly unrelated are illustrated in table number 1. Poison cost has been omitted in estimation of this model.

Table content illustrates that most coefficient

Table 1: Estimation of total cost of production equation for greenhouse cucumber in Khahsh.

Parameter	Coefficient	t-value	Parameter	Coefficient	t-value
C(1)	101.61	8.13	C(15)	-0.14	-2.22
C(2)	-16.4	-8.04	C(16)	-2.79	-1.72
C(3)	-3.74	-1.72	C(17)	4.51	3.50
C(4)	10.05	2.23	C(18)	-0.81	-3.71
C(5)	17.79	4.55	C(19)	-2.59	-1.78
C(6)	-2.01	-2.32	C(20)	0.42	1.70
C(7)	0.31	1.64	C(21)	0.057	0.56
C(8)	0.25	2.47	C(22)	-1.67	-1.04
C(9)	-1.38	-4.47	C(23)	0.43	6.58
C(10)	-1.87	-6.77	C(24)	0.066	0.49
C(11)	0.14	2.16	C(25)	0.21	0.62
C(12)	0.15	1.08	C(26)	-0.65	-2.25
C(13)	-0.48	-1.61	C(27)	-0.13	-1.54
C(14)	-0.2764	-0.99	C(28)	-0.09	-0.65

R^2=0.82 D.W=2.02

in a=0.05 are meaningful and thus the issue that most equation coefficients are meaningful is a suitable result. According to the fact that determination coefficient is 0.82 per cent and shows a good fit of equation, it means that independent variables have been able to explain the reason of total cost change. And Durbin- Vatson data shows that Autocorrelation phenomenon does not exist. (c(1) Labor, c (3) livestock fertilizer, c (4) water, c (5)seeds, c (6) chemical fertilizer).

Based on research goal, share of water cost which is too similar to demand function has been used as follows in order to analyse more.

$$S_w = 10.05-1.38\log(P_L/P_s) - 0.48\log(P_H/P_S) - 2.79\log(P_W/P_S) + 4.5\log(P_B/P_S) - 0.81\log(P_{SH}/P_S) + 0.21\log Q$$

t= (4.4) (0.30) (0.30) (1.62) (1.28) (0.21) (0.34)

As it can be seen, all coefficient of cost-share equation is meaningful based on amount of t. Investigating above coefficients show that cost-share equation of water has a positive relationship with amount of production and price of seeds and negative relationship with price of workforce, livestock fertilizer, water and chemical fertilizer. In other words, increasing the price of seeds as well as increasing production, water cost out of total cost increases and increasing the price of workforce, livestock fertilizer, water, and chemical fertilizer decreases relatively.

Calculating stretching

In this section, minor self and crossing stretching of Allen and self and crossing price stretching of demand using average cost of inputs are being calculated and investigated.

The following results are taken from table 2 and 3:

1- The self-stretching of water has a logical sign(negative) which is compatible with economic theories and show that there is an inverse relationship between prices and amount of inputs and that illustrates the inverse relationship

Table 2: Estimation of replacement stretching of inputs for production of cucumber.

	Labor	Animal manure	water	Seed	Chemical fertilizer
Labor	-0.9	0.3	0.3	0.8	0.2
Animal manure	0.3	-1.6	6.7	19.5	5.4
water	0.3	6.7	-1.6	16.4	4.1
Seed	0.8	19.5	16.4	-3	19.3
Chemical fertilizer	0.2	5.4	4.1	19.3	-0.3

Table 2: Estimation of Demand price stretching of inputs for production of cucumber.

	Labor	Animal manure	water	Seed	Chemical fertilizer
Labor	-0.36	0.12	0.12	0.12	0.03
Animal manure		-0.58	2.27	2.72	0.75
water			-1.69	2.81	0.57
Seed				-2.49	2.69
Chemical fertilizer					0.03

between demand for water and its prices.

2- Absoulute amount of self-stretching for inputs including workforce and chemical fertilizer is less than one which shows none stretching of demand for them. However absoulute amount self-stretching of inputs including water, seeds, livestock fertilizer is more than one which shows stretching of demand inputs.

3- Crossing stretching of demands for water with other inputs shows that it has a replacement stretching in comparison with work force, livestock fertilizer, seeds and chemical fertilizer.

4- Maximum level of replacement of water stretching among inputs is for seeds. In other words, increasing the price of water will increase applying the seeds. And the sensitivity of replacement between these two inputs is less than one and that shows that if the price water goes up , seeds with better quality and modified ones will be used to save water in order to decrease water consumption.

CONCLUSIONS AND RECOMMENDATIONS

1- According to high self-stretching of price , derived demand of water shows that policy in determination of price can be a useful tool to control extreme usage of water in production of green houses.

2- Since amounts of crossing stretching of demand(except work-force input) is more than one, policies to change the price in these inputs will have a considerable effect on planting composition.

3- The amount of self-stretching for water demand is more than one. Thus, changing the price of this input (deep water) may create a pressure on consumption of other kinds of water(surface water).

4- Existence of replacement connection among water, workforce, livestock and chemical fertilizer, and seeds expresses higher usage of these inputs in planting cucumber to save water.

5- Small replacement stretching of workforce in comparison with water lead that relevant policies to factor change affecting in demand of an input not to have considerable effect on composition of other consuming inputs.

As results show demand for water is stretchable. According to being stretchable for water demand, it can be certainly said that price determination of water has an important role in applying this input.

According to obtained results from this study, following solutions are recommended:

1- Possibility to implement the policy of water-price determination with determined water price in this research should be gradually done during the time unless farmers with low efficiency will be omitted.

2- Of course this important issue needs to be taken un consideration that determining water price is not only enough and it is necessary to apply collecting water price and using it in order to improve water resources along with this policy.

3- Creating association to supervise water resources, collecting water cost and expending it to improve the water resources is necessary.

4- Increasing the efficiency of water usage in farms and greenhouses (due to growing up the price of water for irrigation) to farmers and investors is an essential subject.

5- Installing smart counters on agricultural well and controlling the amount of water being taken as well as controlling the irrigation of farms using micro irrigation in greenhouse planting.

REFERENCES

1-Abedi, A. (2005).Using multi-product Translog cost function in simultaneous estimation of supply and demand function of agricultural inputs .case study Isfahan Zayanderoud, fifth two-year conference of agricultural economy, Iran, Sistan and

Baluchistan University
2-Ali, F., & Parikh, A. (1992). Relationships among labor, bullock, and tractor inputs in Pakistan agriculture. American Journal of Agricultural Economics, (4):371-377.
3-Al.Qunaibet, MH., & Johnston, R.S. (1985). Municipal demand for water hodological issues and empirical results. Water Resource. Research. 21(4), 433-438.
4- Antle, J.M. (1984). The Structure of U.S. Agriculture Technology 1910-78. American Journal of Agriculture Economic, 66: 414-121.
5-Bostani, F., & Mohammadi, H. (2006). Investigation of productivity and function of water demand in producing sugar cane in Gelid region. Sugar Cane 23(2), 196-185.
6- Clark, J.S., & Youngblood, C.E. (1992). Estimating duality models with biased technical change: a time series approach. American Journal of Agricultural Economics, (74): 353- 360.
7- Ghafarishirvan, G. (1998). A review on taking water from resources of Iran, collection of ninth conference articles of irrigation and drainage of Iran.
8- Glass, J. C., & McKillop, D. G. (1989). A multi-product multi-input cost function analysis of northern Ireland agricultire, 1955-85. Journal of Agricultural Economics, 40(1), 57-70.
9- Koupayee, M. (2004). Importance and position of underground water in hydrology circle ,collection of second national student conference of water and soil resources, Shiraz University.
10- Mcgeehan, H. (1993). Railway costs and productivity growth: The case of the Republic of Ireland 1973 - 1983. Journal of Transport Economics and Policy. PP, 19 -32.
11- Pakravan, M., Mehrabiboshrabadi, H., & Taherinejad, M. (2008). Investigation of water demand management in production of potatoes: Case study Kerman province, National congress of water crisis management, Islmic Azad university, Marvdasht branch, Iran.
12-Shajari, Sh., Barikani, E., & Amjadi., A. (2008). Management of water demand using determination of water price in Jahrom date farms Case study: Shahani date. Quarterly Journal of Agriculture Economy and Development, 17(65), 55-72.
13- Sherafat, M. (1996). Investigation of production technologic structure and estimation of demands for production inputs; main agricultural products (wheat, oat, cotton, sugar cane), Deputy of Economic Affairs of Economy Ministry, Tehran.
14- Torkamani, J., & Kalani, E. (2001). Multi-product Translog cost function in simultaneous estimating of demand and supply functions of inputs in agriculture: Case study Fars province. Quarterly Journal of Agriculture Economy and Development, 9(34), 101-124.
15- World Bank, (1993). Water resources management, a World bank policy paper. The World Bank, Washington, DC,USA.

11

Investigation of the Factors Hindering and Promoting Tourism Evidences from Villages of Abyaneh Village

Hoda Manafian[1*], Zhila Daneshvar Ameri [2] and Saba Salehi Esfahani [3]

Abstract

Rural tourism, a dynamic industry with a promising future, can play a key role in economic improvement of talented countries such as Iran. Abyaneh village is one of the best-known villages in both Iran and the world. The number of annual visitors of this ancient village is approximately 550,000 that is a considerable number. Within a short radius of Abyaneh there are five villages (Barz, Tare, Komjan, Yarand, Hanjan) with a lot of potentials to develop rural tourism. But only 20% of tourists of Abyaneh population consider the surrounding villages as a separate destination. The large number of tourists visiting Abyaneh village provides opportunity for managing rural tourism in this area is to attract tourists to the surrounding villages around Abyaneh. In the present study, the hindering and promoting factors of fostering tourism in these villages have been compared by both the surrounded villagers' and locals' of Abyaneh village. To this end, 222 questionnaires were distributed among the individuals and the means were compared using Mann-Whitney test. The most important unfavorable factor, from the surrounded villagers' point of view, is lack of enough advertising and marketing and from the locals of Abyaneh village points of view, is the shortage of attractions, compared with attractions of Abyaneh.

Keywords:
Tourism development, Local society, Rural tourism, Abyaneh village

[1] M.Sc Department of Rural Development and Management, Faculty of Agricultural Economic and Development, College of Agriculture and Natural Resources, University of Tehran, Iran.

[2] Assistant Professor of Rural Development, Department of Agricultural Economic and Development, College of Agriculture and Natural Resources, University of Tehran, Iran.

[3] Department of Recreation, Parks & Tourism Management, Kent State University, United States of America (USA).

** Corresponding author's email: hmanafian@ut.ac.ir*

INTRODUCTION

Abyaneh is considered one of the historic and cultural attractions of Isfahan province and a major rural tourism destination in Iran. Within a short radius of Abyaneh, there are well-known villages with plenty of potentials to promote tourism such as Barz-Abyaneh route (Barz, Tare, Komjan, Yarand, Hanjan). These villages take advantage of their capacities in terms of accommodating tourists which result in attracting a lot of visitors from all over the world from which the locals will benefit in particular. Obviously, the more tourists visit a place, the more money the locals would make. However, the aforementioned route has not yet been regarded and recognized as an independent destination by the visitors to Abyaneh village. Center of the Iranian Cultural Heritage and Tourism Organization (ICHTO) (2010) declared that approximately 550000 tourists visit Abyaneh village annually but only 20% of that population consider the surrounding villages as a separate destination of tourism. Although those villages are located close to Abyaneh, they are overlooked by tourists. The main goal of this paper, as a result, is to determine the main factors hindering and promoting of tourism in these villages. In addition of the capacity and the good location of these villages there are two more reasons legitimizing to try on the Abyanian tourists to visit the surrounding villages.

Firstly, Abyaneh is a small village with ancient structure which is vulnerable to the number of visitors larger than its capacity. Nowadays, having too many visitors in Abyaneh has had some negative impacts among which we can name of locals' general discontent (ICHTO, 2010). Other negative outcomes mentioned in the studies are, the accumulation of garbage, pollution of land and water supplies and waste disposal (Stevens, 1993; Ghaderi and Henderson, 2012; Yong, 2012), rise in land and food prices (Stevens 1993; Shackley 1994; Mulligan 1999), some disruptions that local villagers suffered from (lack of respect for their culture and privacy) (Long and Wall 1996; Belsky 1999; Campbell 1999; Saul 1999), obtaining the majority of economic benefits by local elites (Long and Wall 1996; Belsky 1999; Campbell 1999; Saul 1999), congestion in their village as well as on roads and loss of privacy and quietness (Monteiro, 2011). These unfavorable factors caused a strong eagerness to adjustment the population of tourists.

Secondly, based on observations and surveys, the tourists have also proved not to be satisfied with facilities like restaurants, cabins for short lodging, recreational centers and parking lots. Therefore, providing the facilities demanded to promote tourists' satisfaction appears to be logical and essential. Regarding Abyaneh's vulnerability, it would be a good idea to compensate for these inadequacies in Abyaneh by its villages. This has an additional advantage for Abyaneh's tourists because it gives them a chance to experience the beauties of these new destinations.

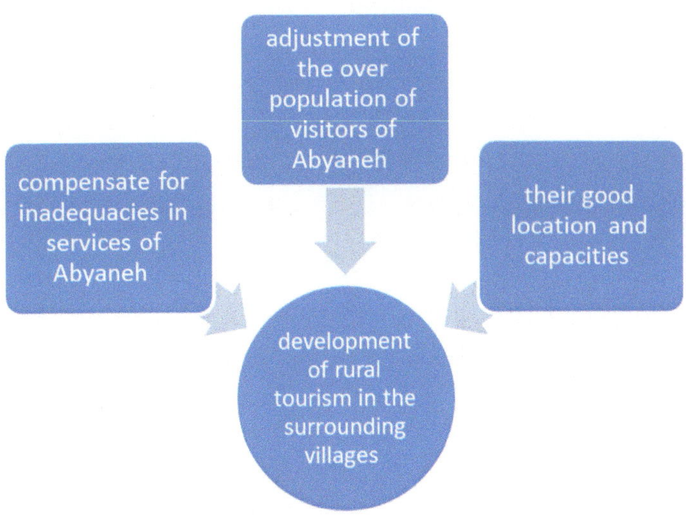

Figure 1: Positive effects of flourishing tourism in surrounding villages on Abyaneh's states.

Figure 1 shows the positive effects of flourishing tourism in surrounding villages on Abyaneh's states.

Generally, any destination can be managed efficiently by improving visitors' satisfaction or introducing a new destination to them (Kozak, 2002). Also it is feasible to extend the market to new tourism destinations by creating variety in current services in different destinations (Jeffrey and Xie, 1995). Because of the vulnerability of Abyaneh village and the necessity of creating new services in this village, introducing a new destination can be the best way to solve the present problems of this historic village. It is possible, as a result, to create a new tourism market (in the five villages) close to the main destination (Abyaneh) by evenly motivating the current visitors of Abyaneh and providing services with higher standards in those nearby places. This is the fundamental ideology of this study. Therefore, it requires investigating the factors that hinder or promote tourism in these villages. In fact it is vital to target prospective customers by being aware of the efficient factors which form bonds between tourism product and tourists (Jeffrey and Xie, 1995). This study tries to investigate these bonds based on the ideas of surrounded villagers' and locals' of Abyaneh. In addition, assessing individuals' perception will help detect areas of strengths or weaknesses within the destination (Kazemi, 2009; Bashar and Puad, 2010; Naidoo *et al.*, 2010). Such knowledge should help governments, tourism agencies, and operators to improve marketing planning and investment decisions in the rural tourism context.

Study area

Barz-Abyaneh route, with its six historic villages at the foot of Karkass Mountain, which overlooks a rocky landscape, is located in the Isfahan province of Iran (Figure 2).

Its attractions are natural scenery and landscape, some historic and cultural monuments, a famous ritual, holy shrine, convince to reach, accommodation facilities (hotel), local products (dried fruits, local sweets) and finally flora and fauna. As it is located near the mountains and desert, the village has a favorable climatic conditions and delicious summer fruits. Abyaneh village is

Figure 2: Abyaneh In Iran.

Table 1: Some tourist attractions in Barz-Abyaneh route.

Village	Natural scenery	Convince to reach	Holly shrine	Local products	Wild life	Famous rituals	Historical and cultural monuments	Hotel	Distance from Abyaneh (Km)
Abyaneh	+	+	+	+	+	+	+	+	0
Tareh	+	+		+	+		+		4
Barz	+	+		+	+		+		8
Komjan	+	+	+	+	+		+		10
Yarand	+	+		+	+		+		15
Hanjan	+	+	+	+	+		+		19

Source: ICHTO (2012)

characterized by peculiar red clay in buildings, traditional costumes, local dialect and a plethora of various handicrafts. From an archeological perspective, this is a village of traditions, architectural styles (all in red clay). Though the attractions of those five villages are not as unique as Abyaneh village, they can be regarded as different destination for tourists. In table (1) their names, distances from Abyaneh and their attractions are mentioned (UNESCO, 2012; ICHTO, 2012).

MATERIALS AND METHODS

This study is descriptive- analytical. The instrument used in this study was a questionnaire which was partly constructed by the researcher and devised in 5-grade Likert scale. The questionnaire focused on promoting and inhibiting factors of tourism development in the villages around Abyaneh. The secondary data were gathered from the internet and the previously published journals. The selection of these independent variables was based on suggestions and empirical testing by a number of researchers (Roknoddin Eftekhari and Mahdavi, 2006; Aref, 2010; Yong, 2012; Ghaderi and Henderson, 2012; Armestrong *et al.*, 1997; Danaher and Arweiler 1996; Richardson and Crompton, 1988; Birdir *et al.*, 2013; Ince and Bowen, 2011; Latkova and Vogt, 2012). Afterwards, through interviews and direct observation, first-hand data were accumulated and finally using Mann-Whitney test, the average opinions of these two groups were compared. Mann-Whitney was chosen as it is used when the data are on ordinal scale. This test measures the differences between the two groups by rating the measures related to the two groups simultaneously regardless of their belonging to a specific group. Use the Mann-Whitney test when the data do not meet the requirements for a parametric test (i.e. if the data are not normally distributed; if the variances for the two conditions are markedly different; or if the data are measurements on an ordinal scale. (The Mann–Whitney test is used to compare two independent samples when data are either interval scale but assump-

Table 2: Some characteristics of respondents.

Characteristic	Groups	Levels of variables	Number	Percent%
	Abyaneh residents	Female	69	61.6
	Other		43	39
Gender	Abyaneh residents	Male	43	38.4
	Other		67	61
Age (year)	Abyaneh residents	18- 40	27	24.1
		41-60	60	53.6
	Other	61-80	25	22.3
		18- 40	36	32.7
		41-60	65	59.1
		61-80	9	8.2

Table 3: Prioritizing promoting factors of tourism in the villages surrounding Abyaneh from the surrounded villagers" and Abyaneh locals' point of view.

Unfavorable Factors	Groups	Priority	CV	S.D.	Mean
Ecological and visual attractions (natural scenery, gardens, farms, moderate climate, Karkas wild life preserve area, snow- covered altitudes)	Abyaneh residents	1	0.33	1.41	4.24
	Other	1	0.14	0.68	4.77
High potential of social participation (locals' inclination to cooperate)	Abyaneh residents	9	0.56	1.19	0.12
	Other	2	0.15	0.76	4.85
architectural, religious, cultural and historical attractions	Abyaneh residents	10	0.57	0.80	1.39
	Other	3	0.16	0.80	4.97
Transport infrastructure (smooth roads, taxi system)	Abyaneh residents	3	0.35	1.35	3.79
	Other	4	0.174	0.80	4.59
The positive effect of population explosion in cities and citizens rush to villages	Abyaneh residents	4	0.38	1.19	3.15
	Other	5	0.176	0.85	4.83
Increasing attention to tourism and its position as a psychological need	Abyaneh residents	6	0.48	1.55	3.25
	Other	6	0.185	0.85	4.58
Rich social capital in these villages (high education of the youth and literacy of the locals)	Abyaneh residents	5	0.36	0.77	2.1
	Other	7	0.186	0.83	4.47
Being close to Abyaneh, Isfahan and Tehran which are likely to send many tourists to these villages	Abyaneh residents	2	0.44	1.35	3.07
	Other	8	0.19	0.80	4.19
Additional attractions (Being close to Kashan, where tourists can enjoy Golab festival*)	Abyaneh residents	7	0.50	1.30	2.59
	Other	9	0.20	0.85	4.35
A chance for these villages to compensate the shortcoming the services in Abyaneh to attract more its tourists	Abyaneh residents	8	0.51	1.43	2.78
	Other	10	0.22	0.95	4.33

*Golab festival: A festival in which locals produce red Rose essence in Kashan city of Isfahan province, near Barz-Abyaneh route (ICHTO, 2012).

tions for t-test (normality) are not satisfied, or ordinal (ranked) scale (Forrest and Andersen, 1986; Bergmanna *et al.*, 2000; Lind *et al.*, 2004; McCrum-Gardner, 2008). Because scale of measurement was ordinal, therefore The Mann-Whitney test was used for comparison. The study was carried out in summer 2010 and the number of subjects was determined by 110 surrounded villagers and 112 locals of Abyaneh.

Preliminary results and discussion

Some characteristics of the subjects in the study are shown in table 2.

Prioritizing promoting factors of tourism in the villages surrounding Abyaneh from the surrounded villagers' and locals' points of view

To develop tourism industry in villages around Abyaneh, promoting factors should be identified. The list of promoting factors from the Abyaneh residents' and surrounded villages' points of view are provided in table 3. Their prioritizing is based on their coefficient variance (CV). The factors in this table were partly constructed by the researcher such as "additional attractions","compensate the shortcoming the services in Abyaneh" and" High potential of social participation". The other factors are extracted from references.

The factor with top priority, from Abyaneh residents' and other points of view is Ecological

Table 4: Prioritizing unfavorable factors of tourism in the villages surrounding Abyaneh from the surrounded villagers' and locals' point of view.

Unfavorable Factors	Groups	Priority	CV	S.D.	Mean
Lack of enough advertisement and marketing (tourists' un awareness of these attractions)	Abyaneh residents	3	0.35	1	2.86
	Other	1	0.36	1.39	3.86
Shortage of architectural, religious, cultural and historical attractions (compared with Abyaneh village)	Abyaneh residents	1	0.31	0.91	2.88
	Other	2	0.37	1.34	3.55
Lack of required facilities and services for tourists	Abyaneh residents	4	0.41	0.99	2.41
	Other	3	0.38	1.29	3.39
Locals' reluctance to attract and accept tourists because of the harms done by them	Abyaneh residents	2	0.33	0.97	2.95
	Other	4	0.40	1.32	3.27

and visual attractions with CV of 0.33 and 0.14 respectively. The least important factors are architectural, religious, cultural and historical attractions with the variance coefficient of 57%. The calculated mean is based on the respondents' answers to the questionnaire on the Likert scale.

Prioritizing unfavorable factors of tourism in the villages surrounding Abyaneh from the surrounded villagers' and locals' point of view

To develop tourism in the villages around Abyaneh, unfavorable factors, as shown in table 4, should be investigated as well.

The most important unfavorable factor, from the surrounded villagers' point of view, is lack of enough advertising and marketing (tourists are not aware of these attractions). This factor has the coefficient variance of 0.36. From the Abyaneh locals' points of view, the most important unfavorable factor is the shortage of architectural, religious, cultural and historical attractions (compared with Abyaneh village). The least important factor, according to surrounded villagers, is the Locals' reluctance to attract and accept tourists because of the harms of tourism industry that are explained before, in introduction part, which has the coefficient variance of 0.40. Locals believe that the least important factor is the lack of facilities and services with the coefficient variance of 0.41. We can conclude that Abyaneh inhabitants think that the villages around Abyaneh are not capable of attracting tourists even if more facilities can be provided due to the fact that Abyaneh locals believe that the nearby villages basically shortage of architectural, religious, cultural and historic attractions (compared with Abyaneh village).

A comparison among favorable factors in creating rural tourism in the villages around Abyaneh from the Abyaneh locals' and surrounded villagers' points of view

As shown in table 5, there is a significant difference between the two groups in all items estimating favorable factors.

The rate averages given to item 9 and 10 by the Abyaneh locals are significantly higher than the rate averages given to them by surrounded villagers. But for other items it is vice versa. Actually, the rate averages given to items 1-8 by surrounded villagers are significantly higher than the rate averages given by Abyaneh locals.

A comparison between unfavorable factors in rural tourism in the villages around Abyaneh village from the Abyaneh locals' and surrounded villagers' points of view

It can be inferred from table 6 that the last 2 items are not significantly different from the notions of the Abyaneh locals and surrounded villagers; so they are not discussed here.

In fact, items 3 and 4 are not significantly different and in this respect, the two groups have similar ideas. But, regarding other items, there are significant differences at the level of 90% and 95%. According to the surrounded villagers, the most important factor is the lack of enough advertising and marketing. While, according to the locals of Abyaneh, the villages around Abyaneh, not having enough tourist attractions, could be considered as the most im-

Table 5: A comparison between favorable factors in creating rural tourism in the villages around Abyaneh village from the Abyaneh locals' and surrounded villagers' points of view.

Unfavorable Factors	Rate Average		U- Mann Withney	p-value
	Other n=110	Abyaneh residents n=112		
Ecological and visual attractions (natural scenery, gardens, farms, moderate climate, Karkas wild life preserve area, snow- covered altitudes)	139.04	110.59	6020**	0.001
High potential of social participation (locals' inclination to cooperate)	115.99	135.97	6549*	0.013
Architectural, religious, cultural and historical attractions	140.29	109.22	5857**	0.001
Transport infrastructure (smooth roads, taxi system)	152.29	96.01	4285**	0.000
The positive effect of population explosion in cities and citizens rush to villages				
Increasing attention to tourism and its position as a psychological need	138.37	111.33	6108.5**	0.003
Rich social capital in these villages (high education of the youth and literacy of the locals, locals' inclination to cooperate)	135.81	114.15	6443.5*	0.014
Being close to Abyaneh, Isfahan and Tehran which are likely to send many tourists to these villages	144.44	106.66	5314**	0.000
Additional attractions (Being close to Kashan, where tourists can enjoy Golab festival***)	115.15	136.89	6538.5**	0.003
A chance for these villages to compensate the	139.37	110.23	5977**	0.001
shortcoming the services in Abyaneh to attract more its tourists	146.06	102.87	5101.5**	0.000

*P < 0.01 ** P < 0.05

*** Golab festival: A festival in which locals produce red Rose essence in Kashan city of Isfahan province, near Barz-Abyaneh route (ICHTO, 2012)

portant factor. Besides, regarding average rate, surrounded villagers' did not consider being close to Abyaneh as an effective factor as locals in Abyaneh.

All in all by comparing the opinions of the 2 groups of subjects, it can be concluded that

Table 6: A comparison between unfavorable factors in creating rural tourism in the villages around Abyaneh village from the Abyaneh locals' and surrounded villagers' view point.

Unfavorable factors	Rate Average		U- Mann Withney	p-value
	Abyaneh residents n=112	Other n=110		
Lack of enough advertisement and marketing (tourists' un awareness of these attractions)	143.64	78.78	2.560**	0.000
Shortage of architectural/ religious/ cultural and historical attractions (compared with Abyaneh village)	97.14	126.12	4.552**	0.001
Lack of required facilities and services for tourists	107.33	115.74	5.694	0.317
Locals' reluctance to attract and accept tourists because of the harms done by them	106.07	117.03	5.552	0.181

** P < 0.05

surrounded villagers were very much optimistic to the development of rural tourism in the areas concerned in the study. As mentioned before, the most effective factor is the role of tourists sending cities like Tehran, Isfahan, and Abyaneh. Of course, in the interview locals rated their village higher than Tehran and Isfahan. Generally it seems that locals of Abyaneh believe that the villages around Abyaneh owe their development to Abyaneh and its remarkable attractions and they don't have any potential on their own. According to surrounded villagers, facilities and transportation infrastructures are more important than other factors.

In fact, it shows that surrounded villagers believe to be in Abyaneh does not necessarily tempt the tourists to visit surrounding villages. Ways of motivating them to be willing to visit these places, such as advertising and marketing, should be explored.

CONCLUSION

Based on the findings of the study, surrounded villagers believed that the most notable as well as the second most important favorable factors are visual and ecological attractions respectively. It is vital to plan and budget properly to further identify, develop and preserve natural and ecological attractions. For example, it can be recommended that a ski slope be constructed and winter sports and mountaineering be encouraged. In addition, there is a good opportunity for professional visits of wild life (to create a situation to visit animals in close range) and flora and fauna of the region and these villages have been identified as suitable sites for bird watching to create variety (especially in comparison to the services offered in Abyaneh) and finally raise money. Of course, it shouldn't be forgotten that all these activities are supposed to be carried out considering environmental principals and the capacity of the region.

Owing to the fact that surrounded villagers estimated the variable of lack of marketing and informing people as the most notable factor, proper measures should be taken to inform public about the attractions and potentiality of the villages as soon as possible. Of course, this necessitates to change potential to practical advantages in the first step and to introduce them to the surrounded villagers.

REFERENCES

1- Aref, F. (2010). Community capacity as an approach for sustainable tourism. E-Review of Tourism Research (eRTR), 8(2).

2- Armestrong, R.W., Mock, C., Go, F., & Chan, G. (1997). The important of Cross-Cultural expectations in the measurement of service quality perceptions in the hotel industry. International Journal of Hospitality Management, 16 (2), 181-190.

3- Belsky, J. M. (1999). Misrepresenting communites: The politics of community based rural ecotourism in Gales Point Manatee, Belize. Rural Sociology, 64(4), 641-666.

4- Bergmanna, R., Ludbrookb, J., & Will P.J.M. Spoorena (2000). different Outcomes of the Wilcoxon Mann WhitneyTest from Different Statistics Packages. The American Statistician, 54(1), 72-77.

5- Birdir, S., Unal, O., Birdir, K., & Williams, A.T. (2013). Willingness to pay as an economic instrument for coastal tourism management: Cases from Mersin, Turkey. Tourism Management, 36: 279–283.

6- Campbell, M. (1999). Ecotourism in rural developing communities. Annals of Tourism Research, 26(3), 534-553.

7- Danaher, P. J., & Arweiler, N. (1996). Customer Satisfaction in the Tourist Industry A Case Study of Visitors to New Zealand. Journal of Travel Research, 35(1), 89-93.

8- Forrest, M., & Andersen, B. (1986). Ordinal scale and statistics in medical research. British Medical Journal, Statistics in Medicine, 292: 537-538.

9- Ghaderi, Z., & Henderson, J.C. (2012). Case study sustainable rural tourism in Iran: A perspective from Hawraman Village. Tourism Management Perspectives, 2(3), 47–54.

10- ICHTO (2010). Booklet of Abyaneh pilot area. Tehran: Deputy of Investment and Projects, Iran Cultural Heritage, Handicrafts and Tourism Organization Published in Persian.

11- ICHTO (2012). About Iran. Destinations. Isfahan. Retrieved from. http://www.ichto.ir/ Default.aspx?tabid=1153.

12- Ince, T., & Bowen, D. (2011). Consumer satisfaction and services: insights from dive tourism. Service Industries Journal, 31(11), 1769-1792.

13- Jeffrey, D., & Xie, Y. (1995). The UK market for

tourism in China. Annals for Tourism Research, 22(4), 857-876.
14- Kazemi, M. (2009). The analysis of citizens, perception role in tourism development in chabahar (case study: zahedan citizens).Geography and Development, 6(12), 81-100.
15- Kozak, M. (2002). Comparative analysis of tourist motivations by nationality and destinations. Tourism Management, 23(3), 221-232.
16- Látková, P., & Vogt, C. A. (2012). Residents' attitudes toward existing and future tourism development in rural communities. Journal of Travel Research, 51(1), 50-67.
17- Lind, D., Marchal, W., & Wathen, S. (2004). Statistical Techniques in Business and Economics, 12e. Retrieved on March 2, 2008, from University of Phoenix, eResource.
18- Long, V., & Wall, G. (1996). Successful tourism in Nusa, Lembongan, Indonesia. Tourism Management 17(1), 43-50.
19- McCrum-Gardner, E. (2008).Which is the correct statistical test to use? British Journal of Oral and Maxillofacial Surgery, 46: 38- 41.
20- Monteiro, Jeronimo g.r. (2011) Dancing to the tune of waves: impact of Maritime history on tourism in Goa. The International Conference on Tourism (ICOT): Tourism in an Era of Uncertainty. 27–30 April. Cyprus University of Technology, Crete, Greece, 525- 538.
21- Mulligan, P. (1999). The marginalization of indigenous peoples from tribal lands in South East Madagascar. Journal of International Development 11(4), 644-659.
22- Richardson, S.L., & Crompton, J. (1988). Vacation patterns of French and English Canadians. Annals of Tourism Research, 15(4).
23- Roknoddin Eftekhari, A., & Mahdavi, D. (2006). Strategies for developing rural tourism using the SWOT Model, Case Study: Lavasan-E Kuchak Rural. Modarres Humanities Journal, 10 (45).
24- Saul, R. (1999). No time to worship the serpent deities: Women, economic change, and religion in norht-western Nepal. Gender and Development, 7(1), 31-39.
25- Shackley, M. (1994). The land of Lo, Nepal / Tibet. Tourism Management, 15(1), 17-26.
26- Stevens, S. F. (1993). Tourism, change, and continuity in the Mount Everest region, Nepal. Geographical Review, 83(4), 410-428.
27- UNESCO. (2012). The historical village of Abyaneh. Retrieve from http:// whc. unesco. org/ en/ tentativelists/5197.
28- Yong, Li. (2012). Tourists' perceptions of ethnic tourism in Lugu Lake, Yunnan, China. Journal of Heritage Tourism, 7(1), 59–81.

Assessment of Land Management Practices in Food Crops Production among Small Scale Farmers in Kwara State, Nigeria

Abdulazeez, Muhammad-Lawal[1], Omotesho, Kemi Funmi [2], Adekola, Olabisi Fatimat [3] and Adekunle, Dayo[1]*

Abstract

The study carried out an assessment of agricultural land management practices in food crops production among small scale farmers in Kwara Sate, Nigeria. Specifically, the study determined the socio economic characteristics of crop farmers in Kwara State; ascertained the cropping patterns common among them; highlighted the soil conservation methods adopted by the farmers; examined the relationship between selected socio-economic characteristics of the farmers and their adoption of major agricultural land management practices; and investigated the constraints to adoption of sustainable agricultural practices among crop farmers in Kwara State, Nigeria A three stage random sampling technique was used in selecting a total of one hundred and forty four small scale food crops farmers. Descriptive statistics, binary logistic regression model and four point Likert-type scale were used to analyse the data for the study. The study revealed that food crops production in Kwara State is dominated by middle aged men who are poorly educated and have poor access to agricultural extension services. Half of the respondents (48.60%) adopted cereal-based cropping systems. 48.6% of the farmers adopted a minimum of three management practices. Crop rotation was mostly adopted by all the farmers while irrigation was the least adopted by only 29.3% of the respondents. Furthermore, the study revealed that farm size, age, education status, number of contacts with extension agents, household size and number of farm plots of the respondents were the significant factors affecting their adoption of land management practices. The study also revealed that the major constraints to the use of sustainable crop management practices among the farmers included inadequate supply of fertilizer, inadequacy of labour and credit, poor knowledge of improved agricultural practices, poor transportation, low produce prices and high cost of production. The study recommended the need for training programme on the use of appropriate sustainable practices that will take into account the cropping systems adopted by the farmers.

Keywords:
Sustainable practices, Land management, Small Scale farmers, Food crops

[1] *Department of Agricultural Economics and Farm Management, University of Ilorin, P.M.B. 1515, Ilorin, Nigeria.*
[2] *Department of Agricultural Extension and Rural Development, University of Ilorin, P.M.B. 1515, Ilorin, Nigeria.*
[3] *Department of Agronomy, University of Ilorin, P.M.B. 1515, Ilorin, Nigeria.*
* *Corresponding author's email: lawaz71@yahoo.com; lawalazeez@unilorin.edu.ng*

INTRODUCTION

A unique characteristic of rural societies in many developing countries is the almost total dependence on agriculture for livelihood. In Nigeria for instance, rural communities are completely agrarian in nature with crop farming, fishing, or livestock keeping being the major occupation of the majority. Since almost all rural households depend directly or indirectly on agriculture, and given the large contribution of the sector to the overall economy, agriculture is a key component of growth and development (IFPRI, 2007). As a major source of income, agriculture plays a significant role in rural and economic development process (Thirtle *et al.,* 2003). As such, agricultural development is an integral part of all rural development approaches. This fact is well captured in the description of agriculture as the engine room of rural development (Ashley and Maxwell, 2001). It is therefore imperative that any attempt at rural development must focus on agriculture. It has however been observed that agriculture has changed dramatically, especially since the end of World War II. Food and fiber productivity soared due to new technologies, mechanization, increased chemical use, specialisation and government policies. Although these changes have had many positive effects and reduced many risks in farming, there have also been significant costs. Prominent among these are topsoil depletion, groundwater contamination, the decline of family farms, continued neglect of the living and working conditions for farm laborers, increasing costs of production, and the disintegration of economic and social conditions in rural communities (Agricultural Sustainability Institute, ASI, 2012).

The Nigerian agricultural sector is made up of four major sub-sectors: crop production, livestock production, forestry and fishery. According to Daramola *et al.* (2007), their contributions to the Nigerian agricultural GDP in 2007 were 85%, 19%, 4% and 1% respectively. The dominance of crop production in the agricultural sector in Nigeria therefore places it in a pivotal position in issues of sustainable agriculture in Nigeria. Food production in Nigeria, as in many developing countries is linked with small scale agriculture. The small scale farmers in this system are responsible for a large share of the total agricultural output, cultivated land and the farming population (Brown and Wolf, 2005). The importance of small scale farmers to sustainable agricultural and rural development is derived therefore from their dominance of the agricultural sector of the country.

Sustainable land management (SLM) has been defined as the adoption of appropriate land management practices that enables land users to maximize the economic and social benefits from the land while maintaining or enhancing the ecological support functions of the land resources (FAO, 2009). It is the key point for improving land resource resilience and productivity within the context of the potentially devastating effects of climate change in Sub-Saharan Africa, bridging the needs of agriculture and environment, with the twin objectives of maintaining long term productivity and ecosystem functions (land, water, biodiversity); and Increasing productivity (quality, quantity and diversity) of goods and services (including safe and healthy food). The major goal of SLM therefore is to develop economically viable agro-ecological system and to enhance the quality of the environment, so that farm lands will remain productive indefinitely. Since SLM includes the maintenance over time of soil productivity, it therefore requires the combination of soil fertility treatment such as application of mineral and organic fertilizers with soil and water conservation measures including implementation of agronomic, soil management and physical measures, such as contour ridging, terracing, tied ridges or providing ground cover through mulching, use of plants and leaving crop residues (Durno, 1992; Woodfine, 2009).

The production systems which follow a predominantly ecosystem approach offer a range of productivity, socio-economic and environmental benefits to producers and to society at large based on a sustainable basis. According to Kassam and Friedric (2011), they are based on five overall objectives: (i) simultaneous achievement of increased agricultural productivity and enhanced ecosystem services; (ii) enhanced input-use efficiency, including water, nutrients, pesticides, energy, land and labour; (iii) judicious use of external inputs derived from fossil fuels (such as mineral fertilizers and pesticides) and preference for alternatives (such as recycled organic matter, biological nitrogen fixation and integrated pest management); (iv) protection of soil, water and biodiversity through use of 'minimum soil disturbance' and maintaining organic matter cover on the soil surface to protect the soil and enhance soil organic matter and soil biodiversity; and (v) use of managed and natural biodiversity of species to build systems' resilience to abiotic, biotic and economic stresses, with an underlying emphasis on improving soils' content of organic matter as a substrate essential for the activity of the soil biota. Furthermore, they noted that the farming practices required to implement these objectives will differ according to local conditions and needs.

Sustainable production practices involve a variety of specific and general approaches. Specific strategies must take into account topography, soil characteristics, climate, pests, local availability of inputs and the individual grower's goals. Despite the site-specific and individual nature of sustainable agriculture, several general principles can be applied to help growers select appropriate management practices: selection of species and varieties that are well suited to the site and to conditions on the farm; diversification of crops and cultural practices to enhance the biological and economic stability of the farm; management of the soil to enhance and protect soil quality; efficient and humane use of inputs; and consideration of farmers' goals and lifestyle choices (Agricultural Sustainability Institute, ASI, 2012).

A common philosophy in sustainable crop management is that a "healthy" soil is a key component of sustainability. This implies that a healthy soil will produce healthy crop plants that have optimum vigour and are less susceptible to pests. On the other hand, crop management systems that impair soil quality often result in greater inputs of water, nutrients, pesticides, and/or energy for tillage to maintain yields. In sustainable crop management systems, the soil is viewed as a fragile and living medium that must be protected and nurtured to ensure its long-term productivity and stability. Methods to protect and enhance the productivity of the soil include using cover crops, compost and/or manures, reducing tillage, avoiding traffic on wet soils, and maintaining soil cover with plants and/or mulches (Agricultural Sustainability Institute, ASI, 2012).

One of the most severe threats to the sustainability of agricultural crops production in Nigeria is declining productivity as a result of the loss of soil fertility. Soils in Nigeria suffer deficiency common to the tropical soil such as low percentage organic matter and nitrogen, shallow depth and high acidity which predispose about 63% of agricultural soils in Nigeria to low productivity (Lekwa and Whiteside 1996). The Federal Ministry of Agriculture and Rural Development (FMARD, 1987) blamed this situation on the inappropriate land and agricultural management practices in many communities across the country. Such agricultural practices have been fingered in natural vegetation transformation causing loss of soil water content, soil structure and porosity. As such, soil depletion in Nigeria can generally be attributed to the following: increase population density and pressure on agricultural lands; over grazing due to increased stocking rate; wide spread deforestation due to reduced fallow period; incompatible and unsustainable agricultural practices; and excessive use of fertilizers and other agro-chemicals (Emma-Okafor *et al.*, 2010).

In spite of the wide spread knowledge (some even indigenous) about cropping patterns such as fallowing and crop rotation known to significantly contribute to soil sustainability and other soil water and nutrient conservative measures which could help to remedy soil condition, the situation has continued to be grim. It is uncertain if farmers in Kwara State, like in many other parts of Nigeria are taking full advantage of the farming practices required for improved and sustainable agricultural production. Studies by Okoye (1998), Marshall (2004) and Rezvanfar *et al.* (2009) have shown that socio-economic characteristics including age, educational background, family size and farm size influence farmers' adoption of land management practices Therefore, the ability of crop farmers in Kwara State to adopt sustainable agricultural practices is probably affected by their socio-economic conditions. As such, this study sets to answer the following research questions:

what are the socio-economic characteristics of crop farmers in Kwara State, Nigeria?

what cropping patterns are common to them?

what soil conservation methods do they practice?

What relationship exist between selected socio-economic characteristics of crop farmers in the State and their adoption of sustainable agricultural practices?

what militates against successful adoption of sustainable agricultural practices among farmers in Kwara State?

Objectives

The main objective of the study was to assess sustainable agricultural practices in food crop production among small scale farmers in Kwara state, Nigeria. The specific objectives of the study were to:

1- Determine the socio economic characteristics of crop farmers in Kwara State;

2- Ascertain the cropping patterns common among them;

3- Highlight the soil conservation methods adopted by the farmers;

4- Examine the relationship between selected socio-economic characteristics of the farmers and their adoption of major agricultural land management practices; and

5- Investigate the constraints to adoption of sustainable agricultural practices among crop farmers in Kwara State, Nigeria.

Literature review

The adoption of land management practices is multidimensional with numerous factors affecting the willingness of farmers to use various conservation practices (Rezvanfar *et al.*, 2009). The effects of most socio-economic variables on the adoption of land management practices have not been conclusive and have been noted to vary with location given the divergent reports available from existing literature.

Higher educational attainment is characteristically linked with increased awareness and hence better adoption. It has been reported that educational attainment has a positive and significant relationship with the adoption of sustainable land management practices among farmers (Okoye, 1998; Deininger *et al.*, 2003 and Pender *et al.,* 2003). While Clay *et al.* (1998) reported educational attainment to be insignificant to the adoption of land management practices, Warriner and Moul (1992) presented education as having a negative relationship with adoption. According to Okoye (1998), age is positively correlated with the adoption of sustainable land management practices; however, Abd-Ella *et al.* (1981) reported a negative relationship between the two. Pender and Gebremadhin (2004) concluded that male headed households were more likely to use contour plowing and manure on a regular basis, Mulat *et al.* (1997) held a contrary opinion from a study carried out in the Ethiopian highlands in which male –headed households were found to most likely use less soil management practices. A high household size could be expected to imply more labour to carry out land management practices, however, Shiferaw and Holden (1998) confirmed a negative correlation between family size and land conservation practices. Holden and Yohannes (2002) however found no relationship between family size and land management practices.

Farm size has been positively linked to the adoption of land management practices (Cramb, 1999; Hagos, 2003; Demeke, 2003 and Teklewold, 2004). Deininger *et al.* (2003) reported a negative relationship between farm size and some land management practices such as tree planting. Land tenure affects adoption of land management practices primarily by influencing transferability of property rights thereby affecting the reversibility of land investments and ability to use land as collateral. Ayalew *et al.* (2005) opined that where land tenure is insecure, chances of sustainable land management practices are slim. Deininger *et al.* (2003) observed that access to agricultural extension services has a positive relationship with the adoption of land management practices. This view was also held by Marshall (2004) and Okunade (2006).

MATERIALS AND METHODS

The study area is Kwara State, Nigeria. With a total of sixteen Local Government Areas, the state has a land area of about 36,825 km^2 (Federal Office of Statistics, 1999) and a population of 2,371,089 (Kwara State Planning Commission, 2007) Major tribes in the State are Yoruba, Nupe, Fulani and Baruba. The state is located between latitudes 7^{0}45'N and 9^{0}30'N and longitude 2^{0}30'E & 6^{0}25'E. The topography is mainly plain to slightly gentle rolling lands. The mean annual rainfall ranges between 1000mm and 1500mm. Average temperature ranges between 30^{0}C and 35^{0}C. The rainy season in the state falls between March and October with a short break in August. The dry season is between November and February. The climatic condition favors the cultivation of arable crops including millet, cassava, yam, cowpea, maize, and rice. As obtainable in other States of Nigeria, Agricultural extension service delivery has been largely public and administered by the Agricultural Development Projects (ADPs) under the supervision of the State Ministries of Agriculture. A number of special agricultural development schemes/projects embarked upon by the Government and some non-governmental organizations (NGOs) over time have also addressed agricultural extension in the State.

Small scale crop farmers in the study area constituted the population for this study. A three stage random sampling technique was used to select the sample for the study. The first stage involved the random selection of four (4) Local Government Areas: Edu, Asa, Kaiama, and Ifelodun out of the sixteen Local Government Areas in the State. The second stage involved the random selection of three (3) villages from each of the selected Local Government Areas using the Kwara State Agricultural Development Programme (KWADP) village listing. The third stage was the random selection of twelve (12) small scale farmers from each of the villages selected. Information on the number of the total farmers in each village was obtained from the village council. This was used in preparing the list of the farmers that form the sampling frame from which the random selection of the farmers was carried out. Every other farmer was thereafter randomly selected from the list. A total of one hundred and forty four (144) respondents were surveyed but responses from only one hundred and thirty five (135) farmers were found to be useful for the study.

Data for the study were obtained by the administration of a structured questionnaire by well trained enumerators. Descriptive statistical tools employed to analyze data collected were frequency, percentages and means. A four point Likert-type scale was employed to investigate constraints to the adoption of sustainable agricultural practices. The scale was graded as follows: Not severe = 1, Less severe = 2, Severe = 3 and Very severe = 4. Respondents rated various possible constraints based on their perception of the severity of the constraints to them. To examine the causal relationship between the use of soil management practices and some selected socio-economic variables, binary logistic regression model was adopted. The regression model in its linear form is presented as follows:

$Y = f(X_1, X_2, X_3, X_4, X_5, X_6, u)$

Where Y_i = use of ith soil management practice (used =1, 0 otherwise)

X_1 = Total farm size (ha)
X_2 = Age of farmer (years)
X_3 = Education level
X_4 = Number of visits by extension agents in the year.
X_5 = Number of farm plots
X_6 = Household size
X_7 = Error term

RESULTS AND DISCUSSION

Socio-economic characteristics of the respondents

The socio-economic characteristics of the respondents including age, household size, sex, marital and educational status are as presented in table 1.

Table 1 shows that with a mean age of 50.72 years and close to 60% of the respondent being 50 years and above, the youth who are known to possess the physical strength required for crop farming are largely not as involved in agriculture as expected. This is particularly significant giving the fact that youths have been reported to constitute about 60% of the population of the Nigeria (Vision 2010 Report, 2005) and that they account for 70% of the State population (Akinyemi, 2011). This may not be unconnected with the fact that agriculture is increasingly being rejected among the youths particularly because of preference for white collar jobs. The Majority of the respondents have household sizes of between 5 and 10. This is of particular importance because of the implication of household size on the availability of farm labour. The dominance of male in crop farming in the study area is expressed in the fact that 92.6% of the respondents were male. Only 28.9% of the farmers had a minimum of primary school education. Education has been known to influence adoption positively. However, the farmers are poorly educated on the average and this may affect their level of adoption of sustainable agricultural practices.

Occupational characteristics

In view of their importance on the use of sustainable crop and soil management practices,

Table 1: Socio-Economic characteristics of the respondents.

Characteristics	Frequency	Percentage
Age of Respondents		
>40	12	8.9
40-49	44	32.6
50-59	69	51.1
60 and above	10	7.4
Total	135	100
Household Size		
<5	4	3
5-10	125	92.6
>10	6	4.4
Total	135	100
Sex of the Respondents		
Female	10	7.4
Male	125	92.6
Total	135	100
Marital Status		
Married	127	94.1
Divorced	2	1.5
Widower	6	4.4
Single	0	0
Total	135	100
Educational Status		
No formal education	51	37.8
Adult education (out of school)	10	7.4
Quranic education	35	25.9
Primary education	24	17.8
Secondary education	4	3
Tertiary education	11	8.1
Total	135	100

Source: Field survey, 2012

this section is based on the discussion of the occupational characteristics of the farmers. Details of such characteristics including the place of farming as a source of income, mode of land acquisition, source of capital for farming, farm size, number of farm plots and contact with extension agents are presented in table 2.

About two thirds of the respondents have farming as their primary occupation. Only 35% of the farmers have full ownership titles by having purchased their farmland. The ownership structure is important as farmers may not be willing to expend effort towards conservation practices on land temporarily held by them. The fact that 83% of the respondents rely on their meager personal savings may have implications for the adoption of agricultural practices that are financially demanding. Mean size of farmland under cultivation was 2 ha. Some agricultural practices such as fallowing are known to require more land area. Inadequate land area may therefore pose a problem in the adoption of such practices. An abysmally poor level of extension contact was recorded among the respondents with only 6.7% of them having extension contact. Thus the use of the sustainable practices might have been affected negatively due to low level of enlightenment on the need for farmers to adopt them by the extension agents.

Cropping patterns among respondents

Certain management practices are dependent on the type of cropping systems adopted by the farmers. This is based on the diversity of cropping potentials in the country at large and the study area in particular. Table 3 therefore presents the distribution of the respondents according to their cropping systems.

The most common cropping pattern among the respondents is the maize/cassava mixture. Table 3 reveals that 19.1% of the respondents followed maize/ cassava intercropping on

Table 2: Distribution of respondents according to occupational characteristics.

Characteristics	Frequency	Percentage
Farming as primary occupation		
Yes	89	65.9
No	46	34.1
Total	135	100
Mode of land Acquisition		
Borrowed	7	5.2
Inherited	63	46.7
Communal	17	12.6
Purchased	48	35.6
Total	135	100
Sources of capital		
Credit	11	8.1
Internal source (friends & family)	12	8.9
Personal saving	112	83
Total	135	100
Farm size (Ha.)		
<2	76	56.3
2-2.99	19	14.1
3-3.99	20	14.8
>3.99	20	14.8
Total	135	100
Contact with Extension Agent		
Yes	9	6.7
No	126	93.3
Total	135	100
vi. No of Farm Plots		
1	47	34.8
>1	88	65.2
Total	135	100

Source: Field survey, 2012

21.87% of the entire farm land. Further analyses revealed that half of the respondents (48.60%) adopted cereal-based cropping systems comprising maize/cassava and maize/melon intercropping as well as maize and, sorghum mono-cropping which cover about half of the entire farm land (51.48%). Furthermore, heavy nutrient drawers such as cereals (maize, sorghum) and tubers (cassava and yam) planted either individually or in various combinations were planted on 58% of the entire farm land by about 3/5 of the farmers (59%). This is in sharp contrast with the cultivation of legumes with only 27.7% of the respondents growing leguminous soil conservation crops (soya beans and melon) on about 24% of the total land area. While the food needs of the farmers is very important in dictating the cropping system adopted, it is also possible that the farmers were not properly informed about the importance of legumes in nutrients fixation as well as their cover cropping characteristics which are beneficial in soil conservation.

Agricultural land management practices adopted by respondents

Various agricultural practices adopted by the farmers which have effects on agricultural sustainability including mulching, use of cover crops, crop rotation with or without legumes, irrigation, composting, use of organic manure, and fertilizer application are presented in table 4.

As shown in table 4, about half of the respondents (48.6%) engaged a minimum of three management practices. Crop rotation was the major single crop management practice adopted by all (100%) of the farmers. On the other hand, irrigation was the least adopted agricultural practice with only 29.3% of the respondents involved in it. This may be connected to the high cost of irrigation. As a sole management practice, crop rotation alone was adopted by 22.2%

Table 3: Cropping patterns among respondents.

Cropping System	Frequency	Percent	Farm Size (ha)	% Farm Size
Maize/ Cassava	53	19.1	67.2	21.87
Maize/ Melon	32	11.5	37	12.04
Cassava	39	14	39	12.69
Maize	34	12.2	36.5	11.88
Soya Beans	31	11.2	26	8.46
Yam	22	7.9	18	5.86
Sorghum	16	5.8	17.5	5.69
Cassava/ Soya beans	14	5	10.5	3.42
Others	37	13.3	55.6	18.09
Total	278	100	307.3	100

R^2=0/82   D.W=2.02

of the farmers. Further analysis however revealed that only 31.20% of the respondents practice crop rotation with legumes. This implies that the majority of the respondents wrongly applies crop rotation with non leguminous plants. The majority of the respondents (52.40%) apply inorganic fertilizer in production while only 16.80% use organic manure and/or composting. Further analysis of the data revealed that only 5.40% (1.6+1.9+1.9) of the respondents apply both inorganic fertilizer and compost or organic manure(The implication is that while it is desirable to apply various management practices, about 94.60% of the respondents who did not include both inorganic and organic fertilizers in their management practices can be said not to apply sustainable land management practices in production. This is because use of both inorganic and organic manure constitute what is generally termed integrated land management system.

Effects of selected socio-economic factors on usage of soil management practices among respondents

This section presents the effect of selected socio-economic characteristics on soil manage-

Table 4: Distribution of respondents according to agricultural land practices adopted.

Management Practices	Frequency	Percentage
Mulching, fertilizer, cover crops, crop rotation	6	1.9
Mulching, cover crops, crop rotation	5	1.6
Mulching, cover crops, crop rotation with legumes	1	0.3
Mulching, crop rotation	13	4.1
Mulching, crop rotation with legumes	3	0.9
Cover crops, crop rotation, crop rotation with legumes	12	3.8
Cover crops, crop rotation, irrigation	2	0.6
Crop rotation	70	22.2
Cover crops, crop rotation, composting	1	0.3
crop rotation, crop rotation with legumes, organic manure	15	4.7
crop rotation, organic manure	5	1.6
crop rotation, organic manure, irrigation	10	3.2
crop rotation, composting, irrigation	5	1.6
crop rotation, irrigation	14	4.4
Fertilizer, cover crops, crop rotation	14	4.4
Fertilizer, crop rotation	57	18
Fertilizer, crop rotation, crop rotation with legumes, composting	5	1.6
Fertilizer, crop rotation, crop rotation with legumes, irrigation	42	13.3
Fertilizer, crop rotation, composting, irrigation	6	1.9
Fertilizer, crop rotation, irrigation	15	4.7
Fertilizer, crop rotation with legumes, composting, irrigation	6	1.9
Fertilizer, crop rotation with legumes, irrigation	9	2.8
Total	316	100

Source: Field survey, 2012

Table 6: Distribution of respondents according to constraints faced in the adoption of sustainable agricultural practices.

Constraints to Use of Sustainable Practices	Frequency of Respondents Based on Perception					Likert Mean Score	Rank
	Very Severe	Severe	Less Severe	Not Severe	Total		
Inadequate Supply of Fertilizer	96	24	9	5	134	3.57	1
Non-availability of Labour	62	63	9	0	134	3.4	2
Non-availability of Credit	57	70	9	0	134	3.4	2
Inadequate Knowledge of Modern Technique	70	45	19	0	134	3.38	3
High Labour Cost	54	73	4	3	134	3.32	4
Transportation Problems	29	76	24	5	134	2.96	5
High Cost of Irrigation Water	28	79	19	8	134	2.95	6
Low Produce Price	40	44	50	0	134	2.93	7
High Cost of Soil Management	17	67	49	1	134	2.75	8
Inadequate Improved Variety of Seed/Planting Material	20	53	52	9	134	2.63	9
Insufficient Land Availability	20	12	47	55	134	1.98	10
Insufficient Extension Services	18	14	22	80	134	1.78	11

Source: Field survey, 2012

ment practices among the respondents. Table 5 shows the result of the binary logistic regression to determine the socio-economic factors that affects the adoption of soil management practices among the respondents.

The results presented in table 5 show that 51.2%, 32.70%, 15.60%, 55.40% and 39.8% of variation in the use of crop rotation, cover cropping, bush fallow, mulching and use of fertilizer respectively were explained by the independent variables as implied by their pseudo R-square (R^2) values. The variables that had significant effects on usage of the soil management practices are farm size, age, education, number of contacts with extension agents, number of farm plots, and household size of respondents.

Farm size positively influenced the use of mulching at 1% level of significance implying that respondents with large farm size had higher probability of making use of mulching. While farm size negatively influenced the use of crop rotation at 10% level of significance which implies that as farm size decreases the farmers may likely increase the use of crop rotation.

The age of the household head was positive and significant at 10% probability level for the use of crop rotation, while the situation was different for the use of mulching which was negatively influenced by age of the household head at 5% probability level. This implies that an increase in the age of the household head will likely increase the use of crop rotation. Hence while the older

Table 5: Effects of socioeconomic factors on adoption of soil management practices.

Variables	Soil Management Practices				
	Crop rotation	Cover Cropping	Bush Fallow	Mulching	Use of Fertilizer
Farm size (X_1)	-0.398***	0.251	-0.012	0.727*	0.069
Age (X_2)	0.05***	-0.024	-0.001	0.20**	-0.0065
Education (X_3)	0.103	0.034	0.001	0.089	0.150**
Contacts with Extension Agents (X_4)	-0.649*	-0.288*	-0.33*	0.569*	0.321**
Number of Farm Plots (X_5)	-1.675*	-1.358**	-0.567	2.612*	-0.58
Household size (X_6)	-0.219***	-0.078	-0.021	0.134***	0.086
Constant	4.569	2.787	-0.356	-4.869	-1.546
R^2	0.512	0.327	0.156	0.554	0.398

Source: Data Analysis, 2012.

* P ≤ 0.01; ** P ≤ 0.05; *** P ≤ 0.1

Note: Logistic regression could not be computed for irrigation, composting and organic manure because there were very few respondents who adopted them in the syudy and were therefore eliminated in the regression.

and more experienced farmers will likely adopt crop rotation, the younger farmers will likely prefer to use mulching. This may be as a result of the fact that younger farmers are more innovative and better adopters than older farmers. Besides, preference for such practice may be due to the fact that it will reduce drudgery associated with highly labour intensive practices.

The educational level of the farmers also positively influenced the use of fertilizer by the farmers. This was found to be significant at 5% level of significance. The more educated the farmers are, the higher their likelihood for the use fertilizer.

While the coefficient of number of contacts with extension agents was negative for crop rotation, cover cropping and bush fallow, it was positive for mulching and the use of fertilizer. This implies that the use of Mulching practice and use of fertilizer will likely increase as the number of contacts with extension agents increases. On the other hand, as the farmers contact with extension workers reduce the farmers' probability of using crop rotation, cover cropping, bush fallow will likely increase. This implies that the farmers will likely rely more on their indigenuos knowledge and practices with reduced level of extension contact.

The number of farm plots cultivated positively influenced the use of mulching. As such, farmers will likely use more of mulching as their farm plots increase in number. This may be due to increased labour requirements in higher number of farm plots which the farmers may not be able to afford. Crop rotation and cover cropping practices were however negatively influenced by the number of farm plots. This indicates that the use of crop rotation and cover cropping is likely to reduce with increase in number of farm plots. Similar findings have been reported by (Aromolaran, 1998).

Constraints faced by farmers in the adoption of sustainable agricultural practices

Farmers are often faced with various challenges in the use of sustainable practices in crop production. An assessment of the constraints faced by the farmers is presented in table 6.

Inadequate supply of fertilizer, with a mean score of 3.57 was reported as the most severe constraint. 89.5% of the respondents identified with this constraint. This often results in low level of application due to inadequate access to fertilizer among the farmers. Other problems considered to be severe are inadequacy of credit and labour (both with mean scores of 3.4); inadequate knowledge of improved methods and high cost of labour were also rated high with average scores of 3.38 and 3.32 respectively. The inadequate knowledge of improved methods may be a result of the poor level of extension contact as earlirer reported. Other constraints included cost of transportation, high cost of adoption, low produce price and poor transportation. Poor transportation may be fingered in a number of other constraints particularly those involving high costs. The low produce price reported may make it difficult for the farmers to afford the various costs associated with sustainable agricultural practices. Contrary to expectation, the least severe problem reported by the respondents was insufficient extension service. Even though, the negative impact of inadequate extension contact is already implied in inadequate knowledge of improved method of land management, response by the farmers may suggest the fact that farmers have not had the opportunity to appreciate the need for extension service in improved and sustainable agricultural production.

CONCLUSION AND RECOMMENDATIONS

The study assessed sustainable agricultural practices in food crop production among small scale farmers in Kwara State, Nigeria. Findings from the study revealed that while farmers engaged in a number of sustainable agricultural practices, the majority who constitute about 95% of the farmers do not apply sustainable practices in food crops production (see above). In order to improve on this situation, this study makes the following recommendations based on its findings:

1- There is the need for better information on the need for complementary use of organic manure and inorganic fertilizer in food crop production among farmers in Kwara State, Nigeria;

2- Increased extension contact is of paramount importance for small scale crop farmers with specific emphasis on sustainable agricultural practices. This may call for a stakeholder conference, including farmers, on the appropriate land management practices that would suit their cropping systems. Besides, there may be the need for regular joint traiing programmes for the farmers on relevant agricultural land management practices;

3- Farmers should make concerted efforts to avail themselves of training opportunities on appropriate use of farm input especially farm land for optimum production that would take account of costs of adoption of the various land management practices adopted by the farmers. Such training programmes should be used as a springboard for updating the knowledge of the of young farmers on appropriate land management practices for their cropping systems;

4- Agriculture should be accorded more priority in youth empowerment scheme of the Kwara State government. This is with a view to encouraging the youth to taking better interest in agricultural food crops production using sustainable practices; and

5- Land fragmentation which often results from high number of farm plots operated by the farmers should be curtailed. This therefore calls for consolidation of land holdings among the farmers which may be achieved through rejuvenation of farmers' cooperatives and other farmers' associations.

REFERENCES

1- Abd-Ella, M., Eric, O.H., & Warren, R.D. (1981). Adoption behavior in family farm systems: An Iowa study. Journal of Rural Sociology, 46(1), 42-61.

2- Agricultural Sustainability Institute ASI, (2012). Sustainable Agricultural Research and Education Program Retrieved on 30th September, 2012 from http:// sarep. ucdavis. edu/ sarep /about/def.

3- Akinyemi, D. (2011). Unemployment: A life-line for youths in Kwara. Vanguard, December 6th Edition Retrieved on 26th April, 2013 from http://www.vanguardngr.com/2011/12/unemployment-a-life-line-for-youths-in-kwara/.

4- Aromolaran, A.B. (1998). Economic analysis of soil conservation practices in South-western Nigeria. Issues in African Rural Development, Monograph Series; Monograph # 11 African Rural Sciences Research Networks

5- Ayalew, D., Dercon, S., & Gautam, M. (2005). Land tenure insecurity and investment in ethiopian agriculture: Evidence from panel data. Mimeo.

6- Ashley, C., & Maxwell. S. (2001). Rethinking rural development. Development Policy Review, 19(4), 395–425.

7- Brown, L.R., & Wolf, E.C. (2005). Reversing africa decline. World Watch Courier (June) Environmental and Development No 87 in Odoemenem, I. U. & V. Adebisi (2011) Sustainable Agriculture For Small Scale Farmers In Niger State, Nigeria Journal of Sustainable Development in Africa, 13(2), 225-235. Retrieved on 30th September, 2012 fromhttp: //www.jsd-africa.com/Jsda/V13No2_Spring 2011 _ B/PDF/Sustainable%20Agriculture%20for%20Smal l%20Scale%20Farmer.pdf.

8- Clay, D., Reardon, T., & Kangasniemi, J. (1998). Sustainable intensification in the highland tropics: Rwandan Farmers' Investments in Land Conservation and Soil Fertility, Economic Development and Cultural Change, 45 (2), 351-378.

9- Cramb, R., (1999). Smallholder adoption of soil conservation technologies evidence from upland Projects in the Philippines, University of Queensland.

10- Daramola, A.G., Ehiu, E., Ukeje, E., & Mclntire, J. (2007). Agricultural Export Potential in Collier, P. and Pattilo, C. (eds) Economic Policy Options for Prosperous Nigerian. London: Palgrave Macmillan, 2007.

11- Deininger K., Jin, S. and Adenew, B. (2003). Market and non-Market transfers of land in ethiopia: Implications for Efficiency, Equity, and Non-Farm Development. World Bank Policy Research Working Paper 2992. World Bank, Washington, DC.

12- Demeke, A.B. (2003). Factors influencing the adoption of soil conservation practices in northwestern ethiopia. Discussion Paper 37. Institute of Rural Development, University of Goettingen.

13- Durno, J. (1992). Integrated farming in the lowlands resources. London: Zed books Ltd.

14- Emma-Okafor, L.C., Ibeawuchi, I.I., & Obiefuna, J.C. (2010). Biodiversity conservation for sustainable agriculture in tropical rainforest of nigeria. New York Science Journal 2010, 3(1), 81-88. Retrieved on 30th September, 2012 from http://www.sciencepub.net/newyork/ny0301/15_1086_Biodiversity _ conservation_ny0301.pdf.

15- FAO. (2009). Country support tool – for scaling-up sustainable land management in sub-saharan africa. Food and Agriculture Organization of the United Nations, Rome, Italy.

16- FMARD. (1987). Federal ministry of agriculture water resources and rural development, Fourth National Development Plan 1981-1985, National Planning Office, Lagos, Nigeria.

17- Federal Office of Statistics (FOS, 1995). Abstracts of Statistics, FOS Abuja, Nigeria .

18- Hagos, F. (2003). Poverty, institutions, peasant behavior, and conservation investment in Northern

Ethiopia. PhD thesis No. 2003:2. Department of Economics and Social Sciences, Agricultural University of Norway, As.

19- Holden, S.T., & Yohannes, H. (2002). Land redistribution, tenure insecurity and input intensity: A Study of Farm Households in Southern Ethiopia. Land Economics, 78: 573-590.

20- IFPRI 2007. The role of agriculture in development: Implications for sub- Saharan Africa Research Report 153.

21- Kassam, A., & Friedric, T. (2011). Conservation agriculture: Principles, sustainable land management and ecosystem services. Società Italiana de Agronomia XL Convegno Nazionale, Università degli Studi Teramo. Retrieved on 26th April, 2013 from http://www.fao.org/ag/ca/CA-Publications/CA_Teramo_Kassam_Friedrich.pdf

22- Kwara State Planning Commission, KWSPC, 2007. Kwara State Year Book. Kwara State, Nigeria.

23- Lekwa, G., & Whiteside E.P. (1996). Coastal plain sand of southeastern Nigeria. Soil science society of American Journals 50: 154-160. In Emma-Okafor, L. C., Ibeawuchi, I. I., & Obiefuna, J. C. (2010). Biodiversity Conservation for Sustainable Agriculture in Tropical Rainforest of Nigeria. New York Science Journal 2010; 3(1), 81-88. Retrieved on 30th September, 2012 from http://www.sciencepub.net/newyork/ny0301/15_1086_Biodiversity_conservation_ny0301.pdf.

24- Marshall, G.R. (2004). From words to deeds: Enforcing Farmers' Conservation Cost-sharing Commitments. Journal of Rural Studies, 20: 157-167.

25- Mulat, D., Ali, S., & Jayne, T. (1997). Promoting fertilizer use in ethiopia: The implications of improving grain market performance, input market efficiency, and farm management. Working Paper 5, Grain Market Research Project, Ministry of Economic Development and Cooperation, Addis Ababa.

26- Okoye, C. (1998). Comparative analysis of factors in the adoption of traditional and recommended soil erosion control practices in Nigeria. Soil and Tillage Research. 45: 251-263.

27- Okunade, E.O. (2006). Factors influencing adoption of improved farm practices among women farmers in Osun State. Journal of Human Ecology, 19 (1), 45-49.

28- Pender, J., & Gebremedhin, B. (2004). Impacts of policies and technologies in dryland agriculture: Evidence from Northern Ethiopia In S.C. Rao (ed.), Challenges and Strategies for Dryland Agriculture. American Society of Agronomy (ASA) and Crop Science Society of America (CSSA) Special Publication 32. Madison: ASA and CSSA.

29- Pender, J., Gebremedhin, B., & Haile, M. (2003). Livelihood strategies and land management practices in the highlands of tigray, presented at the conference on policies for sustainable land management in the East African Highlands, April 24-26, 2002, United Nations Economic Commission for Africa, Addis Ababa (revised). International Food Policy Research Institute (IFPRI), Washington, DC.

30- Rezvanfar, A., Samlee, A., & Faham, E. (2009). Analysis of factors affecting adoption of sustainable soil conservation practices among wheat growers. World Applied Sciences Journal, 6 (5), 644-651.

31- Shiferaw, B., & Holden, S.T. (1998). Resource degradation and adoption of land conservation Technologies in the Ethiopian Highlands: A Case Study in Andit Tid, North Shewa. Agricultural Economics, 18: 233-247.

32- Teklewold, H. (2004). Risk and time preferences on soil conservation decision in the central Ethiopian Highlands of Ethiopia. M.Sc thesis. Addis Ababa University.

33- Thirtle, C., Irz, X. Lin, L. McKenzie-Hill, V., & Wiggins. S. (2001). Relationship between changes in agricultural productivity and the incidence of poverty in developing countries. DFID Report 7946.

34- Vision 2010 Report (2005). Youth, Development and Empowerment in In Adekunle, O. A., Adefalu, L.L., Oladipo, F.O., Adisa, R.S., & Fatoye, A.D. (2009). Constraints to youths' involvement in agricultural production in Kwara State, Nigeria. Journal of Agricultural Extension, 13 (1), 102 - 108

35- Warriner, G.K., & Moul, T.M.(1992). Kinship and personal communication network influences on the adoption of agriculture conservation technology. Journal of Rural Studies, 8 (3), 279-291.

36- Woodfine, A. (2009). Using sustainable land management practices to adapt to and mitigate climate change in Sub-Saharan Africa- Resource Guide Version 1.0, TerrAfrica, Regional Sustainable Land Management.

13

Development of Entrepreneurship in Complementary and Processing Industries in Khouzestan Province, Iran

Ahmad Reza Ommani

Abstract

The main purpose of this study was to identify and analyze factors affecting development of entrepreneurship in complementary and processing industries of citrus products at Khouzestan province, Iran. The Statistical population of this study consisted of all managers in processing and complementary industries of citrus products in Khouzestan province (N=89). By census method all managers were selected for participation in the study. Return rate was 84% (N=75). The main instrument in this study was questionnaire which its validity was confirmed by a panel of experts and its reliability was established by calculating Cronbach Alpha Coefficient (α=0.87). Findings reveal that there were positive and significant relationship between willingness to creativity, risk oriented, responsibility, competitiveness, participation on education programs, attitude to entrepreneurship, income level, educational level and level of entrepreneurship. According to factor analysis, factors affecting development of entrepreneurship in complementary and processing industries were categorized into four groups consisting: development of cognitive and psychomotor domains, development of infrastructure equipments, improving economic policies and development of psychological characteristics that those factors explained 68.82% of the total variance of the research variables.

Keywords:
Entrepreneurship, Complementary and processing industries, Citrus products

[1] *Assistant Professor, Department of Agricultural Extension and Education, Shoushtar Branch, Islamic Azad University, Shoushtar, Iran.*

* *Corresponding author's email: ommani75451@yahoo.com*

INTRODUCTION

Entrepreneurship plays vital role in social and economical development of different societies. From the economic viewpoint, entrepreneurship referred to the production factors that assists the economic growth through discovering and creating new opportunities and from social viewpoint, contributes the improvement of social situations through increasing job or employment chances and social participation (Mohapatra *et al.*, 2007). In general, entrepreneurship is a powerful tool for identifying opportunities, acquiring benefits, and removing such problems as unemployment, the lack of dynamic human resources, low efficiency, quality reduction of products and services, and economic stagnancy (Maclin and Richard, 2004). Economists look at the entrepreneurship from the prospective of profitability, investment, risk, and insight supporting the economical development but it seems that, in modern societies, entrepreneurship`s function is beyond the economical bounds (Hekmat, 2011).

According to Schumpeter (2012), an entrepreneur is a person who is willing and able to convert a new idea or invention into a successful innovation. Entrepreneurship employs what Schumpeter called "the gale of creative destruction" to replace in whole or in part inferior innovations across markets and industries, simultaneously creating new products including new business models. In this way, creative destruction is largely responsible for the dynamism of industries and long-run economic growth. The supposition that entrepreneurship leads to economic growth is an interpretation of the residual in endogenous growth theory and as such is hotly debated in academic economics.

According to Abu-Saifan (2012) Entrepreneurship is a way of thinking, reasoning, and acting that is opportunity obsessed, holistic in approach and leadership balanced.

In general an entrepreneur can benefit from the process of creating your own business several advantages, including (Burdus, 2010):

• Independence and opportunity to achieve the desired objectives, it offers the advantage of not depending on others to implement its wishes;

• Chance to notice a difference in a field they are interested in combining the wishes their social insurance with a win for a better life;

• Opportunity to use its full potential for entrepreneurs because there is much difference between the work of business and recreations, making them find their place of business to obtain satisfaction;

• Opportunities to earn substantial profits, although the reason to start abusiness, an entrepreneur may not be primarily profit;

• Recognition efforts and contribution to the achievement of social objectives, entrepreneurs become very respected person in the community in which it operates;

• Opportunity to do what you love, because most entrepreneurs develop business in areas where they want to work and they get special-satisfaction.

Drucker (2006) believed an entrepreneur is the one who always searches for change, responds to it and exploits it as an opportunity. Innovation is the specific tool of entrepreneurs, the means by which they exploit changes as an opportunity for a different business or different service

Based on the another researchers there is relationship between social factors and development of entrepreneurial (Dodd and Gotsis, 2007; Pages and Markley, 2004).

Also the findings of Mirzaee (2002) and Volery and Muller (2006) showed that allocating the necessary budgets and securing the cost of practical training would have affect on the entrepreneurial development.

Objectives of research:

1- Identifying manager's psychological characteristics.

2- Identifying entrepreneurship development items.

3- Identifying affecting factors on development of entrepreneurship in processing and complementary industries of citrus products.

4- Analyzing the correlation between level of entrepreneurship and other variables.

MATERIALS AND METHODS

The method of research was correlative descriptive. The statistical population of this study consisted of all managers in processing and complementary industries of citrus products in Khouzestan province (N=89). By census method

Table 1: Managers psychological characteristics items.

Subjects	Items	Mean	SD
Willingness to creativity	3	3.34	0.87
Risk oriented	3	2.53	0.89
Responsibility	3	3.31	1.01
Competitiveness	3	3.96	0.93

1: very low, 2: low, 3: moderate, 4: high, 5: very high

all of managers were selected for participation in the study. Return rate was 84% (N=75).

To ensure its content and face validity, the research instrument was reviewed several times by the research group and then implemented in a pilot test to measure its reliability. Questionnaire reliability was estimated by calculating Cronbach's alpha. Reliability of the overall instrument was estimated at 0.87. Data collected were analyzed by using the Statistical Package for the Social Sciences (SPSS). The data were collected between a January and April and July 2013. Questions were generated from the literature review. The instrument consisted of two separate sections according to the purpose and objectives of the study. The first section was designed to gather data on personal characteristics of managers. The second section was designed to gather data regarding the factors affecting development of entrepreneurship in processing and complementary industries of citrus products. Managers were asked to rate their viewpoints on a five point Likert - type scale: 1 = very low,2 = low, 3 = medium, 4 = much and 5 = very much. In order to analyze data, descriptive statistic (mean and standard deviation) and inferential methods (Factor Analysis test used).

RESULTS AND DISCUSSION

Demographic characteristics

The ages of the respondents ranged from 20 to 63. The mean age was 42 (N =75). The majority (31.3%, n=25) of respondent were 31-40 years old. The years ofexperience of respondents ranged from 3 to 30. Only 10.7% of managers had a lower the diploma degree (n = 8). 46.7% of respondents (n = 35) had diploma degree and 42.3% had higher the diploma degree.

Managers' psychological characteristics

The 12 psychological characteristics itemsin four categorize for analyzing psychological behavior of managers wereanalyzed.The results explained in table 1.

Table 2: Rankings of entrepreneurship development items.

Items	SD	Mean	CV	rank
Development of educational programs	0.902	4.61	0.196	1
Development Credit facilities to entrepreneurs firms	0.89	4.46	0.200	2
Control of price fluctuations	0.889	4.43	0.201	3
Marketing management	0.885	4.39	0.202	4
Human development	0.879	4.33	0.203	5
Remove intermediaries	0.88	4.32	0.204	6
Improving the exports	0.868	4.23	0.205	7
Development of managerial skills	0.858	3.31	0.259	8
Facilitate the achievement of the required equipment	0.84	3.09	0.272	9
Development of technical skills	0.864	3.16	0.273	10
Reduction of cost	0.919	3.03	0.303	11
Proper pricing of products	0.995	3.02	0.329	12
Improving psychological characteristics	0.901	2.58	0.349	13
Specialized workshops and scientific conferences	0.939	2.48	0.379	14
Creating a culture of entrepreneurial skills	0.936	2.31	0.405	15
Improving attitude to entrepreneurship	0.929	2.25	0.413	16

R^2=0/82 D.W=2/02

Table 3: Extracted factors, with their eigenvalue, its variance percent and cumulative percent of variance Eigen value.

Factors	Eigenvalue	Variance	Cumulative percent of variance
First	4.567	24.567	24.567
Second	3.890	44.053	19.486
Third	2.980	58.04	13.987
Fourth	1.97	68.829	10.789

Rankings of entrepreneurship development items by the managers

The sixteen entrepreneurship development items have been listed according to their rankings (Table 2). Based on the results most important items were:

1- Development of educational programs regarding entrepreneurship (M= 4.61, SD= 0.902)

2- Development Credit facilities to entrepreneurs firms (M= 4.46, SD= 0.890)

3- Control of price fluctuations (M= 4.43, SD= 0.889)

4- Marketing management (M= 4.39, SD= 0.885)

Factor analysis of items affecting entrepreneurship development

To categorize factors affecting development of entrepreneurship in processing and complementary industries of citrus products, and to determine the variance explained by each factor, an exploratory factor analysis approach was followed. Data revealed that internal coherence of the data was appropriate (KMO =0.87), while and the Bartlett's statistic was significant at the 0.01 level. According to Kaiser Criteria, there were four factors that their extracted Eigenvalues were greater than one. Later, the items were categorized into four factors by using VARIMAX Rotation Method (Table 3). It is worth noting that after Varimax rotation and due to low factor loading (less than 0.5) of some variables therefore insignificance of their correlation with other variables, one variable were omitted from analysis and finally 15 variables were analyzed. Table 4 has summarized the findings.

The correlation between level of entrepreneurship and other variables

To determine the possible relationship between level of entrepreneurship and other variables Spearman correlation coefficient approach

Table 4: Affecting factors on development of entrepreneurship in processing and complementary industries of citrus products.

Items		Factor Loading
Development of Cognitive and psychomotor domains	Development of educational programs	0.78
	Development of technical skills	0.75
	Development of managerial skills	0.89
	Specialized workshops and scientific conferences	0.81
Development of Infrastructure equipments	Facilitate the achievement of the required equipment	0.69
	Improving the exports	0.72
	Development Credit facilities to entrepreneurs firms	0.67
Improving Economic policies	Control of price fluctuations	0.77
	Remove intermediaries	0.73
	Reduction of cost	0.65
	Proper pricing of products	0.66
Development of psychological characteristics	Improving psychological characteristics	0.63
	Creating a culture of entrepreneurial skills	0.76
	Improving attitude to entrepreneurship	0.81
	Marketing and risk management	0.58

Table 5: Relationship between level of entrepreneurship and other variables.

Variable (1)	Variable (2)	r	p-value
Age	Entrepreneurship level	0.015	0.762
Income		0.345**	0.000
Willingness tocreativity		0.473**	0.000
Risk oriented		0.422**	0.000
Responsibility		0.297**	0.000
Competitiveness		0.480**	0.000
Education programs		0.312**	0.000
Attitude		0.343**	0.000
Level of education		0.356**	0.000

** P < 0.01

was utilized (Table 5). Based on the results the correlation between level of entrepreneurship with willingness to creativity, risk oriented, responsibility, competitiveness, participation on education programs, attitude to entrepreneurship, income level and educational level at 0.01 level was significant.

CONCLUSION

It's the people in an organization that carry out many important work activities. Managers and human resources professionals have the important job of organizing people so that they can effectively perform these activities. This requires viewing people as human assets, not costs to the organization. Entrepreneurship in any organization is one of the most effective tools for increasing the productivity of staff both at individual and group levels toward realizing the organizational goals. Based on the results most important items that important for development of entrepreneurship were: development of educational programs regarding entrepreneurship (M= 4.61, SD= 0.902), development Credit facilities to entrepreneurs firms (M= 4.46, SD= 0.890), control of price fluctuations (M= 4.43, SD= 0.889), marketing management (M= 4.39, SD= 0.885). In other researches entitled role of education programs on entrepreneurship, there is a meaningful relationship between entrepreneurial education and the tendency to entrepreneurship (Solomon *et al.*, 2002; Kuratko and Hodgetts, 2004).

According to correlation study, correlation between level of entrepreneurship with willingness to creativity, risk oriented, responsibility, competitiveness, participation on education programs, attitude to entrepreneurship, income level and educational level at 0.01 level was significant. Based on other researches such as Rauch and Frese (2000), this result was convinced.

To categorize factors affecting development of entrepreneurship in processing and complementary industries of citrus products, and to determine the variance explained by each factor, an exploratory factor analysis approach was followed. Data revealed that internal coherence of the data was appropriate (KMO =0.87), while and the Bartlett's statistic was significant at the 0.01 level. According to Kaiser Criteria, there were four factors that their extracted Eigenvalues were greater than one. These factors include: development of cognitive and psychomotor domains, development of infrastructure equipments, improving economic policies and development of psychological characteristics.

REFERENCES

1- Abu-Saifan, S. (2012). Social entrepreneurship: definition and boundaries. Technology Innovation Management Review, (February 2012: Technology Entrepreneurship).

2- Burdus, E. (2010). Fundamentals of Entrepreneurship. Review of International Comparative Management, 11(1), 33-42.

3- Dodd, S.D., & Gotsis, G. (2007). The Interrelationship Between Entrepreneurship And Religion, Entrepreneurship AndInnovation, 8(2), 93-104.

4- Drucker, P. F. (2006). Innovation and entrepreneurship, UK, Elsevier Linacre House, 2006

5- Hekmat, M. (2011). Entrepreneurship approaches in agricultural cooperatives. International Journal of Agricultural Science, Research and Technology,

2011; 1(1), 39-45.
6- Kuratko, D. F., & Hodgetts, R. M. (2004). Entrepreneurship: Theory, Process, Practice (Mason, OH; South-Western Publishers).
7- Mccline, Richard L. (2004). Opportunity recognition: An exploratory investigation of a component of the entrepreneurial process in the context of the health care industry: Theory and Practice. TertiaryEducation Management, 25(2), 81-94.
8- Mohapatra, S., Rozelle, S., & Goodhue, R. (2007). The rise of self-employment in rural China: development or distress?. World Development, 35(1), 163-181.
9- Pages, E.R., & Markley, D. M. (2004). Understanding the environment for rural North Carolina. Center for Rural Entrepreneurship. http:\\www.ruraleship.org
10- Rauch, A., & Frese, M. (2000). Psychological approaches to entrepreneurial success. A general model and an overview of findings. In C.L. Cooper & I.T. Robertson (Eds.), International Review of Industrial and Organizational Psychology (pp. 101-142). Chichester: Wiley.
11- Schumpeter, J. A. (2012). Capitalism, Socialism and Democracy. Routledge. ISBN 0415107628.
12- Solomon, G.T., Duffy, S., & Tarabishy, A. (2002). The State of Entrepreneurship Education in the United States: A Nationwide Survey and Analysis. International Journal of Entrepreneurship Education, 1(1), 65-86.
13- Volery, T., & Muller, S. (2006). Conceptual framework for testing the effectiveness of entrepreneurship education programs towards entrepreneurial intention. Available on the http:// www1. kmu. unisg.ch/rencontres/RENC2006/Topics06/A/Rencontres_2006_Volery-MuellerS. pdf

14

Factors Affecting Nonfarm Income Diversification among Rural Farm Households in Central Nepal

Raju Ghimire[1], Wen-Chi Huang[2*] and Rudra Bahadur Shrestha[1]

Abstract

Nonfarm activities play an important role in the determination of rural household income, consumption expenditure, and household food security. However, much less studies have been devoted to the factors that influence nonfarm income diversification by farm household in developing countries. Using cross-sectional data and a probit model, this study attempts to determine the factors influencing nonfarm income diversification decisions by farm households in Central Nepal. The result reveals that household characteristics such as age, gender and education of the household head, and family size play a significant role in nonfarm work decisions. The households with larger farm size are less likely to participate in nonfarm work than their counterpart. Additionally, for those remains in the rural households, distance to road and market hinders the opportunities for nonfarm work. Finally, regional differences also exist in participating nonfarm activities among farm households. This study suggests that government policy should pay more attention on education, gender and infrastructures such as road and markets, to reduce the entry barriers and facilitate easier access to nonfarm activities. Also, nonfarm activities need to be promoted and incorporated in governmental plans and policies for balanced development between hills and terai areas.

Keywords:
Income diversification, Food security, Rural farm household, Nepal

[1] Doctoral Student, Department of Tropical Agriculture and International Cooperation, National Pingtung University of Science and Technology, Pingtung 91201, Taiwan.
[2] Associate Professor, Department of Agribusiness Management, National Pingtung University of Science and Technology, Pingtung 91201, Taiwan.
** Corresponding author's email: wenchi@mail.npust.edu.tw*

INTRODUCTION

Benefits from nonfarm employment opportunities have influenced the poor directly by raising incomes of households and indirectly by increasing agricultural production through the investment in new technology and high-yielding crop varieties. Therefore, many rural farm households tend to diversify their source of income by involving in nonfarm activities. Income diversification via nonfarm work is associated with higher and stable incomes and food consumption over the years (Reardon *et al.,* 1992). There is an obvious connection between income diversification through nonfarm work and food consumption among farm families in developing countries. For example, Anderson (2002) notes that off-farm income is extremely important to the household livelihood in many developing nations and essential to food security among farm households. Specifically, households with nonfarm work will have a better chance to reallocate their labor and can more efficiently offset the negative price effect on their food security compared to those who have fewer nonfarm options (Chang and Mishra, 2008). Lanjouw and Lanjouw (2001) note that rural nonfarm sector can contribute to economic growth, rural employment, and poverty reduction. Further, Chang and Mishra (2008) find that households with more access to income generating activities or access to higher paying work have higher income and more food secure than households who do not have these benefits.

The poorer, in the rural areas tend to depend more heavily on food-crop production and seasonal wage labor activities for their incomes and are therefore likely to be vulnerable in face of personal (such as illness) and covariate shocks such as droughts. On the other hand, the relatively well-offs have better access to productive assets (such as land and human capital) and use their superior asset endowment to engage in livelihood strategies that offered higher returns and lower risks and are able to escape poverty. Therefore, on-farm investments must necessarily be integrated with nonfarm investments to enable households generate sufficient incomes to escape out of poverty (Karugia *et al.*, 2006). For example, Olale and Henson (2012) provide an evidence of poverty reduction among fish farmers in Kenya by spreading income risk, expanding incomes and relieving the pressure exerted on fish resources. Many researchers (e.g. Abdulai and CroleRees, 2001; Abdulai and Delgado, 1999; Barrett *et al.,* 2001; Reardon *et al.*, 1992; Woldenhanna and Oskam, 2001; Wouterse and Taylor, 2008) have conducted similar studies in the past. With the realization of this milieu, government from developing countries had hammered out more pro-farmer strategies with the support from donors and international support organizations but very little has been achieved so far.

Despite the evidence of the important of nonfarm sector to rural farm households, there has been little systematic study on the determinants of nonfarm work decision by farm households (Reardon, 1997). Additionally, far less attention has been paid to infrastructure related issues that influence nonfarm work opportunities and unknown about how much and why rural farm households diversify their sources of income into the nonfarm sector. In order to fill this research gap, this study investigates the determinants of nonfarm income diversification decisions among rural farming communities in Nepal particularly from two main agro-climatic zones (hills and tropical terai) of Central Nepal. This endeavor is important not only because it serves to bridge a gap in the income diversification and household welfare literature, but also because a better understanding of the situation that facilitates the design and implementation of effective policies to promote the benefits of nonfarm sector and minimize the potential costs. The organization of the paper is as follows: First, we present materials and methods in the next section under which we explained data and variables, and the econometric framework. Next part describes in details the results of the analysis. Finally, we ended up with conclusions and some policy implications.

Non-farm sector in Nepal

Non-farm economic activities of rural households in Nepal have increased over the last two decades (CBS, 2011). As shown in Figure 1 that 37.2 % of the total household income comes from non-farm sector in 2010/11 while it was only 22% in 1995/96. On the other hand, share

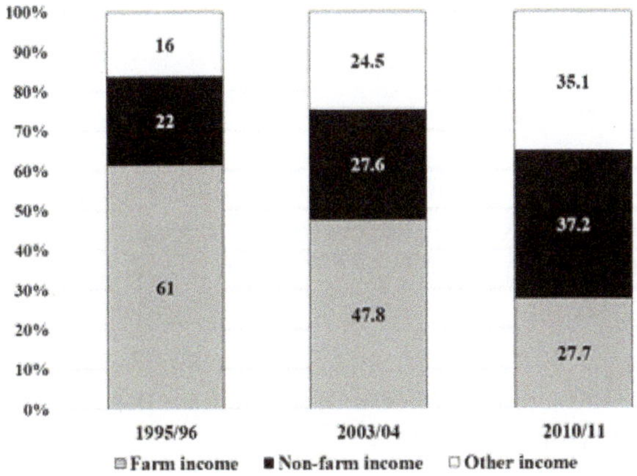

Figure 1: Share of farm, non-farm and other income in total household income. (Source: CBS, 2011)

of farm income in total household income decreased from 61 % to 27.7 % in the same period. The proportion of male wage-earners participated in non-farm activities was 76 % whereas this figure for female was only 45%. Among others, central region has the highest proportion of non-farm participant (39 %) and the percentage share of non-farm income in total household income was 41.5 % which is the highest among regions.

Agriculture sector constitutes 35% of the wage employment and the remaining is accounted by non-agriculture sector. Roughly about 35% of households in 1995/96 had involved in some kind of non-farm activities and this proportion was higher in urban areas relative to rural areas (CBS, 2011). Besides, the highest proportion of households in urban terai areas is operating non-farm activities and the lowest proportion is in the rural mid and far western hills. The majority of the non-farm enterprises in urban areas are in trade sector while manufacturing sector is dominant in rural areas. About 50% of enterprises in rural areas of the central, western and far western hills are in manufacturing sector. Households from the poorest consumption quintile have disproportionately higher share of manufacturing enterprises while those from the richest quintile have trade and services as the dominant type (CBS, 2011).

MATERIALS AND METHODS

Description of the study area

Nepal located in South Asia between 26^0 22' to 30^0 27' N latitude and 80^0 4' to 88^0 12' E longitude, is a landlocked country bordered with China in the north and India in the east, west and south (MOAD, 2011). Nepal covers a total area of 147,181 square kilometers. Topographically, Nepal is divided into three regions: the Himalaya to the north, the Hills in the middle and the Terai to the south. The study was conducted in four districts of Central Nepal namely Kavre, Nuwakot, Chitwan and Rautahat. These districts represent a wide range of agro-ecological variability and are located at an altitude of between 150 and 2500 meters above sea level. Out of eight districts, Kavre and Nuwakot were selected from the hill region; while Chitwan and Rautahat were selected from the tropical plain terai region that represents seven districts (Figure 2).

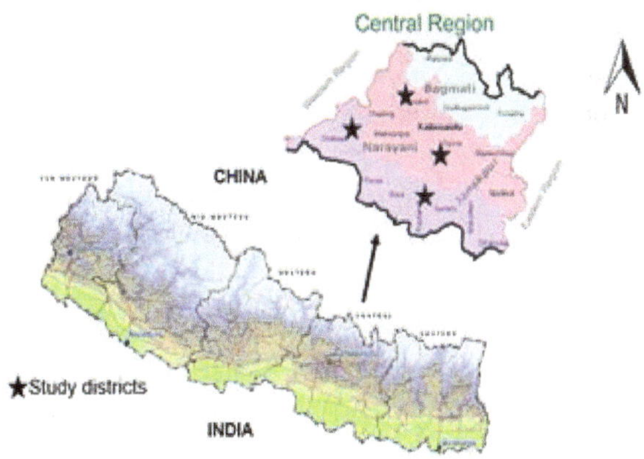

Figure 2: Map of Nepal showing study area (Source: MOAD, 2011)

With the highest percentage of nonfarm work participation, the central region contributes the highest share in total household income from nonfarm sector in comparison to other regions (CBS, 2011). Moreover, selected districts from Central Nepal are benefited with different infrastructures such as road and market networks, business and industry, credit institutions, and potential for agriculture. Because two hilly districts Kavre and Nuwakot are close to the capital city Kathmandu (30 km east and 40 km northwest of the capital city, respectively), there are more nonfarm job opportunities for those residing in the hill districts. Similarly, terai districts (Chitwan and Rautahat) contain mainly of plain areas with easy access to road and markets, and these districts are bordered to India that is an additional benefit to those living in terai in terms of nonfarm job opportunities. Moreover, there was no prior study on nonfarm income diversification conducted in this area.

Econometric model

The farm households are assumed to maximize their utility function subject to constraint resources (Asfaw *et al.*, 2012). The households choose to work nonfarm sector if the utility gained from participating in nonfarm activities is greater than the utility from not participating. Following Ahearn *et al.* (2006) and Gould and Saupe (1989), we used a probit model to determine the probability of choosing nonfarm work by farm households. In probit model, dependent variable is binary which is suitable to determine the probabilities of each households to participate in nonfarm activities. Lass and Gempesaw (1992) and El-Osta *et al.*, (2004) also used a probit model in similar studies in the past.

A household decides to work off the farm only if the market wage (W_i) is higher than its reservation wage (W_i^r) for farm and leisure time. A farm household's decision to work off the farm can be expressed in the framework of a discrete choice model. The specification of this model is:

$$Y_i^* = X_i'\beta_i + \epsilon_i,\ Y_i = 1 \ \text{if} \ w_i > w_i^r;\ Y_i = 0 \ \text{otherwise} \qquad (1)$$

where $\varepsilon \sim N(0, 1)$. Then Y can be viewed as an indicator for whether this latent variable is positive:

$$Y_i = \begin{cases} 1 & \text{if participated in nonfarm work,} \\ 0 & \text{otherwise.} \end{cases}$$

where, Y_i^* is the latent variable which represents the probability of the household to work in the nonfarm sector and takes the value 1 if the household receives higher wage (w_i) than their reservation wage (w_i^r), 0 otherwise. X_i' represent covariates that are associated with the decision to work off the farm, β_i is the coefficient of covariates, ϵ_i is vector of error terms. Variables that increase the nonfarm wage rate increase the probability of nonfarm work and vice versa for variables that decrease the nonfarm wage rate (Ahearn *et al.*, 2006). Marginal effects (ME) assess the influence of each independent variable on the decision of the farm household to participate in nonfarm work as following (Greene, 2012, p. 693):

$$\frac{\partial E[Y_i^*|X_i]}{\partial X_i} = \phi(X_i'\beta)\beta \qquad (2)$$

Based on the theoretical model and experiences of the previous studies (e.g. Ahearn *et al.*, 2006; Chang and Mishra, 2008; Conley and Udry, 2010; Doss and Morris, 2000; El-Osta *et al.*, 2008; Uaiene *et al.*, 2009), we selected our independent variables and specified empirical model as follows:

$$Y_i = \beta_0 + \beta_1 X_1 + \beta_2 X_2 + \beta_3 X_3 + \beta_4 X_4 + \beta_5 X_5 + \beta_6 X_6 + \beta_7 X_7 + \beta_8 X_8 + \beta_9 X_9 + \beta_{10} X_{10} + \beta_{11} X_{11} + \beta_{12} X_{12} + \epsilon_i \qquad (3)$$

where, Y_i = Binary dependent variable, X_1= Age of household head in years, X_2= Male dummy, X_3= Formal years of schooling (years), X_4 = Household size, X_5= Farm size (ha), X_6 = Extension visits (number per year), X_7= Participation in farmers group or cooperatives (dummy), X_8 = Distance to market (km), X_9= Access to seed (dummy), X_{10} = Access to agricultural credits (dummy), X_{11}= Distance to nearest road (km), X_{12}= Region dummy, β_0 to β_{12} = Coefficients to be estimated, and ϵ_i = error term.

Data and definition of variables

The main source of data for this study originates from a survey conducted on a sample of farm households in central region of Nepal during June – August, 2013. A multistage, random sampling technique was used to select the sample. At the first stage, four districts namely Kavre and Nuwakot from hill region, and Chitwan and Rautahat from tropical plain terai region of central Nepal were purposively selected. At the second stage, a total of 8 villages (2 villages from each district) were selected. Subsequently, two wards were randomly sampled in each of the selected village resulting in a total of 16 wards. This was followed by a random sampling of 26 farm households from each ward in proportion to the population size in each ward. A total of 416 farm households in four districts were surveyed using the standardized survey questionnaire. The number of sampled households and their nonfarm work participation status by region are reported in table 1.

In this paper, participants are classified as households who participate any of nonfarm ac-

Table 1: Number of sample households surveyed and their nonfarm work participation categories by region.

Region	Nonfarm work participation status		Regional total
	Group A	Group B	
Hill region	136 (45.18%)	72 (62.61%)	208 (50%)
Terai region	165 (54.82%)	43 (37.39%)	208 (50%)
Total by status	301 (72.36%)	115 (27.64%)	416 (100%)

Note: Group A=Participant in nonfarm work, Group B= Non-participant in nonfarm work.
(Source: Author's survey, 2013)

tivities irrespective of income they earn, and non-participants are those who did not participate in nonfarm work. Many participants did not fully allocate their time to nonfarm work as they also get involved in farming. However, increasing trend of searching nonfarm employment opportunities cannot be undermined as this shares the significant amount of household income in developing countries. Therefore, our main interest in this paper is to see whether farm households' socio-economic and other characteristics affect the decision to participate in nonfarm enterprise. For this, the participation decision is modeled here as binary variable at the household level like other previous studies (e.g. Ahearn *et al.*, 2006; Chang and Mishra, 2008; El-Osta *et al.*, 2008).

The demographic information for each household includes the age, gender and education of the household head and family size were included in the model. Other household characteristics, such as how much land a household owns, whether a household has seed and agricultural credit access, as well as infrastructure and services received such as how far a household is from the nearest road and market and number of contacts or visits with extension officers were additional variables included. Further details of explanatory variables for the regressions are identified in the following section (Table 2). The choice of explanatory variables and its expected signs are explained in more detail as following.

The age of household head is incorporated as it is believed that with age, people accumulate experience as well as more personal capital and, thus, show a greater likelihood of investing in nonfarm enterprise. However, it may also be that younger household heads are more progressive and flexible with situation and hence

Table 2: Definition and summary statistics of selected variables.

Variables	Description	Mean	S.D.
Dependent			
Nonfarm work participation	=1 if household participated in nonfarm work, 0 otherwise	0.72	0.45
Independent			
Age	Age of household head (years)	44.54	10.81
Male dummy	=1 if the head is male, 0 otherwise	0.70	0.46
Education	Years of formal education, head	7.86	3.38
HH_size	Number of household members	5.83	1.98
Farm_size	Area cultivated by household in current year (ha)	0.56	0.27
Ext_visits	Number of extension visits received per year	6.80	6.09
Partici_group	=1 if participated in farmers group and cooperatives, 0 otherwise	0.73	0.44
Dist_mrkt	Distance to input/output market (km)	13.14	6.38
Seed_access	=1 if seed is available at local store, 0 otherwise	0.59	0.49
Agcredit_access	=1 if access to agricultural credit, 0 otherwise	0.45	0.49
Dist _road	Distance to main nearest road (km)	0.40	0.89
Region_dummy	=1 if household reside in terai region	0.50	0.50

(Source: Author's survey, 2013)

likely to take participation decision faster (Olale and Henson, 2012). The expected sign of the coefficient on age is indeterminate. A dummy variable for gender of the household head is included to capture gender difference which takes the value of 1 if the head of the household is male, and 0 if female. It has been argued by some authors that women are generally discriminated against in terms of access to external inputs and information (Doss and Morris, 2000). This hypothesis implies that males are more likely to participate in nonfarm work than females. More educated farmers are typically assumed to be better able to process information and search for appropriate nonfarm work opportunity. The belief is that education gives farmers the ability to perceive, interpret and respond to new information much faster than their counterparts without education (Huffman, 1980; Uaiene *et al.,* 2009). Hence, the expected sign on the coefficient on education is positive.

Size of household is another variable used in model. A large family often has a large number of working members. Generally, an increase in family size is likely to increase the probability of participation. The expected sign for household size is positive. The land area cultivated by the family farm is often argued as an important factor in affecting farm work decision positively and nonfarm work decision negatively (Chang and Mishra, 2008). It is frequently argued that farmers with larger farms are more likely to adopt an improved technology particularly in trying new varieties of seed compared with those with small farmers as they can afford to devote part of their fields to try out the improved technology. Therefore, the expected sign on the coefficient on farm size is negative.

Farmers' contacts with extension agents were measured by number of extension visits per year. Contact with extension agents is expected to have a positive effect on adoption of new agricultural technologies based upon the innovation-diffusion theory. Such contacts, by exposing farmers to availability of information can be expected to stimulate adoption (Polson and Spencer, 1991). Thus, a negative relationship is hypothesized between extension visits and the probability of participating in nonfarm activities. Membership to a farmers group and cooperatives is included because it has been shown that farmers within a group learn from each other how to grow and market new crop varieties. As discussed, the evidence suggests that network effects are important for individual decisions, and that, in the particular context of agricultural innovations, farmers share information and learn from each other (Andrew and Rosenzweig, 1995; Conley and Udry, 2010). The expected sign on the coefficient on membership in farmers group and cooperatives is negative in the context of nonfarm activities rather than farm activities.

Distance to road and market for nonfarm job opportunity is assumed to play an important role in finding nonfarm work opportunities. The hypothesis here is that, the further away a household is from road, the smaller is the likelihood that they will participate in nonfarm employment. Labor markets are also known to influence the probability of participation decision. Thus, the coefficients on the distance of the household to the nearest road and markets are expected to be negative (Beyene, 2008). Access to seed is an another important variable included which takes on a value of 1 if the household has access to seed in local market, 0 otherwise. It is expected that the availability of improved seed is more likely for farmer to engage more on farming. Hence, the expected sign on the coefficient on easy access to seed is also negative. To reflect unobserved regional differences, the location of the farm household is specified as a dummy to indicate whether the respondent lives in terai region.

RESULTS AND DISCUSSION

Descriptive statistics

Descriptive statistics and explanation of the variables used in the study are provided in table 2. As observed in the table, 301 (72%) of the total sample households participated in nonfarm activities. The average household size was 6 persons which is consistent with the national average in Nepal (CBS, 2011). The average age of respondents was 44.5 years and the majority (70%) of the sample households was headed by male which is expected in Nepali context. Similarly, the mean years of formal education of household head was found

Table 3: Characteristics of nonfarm work participants and non-participants.

Variables	Group A	Group B	Difference (A-B)	t-value
Age of Household head (years)	43.75	46.62	-2.87	-2.44**
Gender of household head male (%)	0.74	0.62	0.12	5.43**
Years of schooling of household head				
Household size (number)	8.13	7.16	0.97	2.63***
Farm size (ha)	5.87	5.70	0.17	0.78
Number of extension visits per year	0.45	0.68	-0.23	-4.95***
Participation in farmers group or cooperatives (dummy)	6.59	7.36	-0.77	-1.15
Access to improved seeds of crop varieties and vegeta-	0.72	0.76	-0.05	-0.96
bles (dummy)	0.56	0.69	-0.13	-5.44**
Distance to nearest input/output market (km)	12.14	15.76	-3.62	-5.34***
Access to agricultural credit (dummy)	0.44	0.47	-0.03	-0.26
Distance to nearest main road (km)	0.31	0.65	-0.35	-3.57***

Note: Group A=Participant in nonfarm work, Group B= Non-participant in nonfarm work.
*** P < 0.01, ** P < 0.05 confidence level.

to be 7.8 which is also comparable to national average of 8.1 years for adults (CBS, 2011). The average farm size of 0.56 hectare is comparable to the national average of 0.6 hectare (MOAD, 2011). Amongst the households, majority (73%) had participated in farmers' group and cooperatives, and 45 % had reported to have access to agricultural credits from bank and local saving and credit cooperatives. Furthermore, farmers, on average, had 6.8 contacts or visits with extension technicians per year. The average distance to nearest market and main road head from household is 13.14 km and 0.40 km respectively.

Table 3 presents results of statistical significance tests on equality of means for continuous variables and equality of proportions for binary variables for participants and non-participants. There appear to be a significant difference in age and gender of household head between nonfarm work participants and non-participants. There is also a significant difference in the years of schooling of the household heads among the two groups, with the participants being younger and educated.

Farm size shows significant difference between nonfarm work participants and non-participants, suggesting that higher the farm size, lower the participation in non-farm activities. Further, access to seed by farm households, distance to main road and nearest market also show significant differences between these two groups. In contrast, household size, number of extension visits per year, participation in farmers' group and cooperatives, and access to agricultural credit did not show significant differences.

Nonfarm work decisions by farm households

The results of estimated probit model are presented in table 4 which represents the propensity to work off the farm by rural farm households. Some of the variables have significant effects on the probability to work off-farm and are in agreement with some of the findings in previous studies. For instance, demographic and household characteristics appears to be important in nonfarm work decisions. The results indicate that age and education are statistically significant in influencing the probability of working nonfarm sector. Similar to previous findings by Chang and Mishra (2008); Mishra and Goodwin, (1997) and Huffman and Lange (1989), the results suggests that the more educated the household head is, the more likely the household will participate in nonfarm work possibly because they are qualified and can better process information more rapidly than otherwise. As expected, the probability of working off the farm is negatively influenced by age. This negative relationship shows that older farmers are less likely to participate on nonfarm activities. The households headed by male are more likely to participate in nonfarm employment. This result is supported by positive and statistically significant coefficient of the variable male_dummy.

Table 4: Probit estimates of nonfarm employment decisions.

Variables	Parameter estimates	t-value	Average marginal effects
Age	- 0.019 (0.008)	- 2.23**	-0.005
Male dummy	0.395 (0.163)	2.42**	0.106
Education	0.102 (0.031)	3.34***	0.027
HH_size	0.154 (0.044)	3.45***	0.041
Farm_size	- 1.602 (0.383)	- 4.18***	-0.428
Ext_visits	- 0.002 (0.017)	- 0.12	-0.001
Partici_group	- 0.213 (0.230)	- 0.92	-0.057
Dist_mrkt	- 0.031 (0.013)	- 2.34**	-0.008
Seed_access	- 0.204 (0.185)	- 1.10	-0.054
Agcredit_access	- 0.008 (0.171)	- 0.05	-0.002
Dist _road	- 0.158 (0.088)	- 1.80*	-0.042
Region_dummy	0.494 (0.187)	2.64***	0.132
Constant	1.199 (0.577)	2.08**	
Log-likelihood		- 197.9107	
Pseudo R^2		0.1930	
LR $Chi^2(12)$		94.69	
Correctly classified		0.0000	
		76.68 %	

Note: *** p<0.1, ** p<0.05 and * p<0.1. The number in the parentheses next to the parameter estimate is the standard error.

In order to assess the impact of farm size on the probability of nonfarm work decision, total land area cultivated by households (Farm_size) was included in the regression model. Farm size is negative and significant factor for diversifying nonfarm income sources among rural farm households in Nepal. The result is consistent with Chang and Mishra (2008). Size of household is an important variable that impacts nonfarm work. In this study, household size is found to be positively and significantly associated households' nonfarm work decision. This implies that as the number of household member increases, probability of working off the farm also increases which is consistent with expectations. Further, distance to road (Dist_road) and nearest market (Dist_mrkt) has negative impact on the probability of nonfarm employment by farm household. This implies that farther the household from main road and market, lower the probability to involve in nonfarm work. This result is consistent with Babatunde and Qaim (2010).

Finally, nonfarm work appears to be substantially influenced by regions. Farm households located in tropical plain terai region are more likely to participate in nonfarm work in comparison to those located in hilly areas. The infrastructures (such as road and market network) in terai region possibly capture the opportunities in the non-farm sector available to farm households. Average marginal effects shown in column 4 of table 4 indicate the individual effect of each independent variable for the decisions of nonfarm work by farm households.

CONCLUSIONS

In spite of mounting evidence of the importance of the nonfarm sector to rural farm households, there has been little systematic study on rural nonfarm diversification as a major component of rural nonfarm activity. Using data from a household survey and a probit model, this study investigates the determinants of nonfarm income diversification decisions by farm households, particularly from two main agro-climatic regions (hills and tropical plain terai) of Central Nepal. In our study, we included mainly the household characteristics, institutional and infrastructure related variables.

The empirical results reveal that the household characteristics such as age, gender and education of the household head and household size to be influencing factors for nonfarm work decisions. Farm size is negative and significant factor for diversifying nonfarm income sources, suggesting that larger the farm size smaller the likelihood to participate in nonfarm work. Further, infrastructures related variables play significant role in participating nonfarm activities in rural areas. Finally, nonfarm work appears to be substantially influenced by regions. Farm households living in tropical plain terai region are more likely to participate in nonfarm work in comparison to their counterparts residing in hill region. This possibly captures the availability of nonfarm job opportunities in plain terai areas.

The study, at the policy level, suggests that major attention should be given to build infrastructures such as road networks and market in the rural areas that helps promote nonfarm enterprise, overcome the entry barrier and make it accessible for rural farm households. Also, nonfarm activities need to be incorporated in governmental plans and policies for balanced development between hill and tropical plain terai districts. Finally, programs that encourage women to participate in nonfarm activities should be in place in order to overcome the gender bias, and improve the educational status of farming households are necessary to enhance nonfarm income sources in the long run.

ACKNOWLEDGEMENTS

We would like to thank the anonymous referees for their constructive comments and suggestions. The authors are grateful to District Agriculture Development Offices (DADOs) of four study districts in Nepal for their help and support in conducting the survey.

REFERENCES

1- Abdulai, A., & CroleRees, A. (2001). Determinants of income diversification amongst rural households in Southern Mali. Food Policy, 26(4), 437-452.

2- Abdulai, A., & Delgado, C. L. (1999). Determinants of nonfarm earnings of farm-based husbands and wives in Northern Ghana. American Journal of Agricultural Economics, 81(1), 117-130.

3- Ahearn, M. C., El-Osta, H., & Dewbre, J. (2006). The impact of coupled and decoupled government subsidies on off-farm labor participation of US farm operators. American Journal of Agricultural Economics, 88(2), 393-408.

4- Anderson, A. (2002). The effect of cash cropping, credit and household composition on household food security in southern Malawi. African Studies Quarterly, 6(1-2), 175-202.

5- Andrew, D. F., & Rosenzweig, M. R. (1995). Learning by doing and learning from others: Human capital and technical change in agriculture. Journal of Political Economy, 103(6),1176-1209.

6- Asfaw, S., Shiferaw, B., Simtowe, F., & Lipper, L. (2012). Impact of modern agricultural technologies on smallholder welfare: Evidence from Tanzania and Ethiopia. Food Policy, 37(3), 283-295. doi: 10.1016/j.foodpol.2012.02.013

6- Babatunde, R. O., & Qaim, M. (2010). Impact of off-farm income on food security and nutrition in Nigeria. Food Policy, 35(4), 303-311.

7- Barrett, C.B., Reardon, T., & Webb, P. (2001). Nonfarm income diversification and household livelihood strategies in rural Africa: concepts, dynamics, and policy implications. Food Policy, 26(4), 315-331.

8- Beyene, A.D. (2008). Determinants of off-farm participation decision of farm households in Ethiopia. Agrekon, 47(1), 140-161.

9- CBS (Central Buresu of Statistics). (2011). Nepal Living Standard Survey. National Planning Commission Secretariat, Government of Nepal.

10- Chang, H.-H., & Mishra, A. (2008). Impact of off-farm labor supply on food expenditures of the farm household. Food Policy, 33(6), 657-664.

11- Conley, T. G., & Udry, C. R. (2010). Learning about a new technology: Pineapple in Ghana. The American Economic Review, 100(1), 35-69.

12- Doss, C.R., & Morris, M.L. (2000). How does gender affect the adoption of agricultural innovations? Agricultural Economics, 25(1), 27-39. doi: 10.1111/j.1574-0862.2001.tb00233.x

13- El-Osta, H.S., Mishra, A.K., & Ahearn, M.C. (2004). Labor supply by farm operators under "decoupled" farm program payments. Review of Economics of the Household, 2(4), 367-385.

14- El-Osta, H. S., Mishra, A. K., & Morehart, M. J. (2008). Off-Farm labor participation decisions of married farm couples and the role of government payments. Review of Agricultural Economics, 30(2), 311-332. doi: 10.1111/j.1467-9353.2008.00406.x

15- Gould, B. W., & Saupe, W. E. (1989). Off-farm labor market entry and exit. American Journal of Agricultural Economics, 71(4), 960-969.

16- Greene, W. H. (2012). Econometric Analysis (7th ed.): New York University, Princeton Hall.

17- Huffman, W. E. (1980). Farm and off-farm work

decisions: the role of human capital. The Review of Economics and Statistics, 62(1), 14-23.
18- Huffman, W. E., & Lange, M. D. (1989). Off-farm work decisions of husbands and wives: joint decision making. The Review of Economics and Statistics, 70(3), 471-480.
19- Karugia, J. T., Oluoch-Kosura, W., Nyikal, R., Odumbe, M., & Marenya, P. P. (2006). Access to land, income diversification and poverty reduction in rural Kenya. Paper presented at International Association of Agricultural Economists Conference, Gold Coast, Australia.
20- Lanjouw, J. O., & Lanjouw, P. (2001). The rural non-farm sector: issues and evidence from developing countries. Agricultural Economics, 26(1), 1-23.
21- Lass, D. A., & Gempesaw, C. M. (1992). The supply of off-farm labor: a random coefficients approach. American Journal of Agricultural Economics, 74(2), 400-411.
22- Mishra, A. K., & Goodwin, B. K. (1997). Farm income variability and the supply of off-farm labor. American Journal of Agricultural Economics, 79(3), 880-887.
23- MOAD (Ministry of Agriculture Development). (2011). Statistical Information on Nepalese Aagriculture, Government of Nepal.
24- Olale, E., & Henson, S. (2012). Determinants of income diversification among fishing communities in Western Kenya. Fisheries Research, 125: 235-242.
25- Polson, R. A., & Spencer, D. S. (1991). The technology adoption process in subsistence agriculture: The case of cassava in Southwestern Nigeria. Agricultural Systems, 36(1), 65-78.
26- Reardon, T. (1997). Using evidence of household income diversification to inform study of the rural nonfarm labor market in Africa. World Development, 25(5), 735-747.
27- Reardon, T., Delgado, C., & Matlon, P. (1992). Determinants and effects of income diversification amongst farm households in Burkina Faso. The Journal of Development Studies, 28(2), 264-296.
28- Uaiene, R. N., Arndt, C., & Masters, W. (2009). Determinants of agricultural technology adoption in Mozambique. Discussion papers(67E). National Directorate of Studies and Policy Analysis, Ministry of Planning and Development, Republic of Mozambique.
29- Woldenhanna, T., & Oskam, A. (2001). Income diversification and entry barriers: evidence from the Tigray region of northern Ethiopia. Food Policy, 26(4), 351-365.
30- Wouterse, F., & Taylor, J. E. (2008). Migration and income diversification. World Development, 36(4), 625-640. doi: 10.1016/j. worlddev. 2007. 03. 009

15

Effectiveness of Agricultural Training Centers' Curriculum in Promoting Adoption of Agricultural Technologies: Evidence from Small-scale Potato Farmers in Nyandarua County, Kenya

Ronald O. Nyamwamu[1], Justus M. Ombati[2] and John G. Mwangi[2]*

Abstract

Agricultural Training Centres (ATCs) promote crop and livestock development by providing extension information to farmers and serving as models of improved crop and livestock husbandry practices. Although potato farmers in Nyandarua County were trained on potato production technologies, their average yield was ≤ 16 tonnes per hectare compared to the national potential of 30 tonnes. This study therefore sought to determine the effectiveness of ATCs' curriculum in promoting farmers' adoption of selected potato production technologies in Nyandarua County in Kenya. A survey design was used. Primary data was collected from 136 farmers trained at ATCs on potato production using interview schedules. The respondents were sampled through a five-stage technique. The data was analysed using T-test and ANOVA, then summarized into percentages. Study results showed that adoption of recommended potato production technologies produced high yields compared to traditional technologies. It was therefore concluded that ATCs' curriculum was effective in promoting farmers' adoption of potato seed selection, planting, crop protection and harvesting technologies. It is recommended that ATCs' curriculum prioritise demonstrations that show the influence of clean seed, fertilizer application, timely harvesting and spacing on potato yields. Further, campaigns should be held to educate farmers on the importance of potato dehaulming to promote its adoption. In addition, farmer-based seed production programs should be started in the County aimed at ensuring ready availability of seed.

Keywords:
Adoption, Agricultural Training Centre, Curriculum, Effectiveness, Technology

[1] *Extension Division, Ministry of Agriculture, P.O. Box 30028, Nairobi, Kenya.*
[2] *Department of Agricultural Education & Extension, Egerton University, PO Box 536-20115 Egerton, Kenya.*
* *Corresponding author's email: jombati@egerton.ac.ke, jusmotush@yahoo.com*

INTRODUCTION

The agriculture sector is critical for economic development in most developing countries, for increased incomes, better living standards, poverty reduction and increased food security (Andriese *et al.*, 2007). The agricultural sector provides good opportunities for rural-economic development for most Sub-Saharan countries (Gildemacher, 2012). Agriculture-based economic growth has a stronger effect on poverty reduction than non-agricultural growth (International Fund for Agricultural Development (IFAD, 2010). Kenya's economy relies heavily on agriculture's significant contribution in income generation, employment creation, foreign exchange generation, food security, and provision of raw materials for industries (Ministry of Agriculture, 2010b; GoK, 2007). The economy of Kenya is strongly correlated to agricultural growth and development, as evidenced by the agriculture sector's growth from negative 2.9% in 2009 to 6.3% in 2010, thus increasing the Gross Domestic Product (GDP) from 2.6% by 5.6% in the same period (Kenya Bureau of Statistics, 2011). About 68% of Kenyans living in rural areas rely on agricultural activities for their livelihood (Wambugu and Muthamia, 2009).

Potato is a main tuber crop and third most important food crop in the world after rice and wheat (International Potato Centre, 2008). It is grown in Latin America, Europe, Asia and Africa (Bohl and Johnson, 2010; FAO, 2009). The crop is critical in achieving food security, employment creation and income generation, poverty reduction and economic development (Gildemacher, 2012; Kipkoech *et al.*, 2010). Asia is the world's leading potato producer (48%), second is Europe (33%), third is Africa (7%), with China accounting for more than 20% of the global production (FAO, 2009). Malawi leads in Africa's potato production by producing 20% of the total production in 2007 (FAO, 2009).

The world's average potato yield in 2007 was 16.8 ton/ha, where Africa had 10.8 ton/ha, Malawi 11.9 ton/ha while Kenya had 6.7 ton/ha (FAO, 2009). In Europe and Noth America, the average yield is 40 ton/ha while the developing countries obtain 5-25ton/ha (FAO, 2009). Netherlands produced an average yield of 44.7 ton/ha in 2009 (FAO, 2009; Jansky *et al.*, 2009).

In Kenya, potato is ranked second most important food crop after maize (Kasina and Nderitu, 2010; Abong *et al.*, 2012). About 500,000 Kenyan farmers grow potatoes in over 108,000 ha but obtain low yields due to limited use of high quality seeds and inputs, pest and disease attack (Abong *et al.*, 2012; Kaguongo *et al.*, 2008). The crop is grown mainly for food although it generates employment and income (Lung'aho *et al.*, 2010). This shows that this crop contributes significantly to Kenya's economy. The crop is agro-ecologically suited for high altitude areas (1500-3000 metres above sea level) such as Nyeri, Murang'a, Kiambu and Nyandarua Counties (Lung'aho *et al.*, 2010; Wang'ombe, 2008). In Kenya, potato production is mostly done by small-scale farmers on average land sizes of 2ha (Ogola *et al.*, 2011; Ayieko and Tschirley, 2006). Kenya contributes about 6.5% of Africa's total potato production (Nyagaka *et al.*, 2010). Despite an increase in potato acreage in Kenya, production has not proportionately increased, due to inadequate use of certified seeds, pests and diseases, and poor crop husbandry practices (Mwangi *et al.*, 2008). Potato production involves land preparation, seed acquisition, planting, 1st earthing up, weeding, pest and disease control, 2nd earthing up and lastly harvesting (Bohl and Johnson, 2010). Since Nyandarua County is one of the major potato producing areas in Kenya, it produced 18% and 42% of the total potatoes produced in Kenya in 2009 and 2011 respectively (MoA, 2010a). The county achieved 24% and 25% of the national potato acreage in 2010 and 2011 respectively (MoA, 2012a, b). In Nyandarua, potatoes are produced in two seasons (April-May and October-December) with harvesting being done after 4 months after planting (Ogola *et al.*, 2010).

Agricultural extension started back in the 1900s (World Bank, 2010). In Kenya, much of the agricultural extension is public although it is also provided by the private sector (Anderson, 2008; Christoplos, 2010). According to Abegunde and Ogunsumi (2011), agricultural extension transfers skills between farmers, extension agents and researchers with an aim of promoting technology adoption. The information disseminated in form of advice, skills, and education promotes food security, wealth and employment creation and poverty reduction (ASCU, 2012). The extension service is therefore perceived as a form of non-formal education (Rajalahti and Swanson, 2010), which improves agricultural yield (Maguire *et al.*, 2011). This is because farmer education can influence technology adoption (Anderson, 2008). In Kenya, public extension services are provided by extension agents in the field, at Agricultural Training Centres and parastatals such as Horticultural Crops Development Authority and National Cereals Produce Board (MoA, 2009).

Farmer Training Centres are found in different countries in the world such as Ethiopia (Mengistu, 2009), Tanzania, Kenya, Namibia and Turkey (Nyamwamu, 2013). The Agricultural Training Centers were set up to offer extension services and for durations ranging from 1day to 2 weeks (Desmarais, 2010). Kenya has 28 ATCs distributed in different agro-ecological areas with Olojoro Orok and Njabini found in Nyandarua County (MoA, 2012a). In Kenya the ATCs were formed in 1950s to promote adoption of high yielding technologies. At the start, 11 ATCs were formed in Kenya and were later increased to 28 by the year 2011 (Isinika *et al.*, 2002; MoA, 2012b). Since their formation up to 2006, the ATC training was held without curricula but currently formal curricula are used (MoA, 2006). For crop husbandry, curriculum content includes fertilizer application, seed selection, weed, and pest and disease control, planting and thinning. For potatoes it also includes harvesting, storage and processing (MoA, 2006). Potatoes make a significant contribution to food security, poverty reduction and rural economic development because majority of the producers are smallholder farmers (Gildemacher, 2012).

According to Natumaya (2009), small-scale farmers experience low food reserves due to low yields caused by inefficient crop production practices, poor crop pest and disease management. Low agricultural productivity is reflected by low yields obtained per unit area (Kenya Institute for Public Policy Research and Analysis (KIPPRA, 2009). Adoption of modern farming practices can increase production obtained by small-scale farmers (MoA, 2004). Agricultural extension is an educational process which causes desired changes in people through training (Mengistu, 2009). The extension service shares agricultural knowledge and technology between researchers, extension workers and farmers (ASCU, 2012; Qtaishat and AL-Sharafat, 2012). Since agricultural extension is a form of education, an educated and trained labor-force adapts and easily, utilizes new technologies in production, thus increasing efficiency and productivity (Kenya Institute for Public Policy Research and Analysis (KIPPRA, 2009). The curriculum for the ATCs should therefore motivate the farming population who need skills to chart their destiny out of rural poverty. This paper advocates for farmer training to improve production so that they can earn a productive life through agriculture and regard it as a dignified and profitable occupation. Data for this study was collected in Nyandarua County which is a major potato producing County in Kenya. The aim of this paper was to help in identifying possible areas of intervention in agricultural extension and curriculum development so as to improve potato production in Nyandarua County and also in Kenya through ATCs' curriculum.

MATERIALS AND METHODS

Description of the study area

The study was conducted in Nyandarua County which is served by two ATCs located in Njabini and Oljoro Orok, major potato producing areas in Kenya (MoA, 2010a). Nyandarua County falls within the Aberdare ranges in Kenya, at an altitude of 1,800-3,000 metres above sea level. The average annual temperature is 220c and an average annual rainfall of 1,000 mm, which is bimodal; long rains are received in March-July and the short rains in October-December (Jaetzol *et al.,* 2006). The County has 143, 879 households with 596, 268 persons out of which 82% live in rural areas (KNBS, 2010). Land ownership is freehold with majority of the farmers owning an average of 2 ha (Wang'ombe, 2008). Dairy cattle cattle keeping is the dominant livestock production activity followed by potato production (Nyagaka *et al.*, 2010). Potato production is the major crop production activity in the County (Ministry of Land and Settlement, Ministry of Planning and National Development and United Nations Centre for Regional Development, 2003; Mwakubo *et al.*, 2010).

Other cash crops include wheat and pyrethrum while horticultural crops include cabbage, carrots and garden peas (Obare *et al.*, 2010). The potato is a major crop produced in Kenya with Nyandarua County producing about 18% of the total tonnage (ASCU, 2012; MoA, 2010c).

Sampling and data collection

Primary data for this study was collected from 136 small-scale potato farmers using an interview schedule. To ascertain the reliability of the schedules, they were pre-tested in the field while validity was ascertained through discussions with experts in the Agricultural Education and Extension department of Egerton University. The sample of farmers was obtained using a five-stage technique from farmers who were trained at ATCs between 2009 and 2010. The first stage involved listing all farmers trained at ATCs in Nyandarua County. The second, third, fourth and fifth stages respectively, involved selecting potato farmer groups based on their district, division and location from which they hailed from. Data was collected on farmers' adoption of potato seed selection, planting, crop protection and harvesting technologies after being trained at ATCs. In addition, farmers' socio-economic data was collected on age, gender, education level of potato farmers, reasons for growing potatoes as well as sources of agricultural information for potato farmers.

Data analysis

The proportion of farmers who obtained agricultural information from different sources was expressed as a percentage. The proportion of potato farmers who are knowledgeable and adopting potato seed selection, crop protection, planting and harvesting technologies were expressed as a percentage of the total sample. The independent t-test was used to test the four null hypotheses that stated that: There was no statistically significant difference in yield between farmers adopting and those not adopting the recommended potato seed selection, crop protection, planting and harvesting technologies respectively after being trained at ATCs in Nyandarua County on potato production. The ANOVA was used to determine the difference in yields obtained by potato farmers trained for different number of days at the ATCs. Spearman's rank correlation was used to establish the relationship between the number of times potato farmers were trained at ATCs versus the yield obtained and adoption of certified potato seed.

RESULTS AND DISCUSSION

More males (66%) than females were engaged in potato production in the County. Perhaps this could negatively affect potato production as men participated in agricultural training than women who provided most of the farm labour. This agrees with Kiura (2011) who found that women provide 75% of farm labour while men benefit from most of the agricultural extension services. The United Environmental Programme report of 2008 said that men produce most cash crops while women provide most of the farm labour (United Environmental Programme, 2008). Similarly, the Food Agricultural Organisation of the United Nations report of 2011 also found that agricultural women have limited access to agricultural extension services (FAO, 2011). Of the 136 respondents, 30% of them were aged 40 years and below, while 70% were aged 41years and above. This implied that over two thirds of the potato producers were aging, while the youth were few in potato production. Potato yields may not increase as a result of potato training at the ATCs because most trainees are females whereas it is the males who make most decisions on the farm and also control resources. This means that the knowledge acquired through training may not be put into use on the farm. This also means that if an aging population is involved in training, production may not increase significantly because the individuals are getting less active, may not provide the labour and skill required and acquired through training. In addition, succession may not be possible because the young population is not prepared to take over farming activities by participating in agricultural training. Most of the respondents (97%) reported to have attained at least primary education, while 3% had no formal schooling (Table 1). This means that most potato farmers were educated and literate. Involvement of a literate and educated population in training and farming could increase their understanding of the subject matter, thus increas-

Table 1: Gender, age and education level of potato farmers in Nyandarua County

Variable	Percent
Gender	
Male farmers (n=90)	66
Female farmers (n=46)	34
Age (years)	
Farmers aged 40≤ (n=41)	30
Farmers aged ≥41 (n=95)	70
Education level	
Farmers who had completed at least primary school (n=132)	97
Farmers who had no formal schooling (n=4)	3

Table 2: Farmers' use of potatoes

Variable	Percent
Use	
Farmers who grow potatoes as a business (n=114)	84
Farmers who grow potatoes for home consumption (n=116)	85

ing the adoption of innovations and technologies. This is consistent with KIPPRA (2009) and Nyagaka *et al.* (2010), who found that an educated labour-force easily understands, interprets new information and adopts new technologies such as the use of fertilizers, certified seed and control of crop and insect pests and diseases. A literate farmer is likely to understand easily the technical terms used during training, read the labels on fertilizer bags, follow directions on the operation of machines and tools and other items in order to work out the best system for his farm.

Of the 136 respondents, 84% of them indicated growing of potatoes as a business, while 85% said that they grew potatoes for home consumption as well (Table 2). This could also mean that most farmers who grow potatoes do so for the purpose of providing food for their household members and also as an economic activity. This may also mean that potatoes are critical in the achievement of food security, employment creation and income generation in Nyandarua County. This is consistent with Gianessi and Ashley (2011) and Abong' *et al.* (2012), whose findings show that potato production is done as a food and cash crop, and is significant in Kenyan economy. In order to obtain potato yields enough for food and sale, potato farmers seek training on production technologies that can increase yields. During these training, they are trained on potato seed selection, planting, crop protection and harvesting technologies.

Potato production curriculum

Potato seed selection techniques

Study results showed that 85% of the respondents knew the correct potato seed size to plant while 70% knew the correct minimum number of sprouts per seed tuber, and 96% knew the meaning of 'clean potato seed' tubers of the recommended size that have the correct number of sprouts. However, 47% of the respondents indicated that they obtain potato seed from their own farms, 24% from farmer groups, 14% from ATCs, 9% from Kenya Agricultural Research Institute (KARI), 4% from neighbouring farmers while 20% from Molo Agricultural Development Corporation (ADC) (Figure 1). This meant that majority (75%) of the farmers do not obtain potato seed from recommended sources, despite their good knowledge of potato

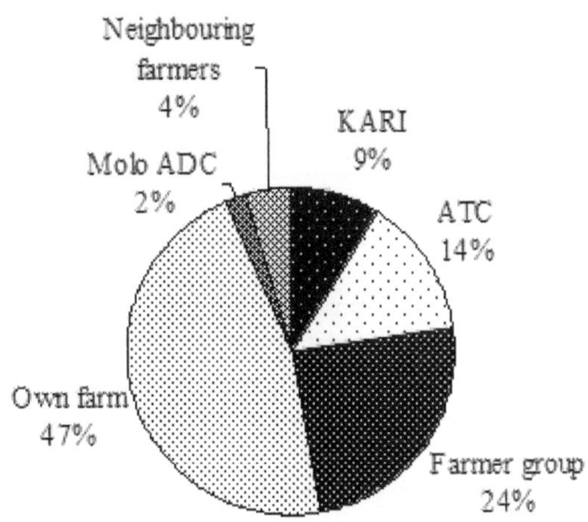

Figure 1: Sources of potato seed in Nyandarua County.

Table 3: Farmers' knowledge of clean potato seed, use of volunteer crop and crop rotation.

Variable	Percent
Clean Potato Seed	
Farmers who understand clean potato seed as being 'disease-free'(n=131)	96
Farmers who understand clean potato seed as being 'free from dirt' (n=5)	4
Use of volunteer crop	
Farmers who eat volunteer potato crop (n=94)	69
Farmers who sell volunteer potato crop (n=22)	16
Farmers who use volunteer crop as fodder (n=7)	5

seed selection technologies. This was against Kirkwyland and Thomas (2012) who recommend the use of certified potato seed when planting potatoes. Gildemacher (2012) observes that potato seed quality is important in sustainably improving potato production. Potato yields can be increased by using the recommended potato seed. If potato farmers are knowledgeable on potato seed selection technologies and practise them, yields obtained from their farms can increase. This is because they will select a seed tuber of correct size, pure, true to type, that is healthy and has at least 3 sprouts, from certified sources.

Potato crop protection techniques

Study results showed that 96% of the respondents understood that clean potato seed was disease-free, 69% eat volunteer potato crop while 89% practice crop rotation (Table 3). This meant that most farmers were knowledgeable on potato crop protection technologies, although about 75% did not use certified potato seed and 69% ate volunteer potato crop. This could imply that majority of potato farmers do not practice most of the recommended potato crop protection technologies. This is consistent with Gildemacher (2009) who found that only 1% of potato farmers in Kenya obtain their potato seed from licensed seed growers. Ogola *et al.* (2010), established that most farmers recycle seed tubers or obtain it from neighbours. Gildemacher (2012) observed that the use of uncertified potato seed significantly reduces potato yields. Potato farmers' failure to use certified potato seed gives them difficulties in controlling potato pests, diseases and weeds. This will be either through introducing new diseases onto their farms or multiplying the already existing diseases. In addition, if the farmers do not practise crop rotation and field sanitation, then they will experience difficulties in controlling insect pests and diseases. This agrees with Wang'ombe (2008) who found that few potato farmers in Nyandarua County use clean potato seed. This is supported by Kwambai and Komen (2012) who found that farmers' use of previous crop or buying potato seed from local markets or from neighbours increases disease build up and spread. Mwangi *et al.* (2008), also established that farmers' failure to practise crop rotation increases disease incidences while the use of clean potato seed and crop rotation helps in controlling weeds, insect pests and diseases (Gildemacher, 2012).

Potato planting techniques.

Study results show that 69% and 54% of the respondents use correct spacing within and between rows for planting potatoes respectively.

Table 4: Potato spacing, ridging and fertilizer application.

Variable	Percent
Spacing	
Farmers using recommended spacing within rows (n=94)	69
Farmers using recommended spacing within rows (n=73)	54
Ridging	
Farmers who practice ridging (n=131)	96
Fertilizer application	
Farmers using recommended fertilizer rates (n=79)	58
Farmers using recommended fertilizer type (n=131)	96

Table 5: Potato maturity indicators and adoption of dehaulming.

Variable	Percent
Potato maturity indicators	
Farmers who check if the plant is dry (n=78)	33
Farmers who check if flowers have dropped (n=19)	67
Farmers who check if the plant has turned yellow (n=31)	57
Farmers who check if tuber skin is hard (n=4)	14
Farmers who use age of crop plant since planting time (n=4)	23
Farmers who do not practise crop rotation (n=15)	3
Dehaulming	
Farmers who do not dehaulm potatoes (n=45)	3
Farmers who dehaulm potatoes (n=91)	11

Further, 96% reported that they prepared ridges, 96% used correct fertilizer rates during planting of potatoes (Table 4). This implied that most farmers used the recommended planting technologies. This was in line with Wang'ombe (2008) who had established that 82% of the farmers in Nyandarua County use fertilizers and FAO (2012), who recommend Di-Ammonium Phosphate (DAP) fertilizer for planting potatoes. Whereas Ogola *et al.* (2011) observed that potato yields can be increased through using correct inputs, Schulte-Geldemann *et al.* (2012), found that application of correct amounts of fertilizers increases potato yields. The spacing used during planting potatoes will determine the plant population which ultimately influences yields. If correct spacing, correct type and amount of fertilizer is used, the correct plant population is obtained per init area, having a healthy crop. This is likely to produce high yields. Further, if a farmer prepares ridges during planting and successively earths up, this practice induces tuber formation as well as prevents pests and diseases. This increases the quantity and quality of potato yield.

Potato harvesting techniques.

Further, study results showed that 57% of the respondents checked whether the plants had dried, 14% if flowers had dropped, 23% checked if the plants had turned yellow/brown, and 3% used age of the crop since planting time, while 3% checked if tuber skin had hardened.

Further, 67% of the respondents dehaulmed potatoes, while 79% used the recommended dehaulming method (Table 5). This implies that majority of potato farmers use the recommended potato harvesting technologies, which is consistent with FAO (2012) recommendation that potatoes should be harvested when stalks are dry, and be dehaulmed 2-3 weeks prior to harvesting (Kirkwyland and Thomas, 2012). Correct diagnosis and interpretation of potato maturity indicators by farmers can significantly affect yields. This is because some maturity indicators resemble pest and disease manifestations or malnutrition for example yellowing and drying of the plant and falling of flowers. Therefore potato farmers need skills for distinguishing between a disease manifestation and a mature potato plant by grasping the maturity indicator for the varieties commonly grown. Mature tubers have higher dry matter content, and therefore are of high quality. If farmers mistaken a disease for maturity and harvest the crop, the total yield may be low and of low quality. Further, the farmers need to prepare for harvest by dehaulming their crop and perform it at the right stage of growth, using the correct method. Dehaulming hardens the tuber skin and therefore reduces bruising and skinning. Harvesting potatoes at the right stage and dehaulming before harvesting can increase yields and quality of tubers.

Sources of agricultural information

A comparison of various sources of agricultural information showed that 83% of the respondents preferred to get information about potato production from ATCs than from other sources (17%). The other sources of agricultural information are; 56% consider the radio, 53% sought from extension officers, 50% got it from field days, and 29% from other farmers (Figure 2). This means that radios, agricultural exten-

Table 6: Independent t-test on yield difference between farmers using and farmers not using certified seed and crop protection technologies.

Variable	Mean Yield (ton/ha)	S.D.	S.E.	t-test	p-value
Crtified seed					
Farmers who plant certified seed after being trained at ATC (n=71)	19	23.89	2.84	5.42	0.00
Farmers who do not plant certified seed after being trained at ATC (n=65)	13	26.36	3.27		
Crop protection					
Farmers using recommended crop protection practices after being trained at ATC (n=113)	17	27.33	2.57	2.05;	0.04
Farmers not using recommended crop protection practices after being trained at ATC (n=23)	13	27.10	5.65		

sion officers and field days are major sources of agricultural information to potato farmers, second to ATCs. This could be attributed to the systematic delivery of subject matter during farmer trainings using radio programs, agricultural extension officers and field days. Further, there is a high likelihood of completing a teaching session in ATCs', radio programs and in farmer training sessions than in other sources. This could mean that more potato farmers listen to agricultural radio programs than they meet agricultural extension officers or attend field days. It could also imply that radio programs are more frequently and regularly used in the County to disseminate agricultural information compared to the frequency of visits by agricultural extension agents and field days. These findings are supported by Mumero (2013) who reported that a radio is a dominant source of agricultural information in Nakuru, Nyanza, Nyeri, Machakos, Makueni and Webuye, although the information provided is inadequate ans sometimes require translation into simple forms usable by local farmers. Mumero (2013) also established that most farmers (44%) trust the public extension service although it is less available as expected, probably due to their ratio of one extension agent to 1,470 farmers agaist the recommended one officer to 400 farmers. Wanzala (2014) also reported that at one time in Kenya's history every village had an extension agent, unlike the situation currently. Consequently, other sources

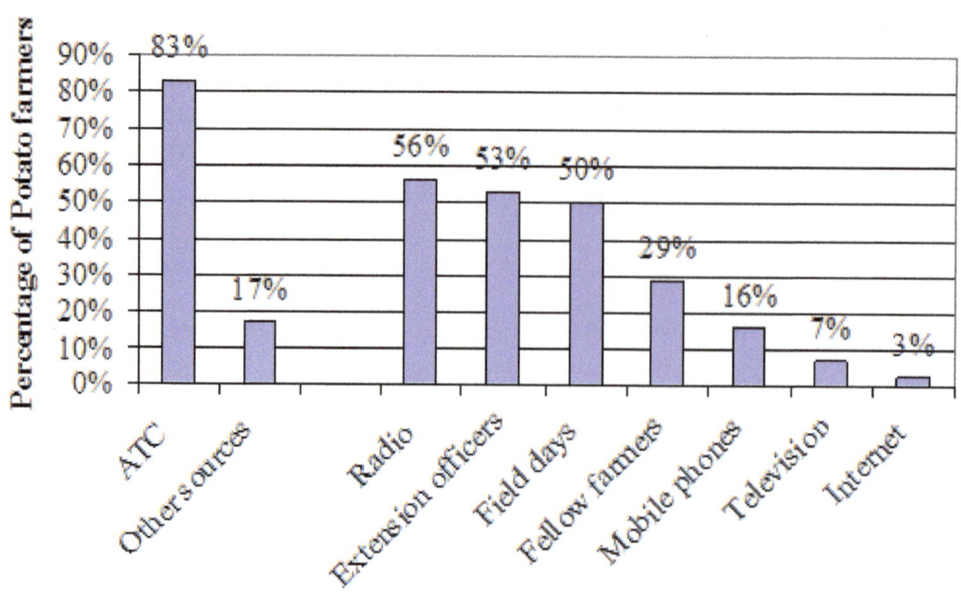

Figure 2: Sources of agricultural information for potato farmers.

Table 7: Independent t-test on yield difference between farmers using and farmers not using recommended harvesting technologies.

Variable	Mean Yield (ton/ha)	S.D.	S.E.	t-test	p-value
Planting technology					
Farmers using recommended potato planting practices after training at ATC (n=108)	17	26.95	2.59	2.73;	0.01
Farmers not using recommended potato planting practices after training at ATC (n=28)	13	27.04	5.11		
Harvesting technologies					
Farmers using recommended potato harvesting practices after training at ATC (n=92	17	26.55	2.77	2.72;	0.00
Farmers not using recommended potato harvesting practices after training at ATC (n=44)	14	27.87	4.20		

Table 8: ANOVA Descriptive statistics on Number of Times Trained versus Potato Yield.

Variable	Mean Yield (ton/ha)	S.D.	S.E.
Descriptive statistics			
Farmers trained on potato production at ATC for one day (n=67)	10	16.41	2.01
Farmers trained on potato production at ATC for two days (n=42)	19	7.63	1.18
Farmers trained on potato production at ATC for three days (n=27)	26	18.05	3.47

of agricultural information such as Farmer Field Schools are used. Mumero (2013) also found that about 11% farmers in Nakuru, Nyanza, Nyeri, Machakos, Makueni and Webuye obtain agricultural information from mobile phones. Mumero (2013) added that some farmers get agricultural information from fellow farmers, through field days and agricultural shows.

The null hypotheses were not supported by the data collected as shown by the t-test results on yield differences between potato farmers adopting and potato farmers not adopting technologies after being trained at ATCs on potato production. The t-test was done on potato seed selection ($p<0.01$), crop protection ($p<0.05$), planting ($p<0.01$) and harvesting technologies ($p<0.01$) as shown in table 6 and 7. The hypotheses were therefore rejected and conclusions drawn that potato farmers who had adopted potato technologies obtained higher yields than those who had not adopted after they were trained at ATCs on potato production.

ANOVA of the number of times trained versus potato yield

The potato farmers trained thrice at the ATC on potato production obtained an average of 26 ton/ha, the farmers trained twice obtained 19 ton/ha, while the farmers trained once obtained

Table 9: ANOVA on the number of times a farmer is trained versus potato yield.

Variable	Sum of Squares	df	Mean Square	F	p-value
Between groups	76061.575	2	38030.787	176.648	0.00
Within groups	28633.829	133	2.01	215.292	

Table 10: Relationship between the number of times potato farmers were trained at ATC versus potato yield and adoption of certified seed.

Variable	rho	p-value
Number of Times trained versus potato yield	0.880	0.000
Number of times trained versus adoption of certified see	0.13	0.13

10 ton/ha (Table 8). The high difference in potato yield between groups than within groups could mean that potato yields obtained by farmers increased with an increase in the number of times that the farmers attended training on potato production (Table 9). This may imply that adoption of potato production technologies by potato farmers may not increase with increase in the number of times farmers attended training at ATCs for example on certified seed and use of fertilizers.

The relationship between the number of times potato farmers were trained at the ATCs and the potato yield was 0.88 (p-value ≤0.00) while the relationship between the number of times potato farmers were trained at ATCs and adoption of potato certified seed was 0.13 (p-value ≤0.13) (Table 10). This means that there was a strong relationship between potato yield and the number of times farmer were trained at the ATCs whereas there was no relationship between the number of times trained at the ATC on potato production and adoption of certified potato seed. This implies that the number of times a potato farmer was trained at the ATC on potato production could influence potato yield, but not the adoption of certified seed. For training to be effective adequate time is required for the farmer trainees to internalize the skills and where possible, when field practicals are carried out.

CONCLUSION

This study concluded that ATCs' curriculum is effective in promoting small-scale farmers' adoption of potato selection, crop protection, planting and harvesting technologies. Adoption of potato technologies increases yields. This therefore implies that ATCs' curriculum significantly contributes towards national development through food security improvement, wealth and employment creation and poverty reduction.

RECOMMENDATIONS

We recommend that ATCs' curriculum prioritize demonstrations that showcase the influence of clean seed, timely harvesting and fertilizer application on potato yields. Further, campaigns should be used to educate on the effect of correct spacing and dehaulming on yields. The County should start farmer-based seed production programs to ensure that clean seed is readily available.

ACKNOWLEDGEMENT

We most sincerely thank the extension workers and farmers in Nyandarua County for their contribution during this study.

REFERENCES

1- Abegunde, B.O., & Ogunsumi, L.O. (2011). Evaluation of agricultural extension and delivery services in South-West Nigeria. International Journal of Agriscience, 1 (4), 581-591.

2- Abong, G.O., Shibairo, S.I., Kabira, N.J., & Aguk, J.A. (2012). Enhancing food security through increased potato production in Kenya. Retrieved from http://www.plantscience. unobi. ac. ke/ node/1039 on 8/7/2012.

3- Agriculture Sector Coordinating Unit. (2012). National Agricultural Sector Extension Policy, (NASEP). Nairobi, Kenya: Kilimo House.

4- Anderson, J.R. (2008). Agricultural advisory services: Background Paper for the World Development Report, 2008. Retrieved from siteresources. worldbank.org/.../ Anderson _AdvisoryServices.pdf on 21/5/2013.

5- Andriesse, W., Giller, K., Jiggins, J., Loffler, H., Oosterveer, P., & Woodhill, J. (2007). The role of agriculture in achieving millennium development Goal 1. A Review of leading Reports. Wageningen International, Wageningen: The Netherlands. Retrieved from http://www. edepot.wur.nl/23281 on 22/7/ 2012.

6- Ayieko, M.W., & Tschirley, D. (2006). Enhancing access and utilization of quality seed for improved food security in Kenya. Tegemeo Institute of Agricultural Policy and Development, Egerton University. Nairobi. Working paper No. 27/2006.

7- Bohl, W.H., & Johnson, B.S. (Ed.). (2010). Commercial potato production in North America. (2nd. Rev. ed). American Potato Journal, supplement (57).

8- Christoplos, I. (2010). Mobilising the potential of rural and agricultural extension. The Global Forum for Rural Advisory Services: FAO, Rome. Retrieved from http:// www.g-fras.org/fileadmin/.../Mobilizing_the_potential_of_extension .pdf on 21 /5/ 2013.

9- Desmarais, P. (2010). Kasisi Agricultural & Training Centre. Retrieved from http://www. canadianjesuitinternational.ca/.../zambia_ka on 17/10/2012.

10- Food Agricultural Organization of the United Nations. (2009). Human Wildlife Conflict in Africa-Causes, Consequences and Management Strategies. FAO Forestry Paper 157. Rome.

11- Food Agricultural Organisation. (2011). The State of Food and Agriculture 2010-11: Women in Agriculture. Rome, FAO. Retrieved from http:// www.feedtheFuture.gov/.../Clinton%20women %20 and the agriculture report…pdf on 10/11/2012

12- Food Agricultural Organisation. (2012). Importance of potato production and consumption in Kenya. Retrieved from http:// www.fao.org/wairdocs/x5420E/x5420e04 on 6/7/2012.

13- Gianessi, L., & Ashley, W. (2011). Fungicide spraying is critical for increasing potato production. International Pesticide Benefits. Case Study No.48. Crop Protection Research institute. Croplife Foundation, Washington DC. Retrieved from http// www. croplifefoun dation.org.

14- Gildemacher, P.R., Demo, P., Barker, I., Kaguongo, W., Woldegiorgis, G., Waguire, W., Wakahiu, M., Leeuwis, C., & Struik, P. C. (2009). American Journal of Potato Research. 86:373-38.

15- Gildemacher, P.R. (2012). Innovations in potato seed systems in East Africa. (Doctoral dissertation, Wageningen University, United Kingdom). Retrieved from http://www. edepot.wur.nl 212110.pdf on 8 /8/2012.

16- GoK. (2007). Kenya Vision 2030; A Globally Competitive and Prosperous Kenya. Ministry of Planning and National Development and the National Economic and Social Council (NESC), Nairobi, Kenya: Government Printers

17- International Fund for Agricultural Development. (2010). Rural Poverty Report 2011: New Realities: New Challenges: New Opportunities for Tomorrow's Generation. International Fund for Agricultural Development, Rome.

18- Isinika, A., Temu, A., Kitalyi, A., & Ngugi, D. N. (2002). Agricultural Education in Kenya and Tanzania (1968-1998). Nairobi; Kenya: Regional Land Management Unit, (RELMA).

19- International Potato Centre (CIP). (2008). The International Year of the Potato. Retrieved from http:// www.cip. org/pressroom /facts_ figures/ 2008 _international_year of the potato.

20- Jaetzol, R., Schmidt, H., Hornetz, B., & Chisanya, C. (2006). Natural Conditions and Farm Management Handbook of Kenya. Central Kenya, Vol. II (2nd. ed). Nairobi; Kenya: Ministry of Agriculture.

21- Jansky, S.H., Jin, L.P., Xie, K.Y., Xie, C.H., & Spooner, D.M. (2009). Potato production and breeding in China. Potato Journal. 52,57-65. doi: 10. 1007 /s/11540-008-9121-2. Retrieved from www.vcru. wisc.edu/spoonerlab/pdf/Potatoes%20in% 20 China. pdf on 22/ 5/ 2013.

22- Kaguongo, W.P., Gildemacher, P., Demo, P., Wagoire, W., Kinyae, P., Andrade, J., Forbes, G., Fuglie, K., & Thiele, G. (2008). Farmer practices and adoption of improved potato varieties in Kenya and Uganda. Social Sciences Working Paper 2008-5. CIP, Lima, Peru.

23- Kasina, M., & Nderitu, J. H. (2010). Policy Implementation and its Economic Impact on Potato Marketing Value chain in Kenya. KARI: Nairobi. Retrieved from http//www. kari.org/ …/pdf.

24- Kenya National Bureau of Statistics. (2010). The 2009 Kenya Population and Housing Census Volume 1. Nairobi; Kenya: Ministry of Planning and National Development

25- Kenya National Bureau of Statistics. (2011). Kenya Economic Survey Highlights presented during Kenya Economic Survey 2011 on 17th May 2011 by Wycliffe Ambetsa Oparanya, Nairobi; Kenya.

26- Kenya Institute for Public Policy Research and Analysis, (KIPPRA). (2009). Kenya Economic Report 2009. KIPPRA. Nairobi, Kenya. Retrieved from http:/ www.kippra.org on 22/7/2012.

27- Kipkoech, D.N., Mburu, J., Ritho, C.W., Ng'ang'a, N.W., & Lung'aho, C. (2010). Technical Efficiency among Small-scale Ware Potato Producers in Kenya.

28- Kirkwyland, M., & Thomas, E. (2012). Production Guide for Organic Potatoes. NYS IPM publication no. 138. New York. Retrieved from http:// www.

nysipm.cornell.edu/organic_ guide/potato.pdf. on 13/ 6/2012.
29- Kiura, L. (2011). Women in agriculture. Retrieved from http://icar.org in/.../12-women %20I N%AGRI...PPT on 15/11/2012.
30- Kwambai, T.K. & Komen, S.K. (2012). Multinational partnership to promote certified potato seed Production and Distribution in the North Rift Region of Kenya. Retrieved from http:// www. kari. org /.... MULTI....pdf on 16/7/ 2012
31- Lung'aho, C., Nderitu, S.K.N., Kabira, J.N., El-Bedewy, R., Olanya, O. M., & Walingo, A. (2006). Yield Performance and Release of Four Late Blight Tolerant Potato Varieties in Kenya. Journal of Agronomy, 5, 57-61.doi:10.3923/ja.2006.57.61.
32- Maguire, C., Rivera, W., Gary, A., & Pierson, M. (2011). Building the base for global food security. Agricultural Education and Training. United States of America: United States Agency for International Development (USAID).
33- Mengistu, F. T. (2009). Problems and prospects of farmers training centers: The Case Study of Ada'a Woreda, East Shewa, Oromia Region (Masters Thesis). Available from http:// www. ipms-ethiopia. org/ Final Thesis
34- Ministry of Lands and Settlement, Ministry of Planning and National Development, United Nations Centre for Regional Development, & University of Nairobi. (2003). Nyandarua District Regional Development Plan (2001-2030). An Integrated Plan for Sustainable Regional Development. Nairobi; Kenya: United Nations Centre for Regional Development, UNCRD.
35- Ministry of Agriculture and Ministry of Livestock and Fisheries Development. (2004). Strategy for Revitalizing Agriculture (2004-2014). Nairobi; Kenya: Ministry of Agriculture.
36- Ministry of Agriculture. (2006). Review of the Status of Agriculture Training Centres' Task-force Report. Nairobi; Kenya: Unpublished.
37- Ministry of Agriculture. (2009). Agricultural Training Centers' 2009 Annual Report. Nairobi; Kenya: Ministry of Agriculture.
38- Ministry of Agriculture. (2010a). National Root and Tuber Crops Policy. Nairobi; Kenya: Ministry of Agriculture.
39- Ministry of Agriculture. (2010b). Agricultural Sector Development Strategy (2010-2020). Nairobi; Kenya: Government Printers.
40- Ministry of Agriculture. (2010c). National Horticulture Policy. Nairobi; Kenya: Ministry of Agriculture.
41- Ministry of Agriculture. (2012a). Economic Review of Agriculture. Prepared by the Central Planning and Monitoring Unit. Nairobi; Kenya: Ministry of Agriculture.
42- Ministry of Agriculture. (2012b). Kilimo News (Jan.-March 2012). Journal of the Ministry of Agriculture. Nairobi; Kenya: Ministry of Agriculture. Available at http://www.kilimo. go.Ke.
43- Mumero, M. (2013, November 12th). Radio: Still the Heartbeat of Farming. The People, online.
44- Mwakubo, S.M., Nguyo, W., Nyagaka, D. O., & Obare, G. A. (2010). Are Kenyan small-holders allocatively efficient? Evidence from Potato Producers in Nyandarua North District. Journal of Agricultural Economics, 2 (3), 078-085.
45- Mwangi, J.K., Nyende, A.B., Demo, P., & Matiru, V.N. (2008). Detection of latent infection by ralstonia solanacearum in potato (Solanum tuberosum) Using Stems Instead of Tubers. African Journal of Biotechnology, 7 (11), 1644-1649. Available at
46- Natumaya, D. (2008). Farmer Training Increases Productivity and Income in Uganda. Retrieved from http://www.acdvoca.org September 2011.
47- Nyagaka, D. O., Obare, G. A., Omiti, J. M., & Nguyo, W. (2010). Technical Efficiency in Resource Use: Evidence from Smallholder Irish Potato Farmers in Nyandarua North District, Kenya. African Journal of Agricultural Research, 5 (11), 1179-1186.
48- Nyamwamu, R.O. (2013). Effectiveness of Agricultural Training Centers' Curriculum in Promoting Small-scale Farmers' Adoption of Selected Potato Technologies in Nyandarua County, Kenya. (Unpublished Masters' Thesis). Egerton University. Egerton, Kenya.
49- Obare, G.A., Nyagaka, D.O., Nguyo, W., & Mwakubo, S. M. (2010). Are Kenyan small-holders allocatively efficient? Evidence from Irish Potato Producers in Nyandarua North District. Journal of Agricultural Economics, 2 (3), 78-85.
50- Ogola, J.B.O., Ayieko, M.W., Orawo, A.O., & Kimani, F.W. (2011). Analysis of fFertilizer use in potato production in Nakuru district, Kenya. African Journal of Agricultural Research. 6 (16), 3672-3677.
51- Ogola, J., Odette, G., & Ondego, N. (2010). Small-holder horticultural marketing programme. Draft report for Value Chain Study on Potato, Cabbage and Garden peas in Nyandarua North. Nairobi; Kenya: Unpublished.

52- Qtaishat, T., & AL-Sharafat, A. (2012). Attitudes of vegetable farmers towards public agricultural extension service. American Journal of Agricultural and Biological Sciences, 7(3), 370-377.doi: 10.3844/ ajabssp. 2012. 370. 377.
53- Rajalahti, R., & Swanson, B.E. (2010). Strengthening agricultural extension and avisory Systems. Procedure for Assessing, Transforming and Rural Development Discussion Paper 45. The World Bank.Washington, D.C. Retrieved from http:// www. g-fras.org /.../ M... /Strengthening-Agricultural-Extension-and-AS.pdfon 21/5/2013.
54- Schulte-Geldemann, E., Gildemacher, P.R., & Struik, P.C. (2012). Improving seed health and seed performance by positive selection in three Kenyan potato varieties: A paper accepted for publication by the American Journal of Potato Research. 13/04/ 2012.
55- United Nations Environmental Programme. (2008).Women Farmers' Productivity in Sub-Saharan Africa. Retrieved from http://www.unep.org / ...Agriculture/.../women farmers.. pdf on 20/11/2012
56- Wambugu, P.W. & Muthamia, Z.K. (2009). The state of plant genetic resources for food and agriculture in Kenya. KARI, Kenya. Retrieved from http:/ www.fao.org/docrep/013/i15 00 e/Kenya.pdf on 6/7/2012.
57- Wang'ombe, J. G. (2008). The potato value Chain in Kenya and Uganda. Maastricht School of Management, Netherlands. Retrieved from http// www.roundtableafricanet/media/uploads/file/ joseph%20 gichuru % wangombe-the %20 potato% 20 value % 20 chain% inkenya 20 and % 20% uganda pdf on 5/7/2012.
58- Wanzala, O. (2014, February 22nd). Information is power and farmers here know exactly that. Saturday Nation, online.
59- World Bank. (2010). A brief history of public extension policies, Resources and Advisory Activities in Kenya: Extension and Advisory Services in Kenya. Retrieved from http://www.worldwide-extension.org/africa/kenya/s-kenya on 19/5/2013.

Examine the Effects of Students' Social Capital Components on Entrepreneurship Intention (Evidences from: University College of Agriculture and Natural Resources, University of Tehran)

***Rasool Lavaei Adaryani* [1], *Mohammad Reza Akbari* [2], *Fatemeh Adel* [1] and *Ali Amiri* [1]**

Abstract

The main aim of this study was to investigate the relationship of students' social capital components and entrepreneurship intention. The statistical population of this study consisted of all students of University College of Agriculture and Natural Resources, University of Tehran (N= 2600) among which 155 students where accessed based on Cochran formula and through convenient sampling. Data were collected through a questionnaire face and content validity of which was approved by faculty members of Agricultural Management and Development Department of University of Tehran. Social capital was measured by four components (*i.e.* social participation, social coherence, social confidence and social communication). Reliability of the research instrument was measured by computing Cronbach's alpha coefficient which ranged from 0.72 to 0.88. Path analysis technique employed to investigate the relationship of social capital components and entrepreneurial intention. Results revealed that social participation and social coherence had a direct effect on entrepreneurial intention while social confidence and social communication had an indirect effect on it.

Keywords:
Entrepreneurial intention, Social participation, Social coherence, Social confidence and communication

[1] *M.Sc. Students in Rural Development, University of Tehran.*
[2] *M.Sc. Student in Agricultural Extension and Education, University of Tehran.*
* *Corresponding author's email: rlavaei@yahoo.com*

INTRODUCTION

In recent years, social capital as a social phenomenon in modern society is creating new opportunities has been a matter of great concern social sciences (Saadat, 2009). Social capital is the raw material from thousands of daily interactions of people. This form of capital between specific individuals or social structures is not certain but also the space between people (Bullen and Ounyx, 2000). Social capital is a set of actual and potential resources that linked networks formed over time as institutional relationships. Finally he emphasized that it is so important to invest in social capital, and provide an opportunity to produce group cohesion as a group activity (Portes, 1998).

Compared to other forms of capital, social capital is intangible and informal relationships between people are hidden (Sedighbanay, 2009). Another definition of social capital is the result of the accumulation potential or actual the property of a durable networks of more or less institutionalized relationships of the people that membership is created in a group (Firuzabadi, 2006). Some of the key elements of social capital can be measured with Include: knowledge of public affairs, political and social motives of those who seek these types of knowledge are Public confidence in each other and informal cooperative participation in volunteer activities and in general it can be said that one of the main criteria for the recognition of social capital is the social relations and practices with one another and how they coexist in society (Adler, 2002). Social capital as a social phenomenon causes creative parenting ideas, facilitate innovation and risk-taking behaviors are (Coleman, 1990). All aspects of general social, economic, cultural and political influences that every society in economics and entrepreneurship that he is not ineffective social capital; So that it can be said that effective as a source of social capital formation and development is an entrepreneur and entrepreneurship.

Entrepreneurship is a social process. The economic and social fabric depends on the circumstances of the: First, entrepreneurs are people who are a product of his social environment, and second, the entrepreneurship is a social activity, resulting in the presence or absence to the nature of entrepreneurship, social relations affects (Anderson, 2003). Therefore, it should be noted that social capital can influential on entrepreneurship. Supposed to understand these relationships need to measure the two main variables are visible in the community. As a structure of social cohesion, social capital variables, social confidence, collaboration and communication has been established to measure the variables to be analyzed. Entrepreneurial intentions is important and hidden aspect of entrepreneurial behavior according to Ajzen model (1991) and In other words, we recognize entrepreneurial behavior by investigate entrepreneurial intentions. Also, entrepreneurial intentions is basic infrastructure for entrepreneurial behavior (Bird, 1988).

It is generally concluded that entering the field of social entrepreneurship could be promoted to provide context, but this is partly responsible for higher education so that Arasteh (2003) believes that one of the goals of higher education centers that prepares graduates for work, and the entrepreneurial success of graduates in the field, depending on their personal and professional skills.

The successful graduates who are not only equipped with the specialized skills but also skills such as creativity, innovation, risk-taking, tolerance of ambiguity, ambition and confidence to succeed in their relative lies and universities develop the skills (Arasteh, 2003).

The Alumni agricultural Estimates indicate that the market needs of agriculture annual maximum of 110 thousand graduates from the fields of agriculture, while 230 thousand people each year in the fields of agricultural universities in the country graduates are on each year by about 25 to 52 percent of graduates are unemployed (Frahani, 2010). Therefore, unemployed graduates increase every year. This issue can be derived of two option, first lack graduates abilities to Creation Business and second it is lack Social potential. Generally, unemployment is as a problem there is unemployment in the community by each reason. In this study, examine the relation between social capital and entrepreneurial intentions in order to explain the effect of social capital on entrepreneurial intentions.

On the subject of entrepreneurship, more research is purely structural characteristics and location have environmental and social capital as the most important factor is not taken into consideration (Ulhoi, 2004). It is also believed by Heydari that due to the growing trend of educated people, on the one hand and economic problems - the state of social entrepreneurship has gained momentum over the past years is the question (Shakiba, 2013).

Therefore this study was to assess the relationship between social capital and entrepreneurial intentions of students as future workers in this country are determined. In the following part of the research done in the field of social capital are mentioned. Rouse (2007) in their study of the factors explaining the national capital, human capital, social capital as factors of business success and Payment And concluded that the national capital, the direct effect of human and social capital and entrepreneurship effects are indirect and complex. Markman and Baron (2003) in their study found that entrepreneurships that higher levels of social capital compared with those with lower levels of social capital are likely to have higher probability of venture capital funding to attract. Other studies suggest that between Social Capital and there was a significant positive correlation with CE In addition, our results suggest that the components of social capital is able to predict changes in corporate entrepreneurship (Gholipur *et al.*, 2008). Ashena (2005) examined the relationship between social capital and entrepreneurship payment and the results indicate a positive and significant relationship between the dimensions of social capital and entrepreneurship. There is positive relationship between social capital and entrepreneurial intentions (Shakiba, 2012). Also it show that the relationship between social capital and gender is not significant; between male and female students is not significant in terms of entrepreneurial intentions. The results of this study indicate that social capital variables among students in rural areas is higher than urban residents, but this difference was not significant. This conclusion also applies to the variables of entrepreneurial intentions. The main purpose of this study was to investigate the relationship of social capital components and entrepreneurial intension of students and of particular interest were to:

1- Analysis of the relationship of social capital components and entrepreneurial intention in terms of gender.

2- Determining the most effective social capital component on entrepreneurial intension. In the figure 1 shown theoretical frameworks.

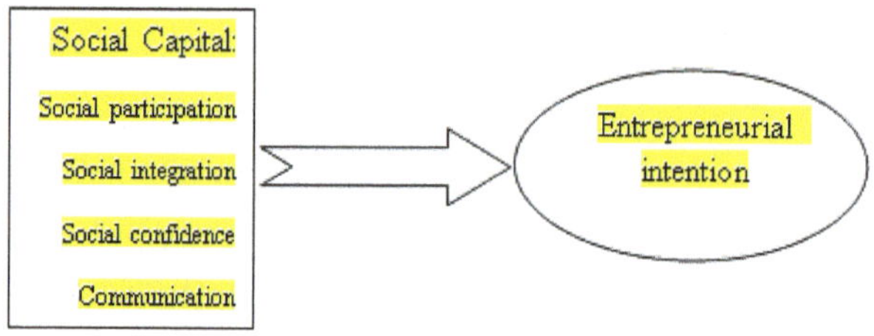

Figure 1: Theoretical frameworks.

MATERIALS AND METHODS

The present study was a correlation research that for data collection use of survey. The study's subject involves the factors influencing the entrepreneurial intention of students within the university college of Agriculture and Natural Resources, University of Tehran., in the fall and winter of 2013-2014. The survey includes 2600 students of College of Agriculture and Natural Resources, Tehran University. Convenient sampling method was used for collecting data and the sample size, which was identified using Cochran formula, was estimated as 155 subjects. To measure the entrepreneurial intentions as dependent variable, as well as social integration, social participation, communication and social confidence of the students as independent variables, a questionnaire with Likert scale was

Table 1: Chronbachs' alpha coefficient for different part of questioner.

Variable	Number of Items	α
Entrepreneurial intention	5	0.88
Social participation	5	0.80
Social coherence	5	0.84
Social confidence	8	0.72
Communication	5	0.82

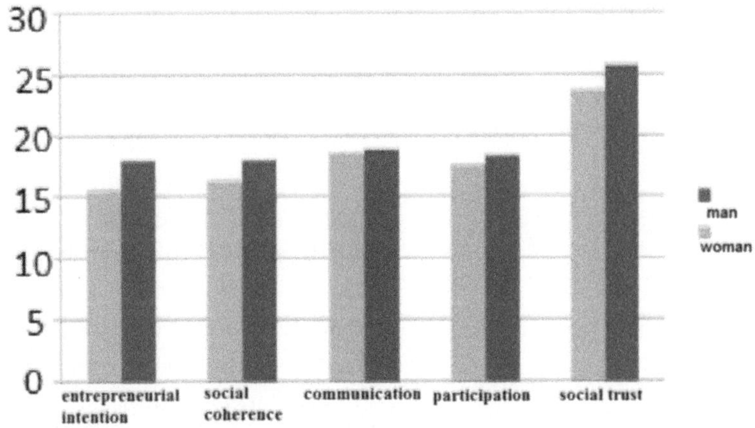

Figure 2: Comparison of studied variables in terms of gender

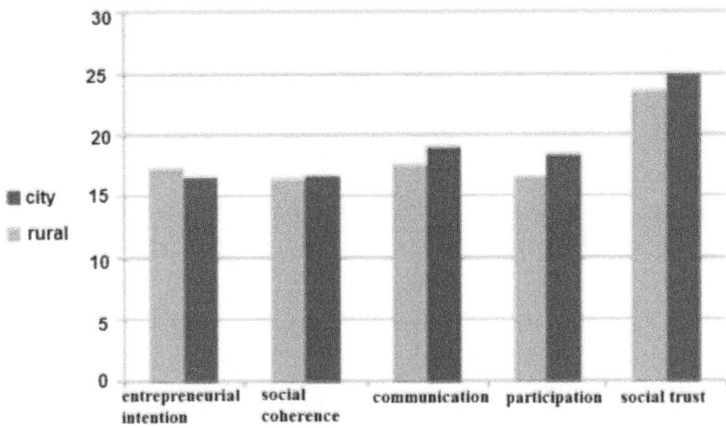

Figure 3: Comparison of main variables in terms of habitat.

applied. Formal methods were used to assess the questionnaire's validity, approved by the professors of Agriculture Development department of College of Agriculture and Natural Resources, Tehran University. To determine the reliability of the instruments used, a pretest was conducted among 30 students of the studied sample which based on Cronbach's coefficient alpha. The value of this coefficient, presented in table 1, showed that all parts of questionnaire have appropriate reliability to measure the studied variables. Other variables were presented by closed and open questions. SPSS software was used for data analysis. To describe the sample, frequency tables, central indices and descriptive figures were used. In the analytical study, the predominant statistical method is path analysis.

RESULTS

The instrument used in the study, to measure the entrepreneurial intention, social integration, social participation, communication and social confidence variables of the students, was a questionnaire with Likert scale whose reliability was confirmed by Cronbach's coefficient alpha.

The average age of students participating in the study is 23, ranging from 18 to 35 years.

Studying the frequency of gender variable, it was determined that 47 percent of the respondents in the questionnaire were male students and 53 percent of them were female students. Figure 2 was used for the descriptive comparison of two groups of respondents and it has been plotted based on the average of studied variables according to the gender variable.

Table 2: Comparison of entrepreneurial intention in terms of gender.

Entrepreneurial intention	N	Mean Rank	Z Statistics	Sig.
male	73	88.8	-2.832	0.005
female	82	68.38		

Table 3: The frequency of respondent in terms of major.

Case	Field	Percent
1	Irrigation	7.7
2	Pedology	11.6
3	Agronomy	11.6
4	Gardening	5.2
5	Agricultural Extension	3.9
6	Agricultural Development	3.2
7	Animal Sciences	6.5
8	Economy	2.6
9	Fisheries	2.6
10	Woods	12.3
11	Plant protection	5.2
12	Wood Industry	7.7
13	Environment	5.8
14	Watershed	3.9
15	Wood and Paper Industry	8.4
16	Agricultural Machinery	1.9

Figure 2 has showed that in most of the study variables except entrepreneurial intention one, both male and female students are equal. Mann White Cane test was used for compare of two groups in the study's dependent variable (The results of this test shown in table 3). There is significant difference between the two groups and male students are more interest to entrepreneurship.

Another variable that has been examined in this study is the students' residence variable. Overall, the residence as the culturing environment plays an effective role in one person's career and life. It's important to note that the respondent frequency in the two groups does not have a reasonable balance, so that only 30 of 155 respondents were from rural communities, so in this part we only relied on the descriptive statistics. According to figure 3, it was revealed that the participants who'd live in a rural environment had higher levels of entrepreneurial intentions. The important point is that with respect to the fact that rural communities have more traditional norms than rural communities, social integration and social confidence are expected to be higher among the students inhabiting villages. This unexpectedly contradictory observation could be explained by the fact that university is generally located in an urban area and when both rural and urban students are registered in the university, as the students inhabiting cities have an urban background and their education place where is presumably in cities, they will be better integrated. This situation will take place, if the students living in village, due to the differences in personal life and education place, cannot integrate adequately with the environment. These conditions are the same as entrepreneurial intention.

Other variable is investigate the students' field of study in this study. The respondents to the questionnaire would consist of different departments of Agriculture and Natural Resources College, Tehran University (Table 3).

Based on the mean entrepreneurial intention variable, it was determined that among different groups, the Agricultural Machinery students had the highest level of entrepreneurial intention. Three undergraduate, graduate and PhD students were surveyed in the study. The combination of participants includes 51%, 41% and 8% for undergraduate, graduate and PhD students, respectively. The mean entrepreneurial intention among undergraduate, graduate and PhD students is 17.35, 16.03 and 16.83, respectively. These average values indicate that descriptively there is no significant difference between different degrees on entrepreneurial intentions. Of course, it should be noted that due to the imbalance in frequency of the three degrees, this comparison cannot be completely reliable. The variables which have been assessed in the present study by Likert scale would include the entrepreneurial intention, social integration, social participation, communication and social confidence of Agriculture and Natural Resources students. The scale of the variables is pseudo- parametric or pseudo-

Table 4: The frequency of students in terms entrepreneurial intention.

Entrepreneurial intention Group	Lower entrepreneurial intentions	Relatively lower entrepreneurial intentions	Relatively high entrepreneurial intentions	High entrepreneurial intentions
Frequency	25	46	53	31
Frequency Percent	16.1	29.7	34.2	20

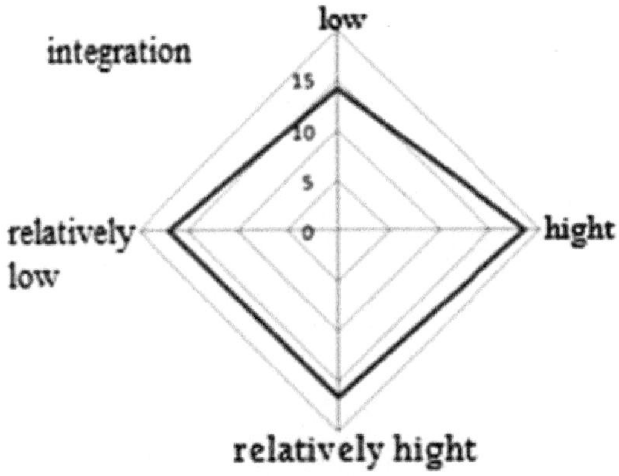

Fig 4: Social coherence in four level of entrepreneurial intention.

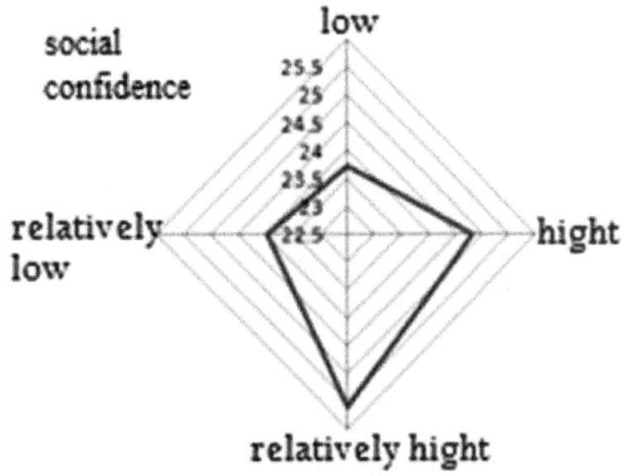

Fig 5: Social confidence in four level of entrepreneurial intention.

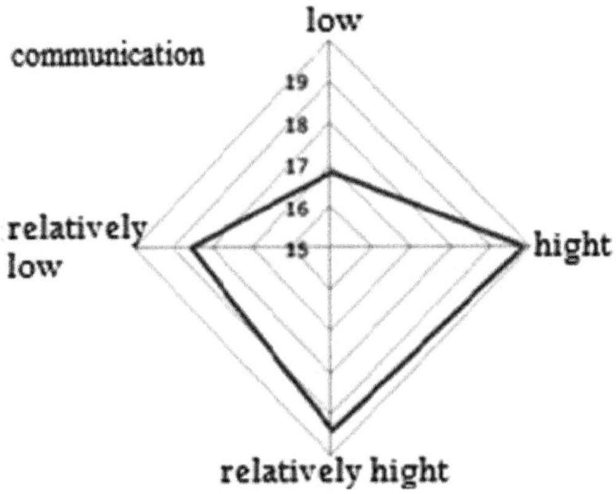

Fig 6: Social communication in four level of entrepreneurial intention.

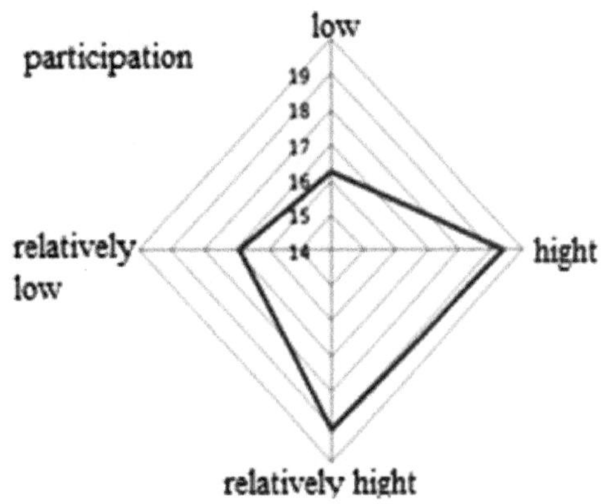

Fig 7: Social participation in four level of entrepreneurial intention.

distance. To describe the sample, firstly the students by means of mean and standard deviation are divided into four groups in terms of the dependent variable, i.e. entrepreneurial intention and then, they'll be compared with other variables. First group are the students who have low entrepreneurial intention. This group constitutes the students whose intention scores are less than one unit of SD. The second group are the students who have relatively low entrepreneurial intention. This group's entrepreneurial intention scores are between the mean and one standard deviation below. The third group of students whose entrepreneurial intention score is between the mean and one standard deviation higher than the average. The fourth group who enjoy a high level of entrepreneurial intention have a score which is one standard deviation relatively higher than the average.

Table 4 presents the frequency and frequency percentage in each group. According to this table, the highest frequency belongs to the third group, i.e. the students with high entrepreneurial intention. In other words, based on grouping, entrepreneurial intention mode in the studied samples is related to the third group.

According to this grouping, a comparison was made among these four groups using the descriptive figures, on the basis of the variables of social integration, social participation, communication and social confidence. In figure 4, social integration variable has been studied among the four entrepreneurial intention groups; the students placed in the low entrepreneurial intention group have the lowest social integration. The same procedure is true for social participation and communication variables as well. As it can be observed in figure 5, however, the lowest social confidence is within the first group with low entrepreneurial intention and the highest social confidence is seen among the third group with relatively high entrepreneurial intention.

Analysis of causal relations by path analysis technique

After describing the study sample, path analysis technique was used to test hypotheses and explore the causal relations among the variables. Path analysis is considered a suitable technique for the analysis of direct and indirect effects of independent variables on the dependent variable. In the present study, the entrepreneurial intention variable as the dependent variable and all other variables were considered as independent variables. Then, the path of the causal relations between these variables was drawn by the diagram in figure 8. The F test used for determining the model fit with degrees of freedom 3 is equal to 8.97, which is significant at the 99% level. Furthermore, Adjusted R square for the model is 0.134. That is, the independent variables in the study explain 13.4% of the variance of the students' entrepreneurial intention. The coefficients used in the path diagram are the regression beta coefficients. Being standard, these coefficients would be able to be compared within different variables. Therefore, it can be concluded that among the variables of age, social participation and social integration, which influence directly on entrepreneurial intention, the participation variable has the greatest role in explaining the variance in entrepreneurial intention.

To interpret the beta coefficients in path analysis, the changes in the standard deviation (SD) of dependent variable and independent variables need to be taken into account. Thus, it can be said that for each unit of increase in the standard deviation of age, standard deviation of entrepreneurial intention will experience 0.164 unit of decrease. In other words, as the students get older, they will be less intended toward entrepreneurship. Also, for each unit of increase in the standard deviation of the participation variable, standard deviation of entrepreneurial intention will experience 0.212 unit of increase. With respect to the social integration variable, it should be noted that one unit of increase in social integration will lead to 0.99 unit of increase in the standard deviation of entrepreneurial intention. In other words, the students who enjoy high levels of social participation and integra-

Table 5: Direct, indirect and total effects of independent variables on entrepreneurial intention.

Independent variable	Direct effect	Indirect effect	Total Effects
Age	-0.164	-	-0.164
Social participation	0.212	-	0.212
social coherence	0.199	-	0.199
communication	-	0.257	0.257
social confidence	-	0.145	0.145

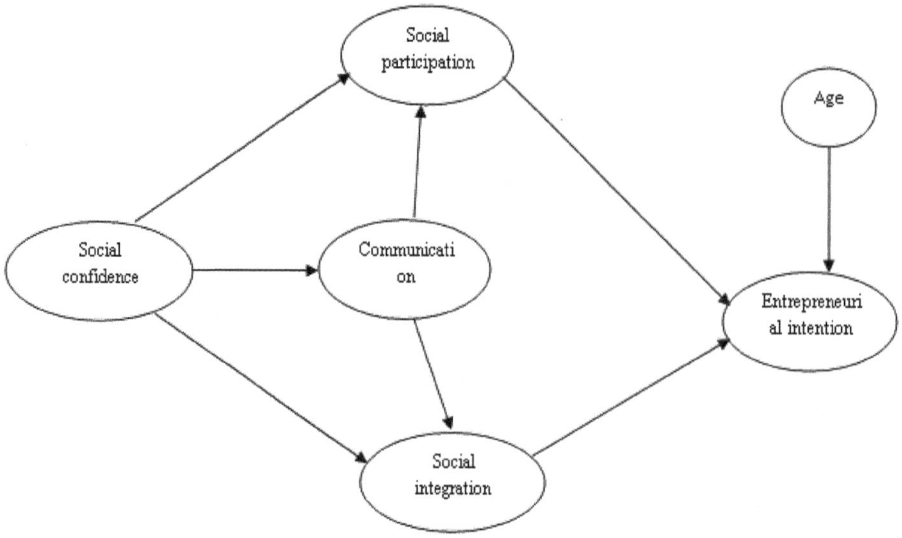

Figure 8: Path analysis of entrepreneurial intention.

tion are more likely to become entrepreneurs.

With respect to the other variables, it should be stated that communication variable has the greatest impact on the social participation and integration variables. In other words, the students who have a higher level of communication capabilities have higher social participation and integration, thereby having a positive impact on entrepreneurial intention. Communication as an important variable is affected by the social confidence variable. This relationship is direct, positive and significant; that is, the students with higher social confidence have more favorable communication capabilities. In the following in table 5, the direct and indirect effects of independent variables on the dependent variable have been presented. According to table 5, the communication variable has the most contribution to explaining the variance in entrepreneurial intention variable. Social confidence has the lowest contribution to the entrepreneurial intention, figure 1, however, indicates that the influence of this variable occurs in different ways.

DISCUSSION

In the present study, the relationship between social capital and entrepreneurial intention was studied. Social capital can be investigated as a structure with the variables of social integration, participation, communication and of social confidence. Thus, by measuring the variables and their direct and indirect effects on entrepreneurial intention, the relationship between social capital and entrepreneurial intention can be analyzed.

In this study, firstly we identified and introduced the study sample using descriptive statistics and then on the basis of entrepreneurial intention, the students were divided into four groups and the social capital variables were analyzed within the four groups. Results showed that students with higher levels of entrepreneurial intentions benefit obviously more from the social capital variables. In other words, the more acceptable social capital, the more favorable the students' entrepreneurial intention is. This finding is consistent with the results from Shakiba's (2013) study. Finally, the causal relations were examined using the path analysis technique and results showed that age, participation and social integration variables directly affect entrepreneurial intention and communication and social confidence variables indirectly affect entrepreneurial intention, where communication variable has the greatest role in explaining the variance in entrepreneurial intention. Thus, the following recommendations are offered for the development of social capital variables, so that with doing this among the students, the conditions for development and pushing students to entrepreneurship are provided.

1- It is of great importance to pay attention to the students' communication capabilities, because this variable plays the most significant role in explaining the entrepreneurial intention. There are different ways to improve the communication capabilities, but training workshops as a short-term action and introducing the subjects associated with communication and social capital in the courses taken by students can be represented as a long-term strategy.

2- Students with lower ages are more favorably tended toward entrepreneurship. Thus, for entrepreneurship development, more attention should be paid to this group and using the field researches, the reasons for lack of interest in entrepreneurship among older people should be identified and attempts need to be made to resolve them.

3- Participation has an effective contribution to improving social capital and subsequently, development of entrepreneurship.

REFERENCES

1- Adler, Paul, S. (2002). Social capital prospect for a new concept. Academy of Management Review. 12(1), 17-18.

2- Ajzen, I. (1991). The theory of planned behavior: Organizational Behavior and Human Decision Processes. 50(2), 179-211.

3- Anderson, A. R., & Miller, C. J. (2003). "Class matters": Human and social capital in the entrepreneurial process. The Journal of Socio-Economics, 32(1), 17-36.

4- Arasteh, H. (2003). Entrepreneurship and higher education, principals and obstacles, Rahyaft Magazine. (29), 6-11. (In Farsi).

5- Ashena, M. (2005). Investigation the relationship between social capital and organizational entrepreneurship in pars Arak wagon, M.Sc. Thesis, faculty

of management, university of Tehran. (In Farsi).
6- Bird, B. J. (1988). Implementing entrepreneurial ideas the case for intention. Academy of management Review, 13, 442- 453.
7- Coleman, S. (1990). Foundations of social theory. Cambridge, Harvard University press.
8- Farahani, M. (2010). The Number of Graduates in Agricultural Sector. Retrieved from http:// www.ilna.ir\newstext.aspx?ID=122172. (In Farsi).
9- Firooz abadi, A. (2006). Review of social capital and the ways in promoting it. Tehran: Supreme Council of Cultural Revolution. (In Farsi).
10- Gholipur, R., Madhushi, M., & Jafarian, V. (2008). Analysis of the relationship and impact of social capital on organizational entrepreneurship: (Case Study in Sadid Industrial Group). Journal of Management Culture, 6, (17), 111-129. (In Farsi).
11- Markman, G.D., & Baron, R. A. (2003). Person–entrepreneurship fit: why some people are more successful as entrepreneurs than others. Human Resource Management Review, 13(2), 281-301.
12- Ounyx, J., & Bullen, P. (2000). Measuring social capital in five communities. Journal of Applied Behavioural Science, 36: 23- 42.
13- Portes, A. (1998). Social Capital: Its origins and applications modern sociology, Annual Review of Sociology, 24(1), 1- 24.
14- Rouse, J. (2007). Path ways to entrepreneurship across the life course: An Innovation model. Institute for Small Business and entrepreneurship (ISBE). 7-9.
15- Saadat, R. (2009). Estimation of trends in social capital in Iran. (Using fuzzy). (In Farsi).
16- Seddighbannay, H. (2009). Mass media and social capital. Aftab Magazine, 34(3), 22-25. (In Farsi).
17- Shakiba, H. (2013). Study of Relationship between social Capital and Entrepreneurship Intentions Agri-Science students, university of Tehran, A thesis Submitted to the Graduate Studies office in partial fulfillment of the requirements for the degree of Master of Science in Agriculture Education, University of Tehran.
18- Ulhoi, J. P. (2005). The social dimensions of entrepreneurship. Technovation, 25(8), 939-946.

The Law of One Price and the Cointegration of Meat Price in the Global Market: the Case of Iran's Market

Behrouz Eskandarpour [1*], Parisa Khaligh Khiyavi [2] and Mohammad Kavoosi Kalashami [3]

Abstract

In view of the importance of the meat imports and the exports of meat in Iran, this study focuses on the status of the global market of meat imports the unit price test in these markets between the years 1961 and 2011 using the market cointegration approach. The results of this study are indicative of prices co-integration in the international market of meat as well as among such major importers as the US, Germany, Italy, Japan, France and Russia. However, the important point is that no price co-integration has been observed between the global prices of meat and the import price in Iran's import market. Moreover, no law of one price has been established in Iran's import market in the long term; and in this respect, these markets lack the appropriate efficiency.

Keywords:
Cointegration, Iran, Meat market, The law of one price, World market

[1] *Faculty Member of Payam Noor University, Iran.*
[2] *Lecturer of Economics, Rasht Branch, Islamic Azad University, Rasht, Iran.*
[3] *Assistant Professor, Department of Agricultural Economics, Faculty of Agricultural Sciences, University of Guilan, Iran.*
* *Corresponding author's email: skandarpur@gmail.com*

INTRODUCTION

The influence of such political and economic issues as joining World Trade Organization (WTO), removing tariff barriers and trade restrictions, considering the principles of the General Agreement on Tariffs and Trade, food security etc. on the international trade of agricultural commodities and food all point to the importance of the law of one price and the co-integration of the global and domestic prices. Two markets in different locations are joined only when the prices of the traded goods in the importing market equals such prices in the exporting market plus transportation costs and other delivery expenses. In other words, the market is integrated for those products in which the prices of the goods move together over the time; implying that the law of one price is preserved.

Existence of distance between selling markets of different products and manufacturers leads to the price differences between such markets. The efficiency of such markets is assessable using the relationship and integration criterion. The law of one price suggests that when there are no unusual transaction costs, the price of similar goods in different countries will be the same as long as they are calculated based on a common currency. The largest meat importers in the global markets are countries such as US, Germany, Italy, Japan, France and Russia. Iran too has always been one of the importers of meat from the global market, too.

Several fluctuations can be observed in the quantity and value of imports in the global market of meat. As for the value of the meat imports from the global market, France had the highest rate in the early years, whereas in more recent years the US has taken over. In recent years, the import trends of the US, Russia and to some extent Japan have been negative, while such trends are still positive for the other countries.

So far, several studies have been done in the area of market integration and the law of one price. Baffes (1991) examined the relationship of seven products among four countries using the law of one price. The findings suggest a convergent relationship among thirteen markets out of the available sixteen markets. Mohanty *et al.* (1998) examined the law of one price in the international market of wheat, sugar, wool and tea in Canada, Australia, US, New Zealand and England. In this article, the two approaches of general and detailed co-integration are employed and their results are compared. In three cases, the general convergent approach has confirmed the law of one price while the detailed convergent approach ratifies long term relations in 8 pairs. Kuiper (1999) investigated six regional corn markets using Johansen's integrative approach. The results obtained point to the long term relations among such markets. Zanias (1993) studied the stability of the corn, milk, and meat and potatoes market among England, Belgium, Denmark, France, Germany, and Italy. The results indicated that the milk and corn markets show the least and most stabilities respectively. Delpachitra and ST Hill (1994) studied the law of one price in New Zealand through investigation of the data provided by 15 agricultural companies. The results showed that market stability is not feasible over short periods of time. Vinuya (2007) investigated the stability of the law of one price in the shrimps market using data on import price provided by Japan, US, and the European Union. The findings show that there is strong stability between Japanese, American and European markets, and that the law of one price is in effect in the shrimps market. Toosi (2013), used Error Correction Model and Engle-Granger Method to investigate the price linkages between domestic and world cotton market. The variables of the study are domestic and world cotton monthly prices. The results showed that any price shock induced by world market causes the domestic cotton price to get out of equilibrium and it takes more than approximately three months to eliminate the disequilibrium. Katrakilidis (2008), explored the long run linkages among milk prices of five European markets in a dynamic framework by employing multivariate cointegration analysis and appropriate Vector Error Correction (VECM) specifications. The detection of causal effects and the identification of possible dominant markets that drived the prices of other markets was carried out by means of Granger causality testing and exogeneity tests. Finally, the short nm dynamics of the milk markets are explored by applying variance decomposition analysis. Fossatti *et al.* (2007), studied the relationship between a set of commodity prices in a small open economy like Uruguay and the corresponding international and regional prices. The empirical methodology used is the multivariate cointegration procedure based on maximum likelihood methods introduced by Johansen. In the case of cereals, the evidence suggested strong market integration between domestic and regional markets and, to some extent, also to international markets. Therefore, directly or indirectly, domestic prices were connected with the efficient price signal. Results for beef indicate strong market integration between the domestic market and the regional market, which is not so well connected with international markets. Thus, domestic price appears to be linked to a regional price that is not linked to the efficient price signal.

This study investigated whether the law of one price can be established in a market with clear fluctuations in the quantities and values of imported meat in major importing countries? Is Iran's import market competitive with the world market? This study too, will also investigate the relationship of the meat import markets. The overall aim was to investigate the world market relationship and the status of Iran it.

MATERIALS AND METHODS

In this study, at first, stationary of variables was examined and then Engel-Granger's as well as Johansen's co-integration tests are used to assess market integration. Integration and Co-integration analysis entered economic literature in 1980, and is considered by many economists as one of the most important developments in experimental modeling. If two or more variables possess varying means of variance and covariance over time, they will be used in regression analysis. The obtained regressions will be false. However, this does not hold true if the same variables converge. In essence, co-integration means studying the stability of the relationship between variables in the long term. This concept enables researchers to look into the possible existence of an equilibrium or stability among two or more time series variables which are by themselves non-stationary. We employed Engle-Granger's co-integration test. Based on this test, if there is long term equilibrium between two non-stationary variables, they converge.

Before conducting the above test, examination of the stationary and non-stationary time series variables is necessary as a pre-condition for a co-integration test are for the variables to be non-stationary and co-convergent. This is to say if a convergent variable is compounded d times to become stationary, it then has d roots of grade d, as shown with I (d).

To examine the stationary of the variables, the unit root test is utilized. In the present study, the Augmented Dickey – Fuller (ADF) Test is used for this purpose.

In the next stage and in the case of nonstationary of variables, their co-integration will be studied. Engel- Granger approach proceeds as follows: first a regression is obtained using OLS to remove the false statements ($U_t = y_t - Bx_t$). Then using ADF, we investigate the stationary of the false statements. If they are stationary, we conclude that the variables are co-integrated. To investigate market relatedness, we use the following:

$$p_{1t} = a_0 + a_{jt} + E_t \qquad (1)$$

Where p_j, p_i represents price series for a commodity in two markets and E is the false statement. If E_t is stationary, we conclude that the price series have long-lasting relatedness. The most common and important test used for assessing long term equilibriums among time series variables is what was presented by Johansen in 1990. The advantage of this approach over the others is that the trends and kinds of the long term relationships are also shown. Determining such trends among variables is done using VAR.

$$\Delta Y_t = \beta Y_{t-p} + \sum_{i=1}^{p} \Gamma_i \Delta Y_{t-i} + \varepsilon_t \qquad (2)$$

Johanesen proposes two criteria of λ_{trace} and λ_{Max} for testing long run relation. (Johansen and Juselius, 1990).To obtain the intended results, price data from imports of meat by US, Germany, Italy, Japan, France, Russia and during the years 1961-2011 available at FAO website were used. Eviews7 Software was used to analyze the data.

RESULTS AND DISCUSSION

As shown in the results obtained from the validity of variables of the Augmented Dickey

Table 1: Analysis of the validity of the variables under investigation.

Variable	Statistic	Critical values at 0.01
France import price	-5.21	-4.17
Germany import price	-4.23	-4.17
Iran import price	-4.30	-4.17
Italy import price	-6.23	-4.17
Japan import price	-7.45	-4.17
Russia import price	-4.80	-4.17
US import price	-4.87	-4.17
Average Global import price	-4.68	-4.17

Table 2: Determining the optimum stoppage in the VAR model for meat import.

Stoppage	LogL Statistic	LR Statistic	FPE Statistic	AIC Statistic	SC Statistic	HQ Statistic
0	-75.38	NA	2.26×10-6	6.86	7.21	6.96
1	32.18	143.42*	2.07×10-8	1.98*	6.73*	2.71*

Source: current study

Table 3: Trace test in Johansen Juselius.

Null hypothesis	Eigenvalue parameter	Trace statistic	Critical values at .05
Nonexistence of long-term relationship	0.89*	164.68	150.56
More than one long-term relationship	0.78	109.70	117.70
More than two long-term relationships	0.67	73.73	88.80
More than three long-term relationships	0.56	47.36	63.88
More than four long-term relationships	0.46	27.71	42.92
More than five long-term relationships	0.32	13.11	25.87
More than six long-term relationships	0.15	3.86	12.52

Source: current study

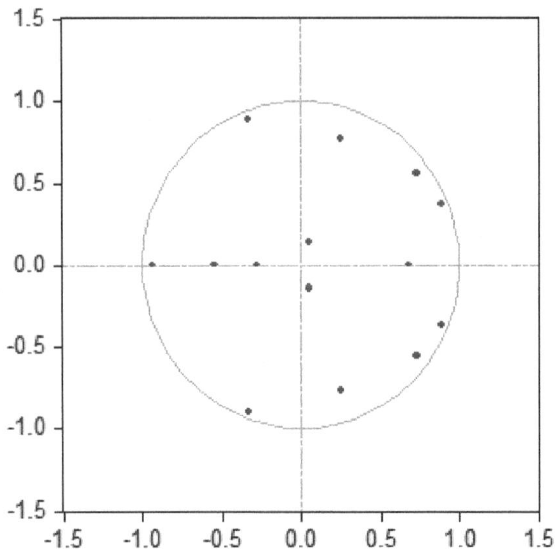

Figure 1: Inverse Roots of Auto Regressive Characteristic Polynominal

Fuller test, all the variables under investigation are of the validity level I (1).

To assess the law of one price in this case, it is necessary to first determine the amount of optimum stoppage in the co-integrated model. To this end, a VAR model was prepared, after which the optimum stoppage was elicited. As shown in table 2, the optimum stoppage in this model is equal to 1. Thus, such convergent models have lower stoppage rates.

To ensure the validity of the VAR model, it is necessary to run tests such as the inverse root. As can be seen in Figure 1, the present model lacks self-regressive inverse root, and as such is appropriate for the case at hand. Also the normality tests of existence or nonexistence of correlation and heterogeneity of variances was also studied, which also proved to be effective.

Ultimately, to study the law of one price among big importers of meat in the global market, integrative models have been used. The results indicate that the import prices in the global market of meat tend to be similar in the long

Table 4: Max-Eigen test in Johansen Josilus models.

Null hypothesis	Eigenvalue parameter	Trace statistic	Critical values at .05
Nonexistence of long-term relationship	0.89*	54.98	50.60
More than one long-term relationship	0.78	35.97	44.49
More than two long-term relationships	0.67	26.37	38.33
More than three long-term relationships	0.56	19.65	32.12
More than four long-term relationships	0.46	14.60	25.82
More than five long-term relationships	0.32	9.25	19.39
More than six long-term relationships	0.15	3.86	12.56

Source: current study *P< 0.05

Table 5: Trace Test in Johansen Juselius.

Null hypothesis	parameter	Trace statistic	Critical values at .05
Nonexistence of relationship	0.65*	96.13	95.75
More than one relationship	0.54	64.89	69.81
More than two relationships	0.47	41.76	47.85
More than three relationships	0.29	22.52	29.80
More than four relationships	0.21	12.22	15.49
More than five relationships	0.12	3.29	3.84

Source: current study *P< 0.05

Table 6: Max-Eigen test in Johansen Juselius models.

Null hypothesis	parameter	Trace statistic	Critical values at .05
Nonexistence of relationship	0.65*	41.24	40.08
More than one relationship	0.54	23.12	33.87
More than two relationships	0.47	19.24	27.58
More than three relationships	0.29	10.30	21.13
More than four relationships	0.21	7.32	14.26
More than five relationships	0.12	3.29	3.84

Source: current study *P< 0.05

term; and as such, one can conclude that such a market tends to be competitive and that the prices are convergent. In fact, considering tables 3 and 4, and with reference to both Trace and Max-Eigen tests, the null hypothesis stating that there is no long term relationship among import prices in big countries is rejected, and this relationship is well proved.

In the next section, Iran's status has been separately considered in terms of long term relationships. To his end, such relationships have been studied on meat import prices in comparison to the (average) world prices. The results of both statistics of Trace and Max-Eigen reveal that in all 5 cases, existence long term relationship between world prices and the import price of meat for Iran is not confirmed- hence lack of the law of one price in this case. This could be

Table 7: Trace Test in Johansen's model.

Null hypothesis	Eigenvalue parameter	Trace statistic	Critical values at .05
Nonexistence of along term relationship	0.32	18.07	25.87
More than one long term relationship	0.19	6.63	12.52

Source: current study *P< 0.05

Table 8: Max-Eigen test in Johansen model.

Null hypothesis	Eigenvalue parameter	Max-Eigen statistic	Critical values at .05
Nonexistence of along term relationship	0.32	11.45	19.39
More than one long term relationship	0.19	6.63	12.52

Source: current study *P< 0.05

due to variable tariffs drawn up on meat import. The law of one price indicates that prices could be similar based on a common currency when there are no business impediments or unusual expenses. Considering the existence of such variable tariffs on meat import and their temporary lifting in some years (as in 2010), the law of once price will not be observed in the Iranian market. The results from one of the existing five approaches are shown in tables 7 and 8, which are confirmed by other approaches as well.

CONCLUSION

As can be seen in the results of the present study, the global market of meat import is generally competitive market over the long run; however, Iran's market of meat import, contrary to its, is not competitive with the global market. In fact, global meat market has been very efficient, whereas the same cannot be said for its Iranian counterpart. This lack of competitveness can be, as mentioned earlier, due to drawing up variable tariffs on meat import. To support domestic producers of meat, such tariffs have been varied by the government continuously, which indicates that such producers cannot compete with their off-shore counterparts. However, if in this particular case, instead of varying tariffs, the government attempts to reduce the finished cost of meat at home and reduce its price, then such local businesses would be able to compete with their international rivals, and the efficiency of the meat market will be improved as well.

REFRENCES

1- Baffes, J. (1991). Some further evidence on the law of one price: the law of one price still holds. American Journal of Agricultural Economics, 17, 1264-1273.

2- Delpachitra, S. B., & St Hill, R. L. (1994). The law of one price: a test based on prices for selected inputs in New Zealand agriculture. Agricultural Economics, 10(3), 297-305.

3- FAO database, (2011). Available at: www.FAO.org

4- Fossati, S., & Lorenzo, F., & Rodriguez, C.R. (2007). Regional and international market integration of a small open economy, Journal of Applied Economics, 1, 77-98.

5- Johansen, S. (1988). Statistical analysis of cointegrating vectors. Journal of Economic Dynamics and Control, 12, 231 - 254.

6- Johansen, S., & Juselius, K. (1990). Maximum likelihood estimation and inference on cointegration – with application to the demand for money. Oxford Bulletin of Economics and Statistics.

7- Katrakilidis, C. (2008). Testing for market integration and the law of one price: An application to selected eurpean milk market, International Journal of Economic Research, 5(1), 93-104.

8- Kuiper, W.E. (1999). Testing for the law of one price and identifying price leading markets: an application to corn markets in benin. Journal of Regional Science, 39, 713-739.

9- Lütkepohl, H., & Reimers, H. E. (1992). Impulse response analysis of cointegrated systems. Journal of economic dynamics and control, 16(1), 53-78.

10- Mohanty, S., & Peterson, E., & Smith, D.B. (1998). Fractional cointegration and the false rejection of the law of one price in international commodity markets. Journal of Agricultural and Applied Economics, 30: 267-276.

11- Toosi, M. (2013). The price linkages between domestic and world cotton market. American-Eurasian J. Agric. & Environ. Sci., 13 (3), 352-356.

12- Zanias, G. P. (1993). Testing for integration in European Community agricultural product markets. Journal of Agricultural Economics, 44(3), 418-427.

Assessment of Rural Farming Households WTP for Fertilizers and Agrochemicals in Kwawara State, Nigeria

Ademoye Emmanuel Akinboboye, Fadipe, A.A., Adigun and Jubril Olayinka, Animashaun*

Abstract

Given the specific geographic and spatial location of rural areas in developing countries, to bring agrochemical to the rural farming households, it is argued, may have to come at a cost over and above the normal price it is sold in market. To this end, this work focuses on the willingness of rural farming households to pay more than the mean average regional retail price for agrochemicals in Kwara state, North Central Nigeria. Questionnaire was administered to 100 randomly sampled in the two agricultural development zones (ADP) in the study area. Descriptive statistics and the Logistic Binary Regression model were fitted to examine factors that influence respondents' decision to pay more than the prevailing average prices for access to agrochemicals in the study area. Findings indicated a high level of awareness of agrochemical use and modal responses to quantities used include: fertilizers (41-50Kg/ha), herbicides (<10liters/ha), and pesticides (<10liters/ha) on the cultivation of yam, cassava and maize, which were the dominant arable crops in the study area. About 90% of the respondents purchase agrochemicals from their personal savings and less than 10% of the respondents got their agrochemicals from other sources (governmental, developmental agencies, ADP and farmers' cooperative). Furthermore, 88% ($p<0.01$) of the respondents indicate willingness to pay more than the current average price to have access to agrochemicals and the logistic regression reveals that level of education ($P<0.01$), had a positive relationship with respondents' to willingness to pay more for agrochemicals in the study area. Recommendations were made towards encouraging expanded use of agrochemicals throughenhanced marketing strategies that will facilitate contact of marketing agents to prospective customers located in the rural communities.

Keywords:
WTP, Agrochemicals, Fertilizers contingent valuation

Department of Agricultural Economics and Farm Management, University of Ilorin, PMB 1515, Ilorin, Nigeria .
** Corresponding author's email: emmanfad@yahoo.com*

INTRODUCTION

Agrochemicals refer to substance use to help manage an agricultural ecosystem or the community of organism in a farming area. They are important agricultural input useful for sustaining and increasing yields of agricultural products. Agrochemicals include fertilizer, pesticide, herbicide and rodenticide to eliminate the presence of living things that causes injury or disease to crops and also to increase soil nutrient (fertilizer addition) for improve crop production (Ayoola, 1990; Morris *et al.,* 2007).

However, low fertilizer use is a serious constraint to agricultural productivity growth in Nigeria. In Nigeria, consumption of fertilizer was only seven kilograms per hectare (kg/ha) of arable land in 2005, significantly lower than India's rate of about 121 kg/ha or Pakistan's 184 kg/ha (World Resources Institute, 2010). In addition, rural-based farmer demand for and utilization of fertilizer and agro-chemicals is still lower than what is experienced in advanced countries. This low use could be as a result of rural small scale farmers not being aware, unable to afforddue to higher cost attributed to longer distances or lack of access resulting from geographic and locale specific imposed constraints. Therefore, to effectively get agrochemicals to rural farmers it is argued, will have to come with an additional cost that will add up to its composite cost.

As agrochemical suppliers and marketers constantly seek for market outlets for the sales of their commodities, however, the geographic and spatial location of small-scale rural based farmers, who constitutes a significant portion of the agrochemical market, suggest that getting these products to them may come at a cost over and higher than the prevailing regional retail price. To estimate potential producers' demands, suppliers and marketers often rely on stated preference methods such as Contingent Valuation.

Initially, contingent valuation, a survey-based methodology, was developed to elicit the WTP for non-market goods and services like environmental services (Boyle, 2003; Carson and Hanemann, 2005; Zapata *et al.*, 2012), health economics (Diener *et al.*, 1998; Krupnick *et al.*, 2002), real estate appraising (Banfi *et al.*, 2008; Lipscomb, 2011), art valuation (Thompson *et al.,* 2002), and in agribusiness (Lusk and Hudson, 2004).

While majority of studies in the study area have holistically focused more on the adoption of fertilizers and agrochemicals, none of such studies critically examined farmers willingness to pay an extra cost that may be associated with getting the agrochemical inputs to them (Owusu *et al.*, 2007; Lagat *et al.*, 2007; Irene, 2012; Moser and Barret, 2003). In view of this, the purpose of this study was to identify the factors that determine rural farming households' willingness to pay an additional cost to have access to agrochemical in the Kwara State, Nigeria.

Specifically, it:

1- examines the rural farmers level of awareness about knowledge of the benefits of agrochemicals;

2- examines the quantity and cost of fertilizers and agrochemicals used by the rural farmers;

3- investigates the level of the farmers willingness to pay more for the access to the use of agrochemicals; and

4- identifies determinants of rural farmers willingness to pay to have access to agrochemicals.

Findings emanating from this study will add to the body of knowledge and assist governmental and non-governmental agencies with sound policy formulation particularly as its relates to agricultural development in Nigeria.

MATERIALS AND METHODS

The study was conducted in Kwara State of Nigeria. It is located in the guinea savannah region of Nigeria and falls within latitudes 7^0 45! N and $9^0$30!N and longitudes $2^0$30!E and 6^0 25!E. Farming is the predominant occupation of residents in Kwara while some engage in craft activities such as weaving, blacksmithing, bricklaying, carpentry, welding etc. Fishing is also prominent along the lower River Niger Basin. The target respondents were rural small-scale farming households. A three stage sampling procedure was used in selecting our respondents. In the first stage, the list of four ADP zones in Kwara state was collected from ADP (Agricultural Developmental Program) of which two out of the four zones were randomly selected.At the second stage, 50 respondents were randomly sampled from each of the two zones. At the end, a total of 100 respondents were randomly selected and interviewed with a structured questionnaire.

Descriptive statistical analysis and the Logistic regression was fitted for to assess determinants of farmers' willingness to pay for access to agrochemicals, using the function:

$Y= \alpha+\beta X_1+ \beta X_2+ \beta X_3+ \beta X_4+ \beta X_5+\beta X_6+ \beta X_7+ \beta X_8+ \beta X_9+ \beta X_{10}+e$

Where

Y is a dichotomous response variable (1 for farmers that are willing to higher than the current determined market price for agrochemicals and 0 for otherwise). The study determined the hypothesized independent variables to include in the model after a review of relevant literatures on factors that influence WTP. The independent variables include:

X_1= age (years)

X_2= Yield of yam (kg/ha)

X_3= farming experience (years)

X_6= educational status (formal=1 and informal= 0)

X_7= income (Naira)

X_8= mode of action of pest

X_9= cost of agrochemicals (Naira)

X_{10}= yield of cassava (tonnes/ha)

α= constant; e=Error term, it is assumed to be normally distributed with zero and constant variance and β is the parameter to be estimated.

Socio-economic characteristics of the respondent

As shown in the Table 1 86% of our respondents are male compared to the 14% female. Accordingly, from the Table it reveals that 14% of the respondents are single, 79% are married, and 5% are widows while 2% are widowers. This could indicate that married people are more engaged in farming probably because they have support from their family members, in terms of labour unlike the rest of the respondents. The modal household size of our respondent was 6-9 (51%). Households with smaller sizes were more engaged in farming activities than those with larger farm sizes. The Table reveals that greater percentage (90%) of the respondents has one form of formal education and that about 10% that are with informal education. This makes it easier to conclude that the farmers in this area could improve on their ability to embrace new innovations if all their problems of availability and affordability of agrochemicals are solved.

Furthermore, the survey revealed that 25% of the respondents were engaged in farming alone while 75% of the respondents combine farming with other businesses (Table 1). Rural farmers always find it difficult to get capital required to buy inputs such as fertilizer and agrochemicals, and as such, they prefer to combine farming with other businesses so that they can source for funds. Capital earned by farmers in current production season would determine the extent to which the farmers would invest in the next session. The modal income class of the respondents ranges between 51,000-60,000 and the average income of the respondents is N 73,530 annually. This will enable the farmers to invest more in the next farming season (Table 1).

The survey showed that those that had experience less than 10 years were more in the study area which shows that they will be willing to accept or embrace new ideas ,since they will be looking for ways to improve their productivity (Table 1).

Respondents Perception of the Benefits of Fertilizers and Agrochemicals

The result of the respondents' awareness of the perceived benefits of agrochemicals was presented as their percentage response on the strongly agree, agree, disagree and strongly disagree.

The respondents' awareness of the benefits of agrochemicals would to an extent determine their willingness to pay for agrochemicals and their utilization of these agrochemicals. As shown in the survey (Table 2), about 69% of the respondents strongly agree that agrochemicals help to increase their yield while 27% agree, 2% disagree and 2% strongly disagree to this.

Also according to the Table, 62% of the respondents strongly agreed that control of pest and diseases is a benefit that can be gotten from the use of agrochemicals, while 35% agreed and 3% disagreed to this fact (Table 2).Majority of the rural farmers don't believe that use of agrochemicals provides for recreational areas, which is clearly shown from the response of the respondents with just having 1% of the respondents strongly agreeing and 33% agreed and another 33% disagree, with

Table 1: Socio economic distribution of respondents.

Socio Economic variables		Frequency	Percentage
Educational status	No formal	10	10
	Quranic	9	9
	Adult	15	15
	Primary	30	30
	Secondary	30	30
	Tertiary	6	6
Occupation	Farming alone	25	25
	Farming &others	75	75
INCOME(N)	≤ 30,000	6	6
	31,000-40,000	16	16
	41,000-50,000	18	18
	51,000-60,000	22	22
	61,000-70,000	10	10
	71,000-80,000	12	12
	81,000 >	16	16
Experience (Years)	1-10	33	33
	11-20	19	19
	21-30	16	16
	31-40	10	10
	41-50	8	8
	51 >	14	14
Sex	MALE	86	86
	FEMALE	14	14
Marital status	Single	14	14
	Married	79	79
	Widow	5	5
	Widower	2	2
Household size	≤ 5	25	25
	6-9	51	51
	10-13	21	21
	14-17	2	2
	≥ 18	1	1
Age (Years)	21-30	11	11
	31-40	16	16
	41-50	18	18
	51-60	28	28
	61-70	13	13
	71-80	14	14

Source: Field Survey

24% strongly disagreeing (Table 2). The survey showed that 45% of the respondents were of the opinion that the use of agrochemicals eliminates storage pest, 31% agreed, 11% disagreed and 13% strongly disagreed probably due to the fact that some of these farmers are not accessible to the use of these agrochemicals which could be used for the elimination of pest that destroys their stored crops (Table 2).

Estimation of Quantities and Cost of Fertilizers and Agrochemicals used by Respondents /ha

As revealed in the Table, fertilizers remains the most widely used as shown by its spread among the respondents. Quantity of fertilizer (Table 3) purchased and used could be dependent on the farm size of the respondents as well as affordability and availability .The table shows that a larger percentage of the respondents use

Table 2: Respondents' awareness of perceived benefits of agrochemicals.

Perceived benefits	Strongly agree (%)	Agree (%)	Disagree (%)	Strongly disagree (%)
Yield increase	69	27	2	2
Control of pest and diseases	62	35	3	0
Decrease in cost	49	32	17	2
Saves farmers money	46	32	21	1
Reduces drudgery	28	35	25	12
Helps farmers to grow more per unit area	21	44	28	7
Safe guard public health	13	4	32	15
Provides for recreational areas	1	33	33	24
Eliminates storage pest	45	31	11	13

source: Field survey

Table 3: Estimation of total cost of production equation for greenhouse cucumber in Khahsh.

Agro chemicals	Qty (kg, litre/ha)			Cost range (N/ha)		
		Freq	%		Freq	%
Fertilizer	≤10	33	33	≤1000	32	32
	20-Nov	8	8	1100-2000	1	1
	21-30	7	7	2100-3000	5	5
	31-40	9	9	3100-4000	12	12
	41-50	22	22	4100-5000	11	11
	51-60	3	3	5100-6000	10	10
	≥61	18	18	6100-7000	3	3
				7100-8000	1	1
				8100-9000	3	3
				9100-10,000	5	5
				10,000 >	17	17
Total	100	100			100	100
Herbicide	≤10	77	77	≤1000	57	57
	11-20	16	16	1100-2000	16	16
	21-30	6	6	2100-3000	12	12
	31-40	1	1	3100-4000	9	9
	-	-	-	4100-5000	6	6
Total		100	100		100	100
Insecticide	≤10	93	93	≤1000	93	93
	11-20	7	7	1100-2000	2	2
				2100-3000	2	2
				3100-4000	3	3
Total		100	100		100	100

Source: Field Survey

less than 10kg of fertilizer per hectare and a larger percentage of the respondents spend less than N1000 on the use of fertilizer per hectare which could largely be due to the fact that the fertilizer are either expensive or are not accessible to the respondents. The mean or average usage of fertilizer per hectare is 36.65kg and the average cost per hectare is N5148.07.The average use of herbicide and insecticide per hectare by the respondents is 6.28 litres and 2.58 litres respectively. While the average cost of usage is N1261.64 and N316.49 respectively (Table 3).

Sources and accessibility of agrochemicals by respondents

Table 4 shows the sources and accessibility of agrochemicals to the respondents'. From the table it shows that 58% of the respondents' re-

Table 4: Perception of accessibility of sources of agrochemicals by farmers.

Accessibility	Very accessible (%)	Accessible (%)	Poorly accessible (%)	Not accessible (%)
Savings accessibility	58	33	7	2
Government accessibility	1	6	1	92
Developmental agencies accessibility	0	0	12	88
ADP accessibility	1	3	1	95
Farmers' cooperative	4	11	1	84

Table 5: Statistical test to determine significance of respondents willingness to pay more.

Willingness to pay	Frequency	Percentage	Standard error
Yes	88	88	0.32
No	12	12	0.32
Total	100	100	

vealed that agrochemicals gotten from their personal savings were very accessible, 33% said it was accessible, 7% said it is poorly accessible and 2% said their personal savings were not accessible at all to get the agrochemicals. The table also reveals that accessibility of agrochemicals from other sources are poor, majority of the respondents were of the opinion that they are not accessible at all.

Assessment of farmers willingness to pay for more for access to fertilizers and agrochemical

Respondents were asked to state the price they bought their agrochemicals and fertilizers outside the rural areas. They were then asked to indicate if they would be willing to pay more to buy the inputs at their bases. About 88% indicated the affirmative and 12% of the respondents were negative. A test was carried out to examine the significance of there response. The result is presented in Table 5.

The Logistic regression was used to determine the respondents' willingness to pay for agrochemicals. The model parameters as shown by the Pseudo R^2 and the significance of the model chi ^{2} (<0.01) affirms the relative strength of the model as a good fit. As shown in the Table, possession of education is positively related to willingness to pay and it is significant at 1 % which indicates that the more educated

Table 6: Logistic regression analysis for the determinants of farmers willingness to pay.

Predictor variable	Average Marginal Effect	Standard error	p>/z/
Formal Education	14.76	5.40	0.01***
Income	0.00	0.00	0.07
Farming experience	-0.34	0.19	0.07*
Control of pest and disease	14.65	8.55	0.09*
Saves farmers money	2.36	1.29	0.07*
Drudgery	2.42	1.62	0.14
Cost of herbicide	0.00	0.00	0.13
Cassava yield	0.00	0.00	0.06*
Yam yield	0.00	0.00	0.10

Note: single asterisk refers to 10% significance and three asterisks refer to 1% significance.
Prob>chi=0.0023
Pseudo R^2=0.6326
Source: Field Survey

the farmers, the more they are willing to pay for agrochemicals. Income and knowledge of the benefit of control of pest and disease function of agrochemical are positively related to willingness to pay and significant at 10% which indicates that the higher the income and knowledge of benefits of agrochemicals, the more willing they will be to pay for it. Equally, respondents who record a higher yield from cassava are more likely to pay more for access to agrochemicals.

Conversely, farming experience shows a negative relationship with willingness to pay ($p<0.1$). This indicates that the higher the farming experiences of the farmer, the lower their willingness to pay.

CONCLUSION AND RECOMMENDATION

The study was carried out to assess rural farming households' willingness to pay more than the central market price for agrochemicals. It was conducted in two ADP zones (Zone C and Zone D) in Kwara state. Findings show that majority of the respondents are aware of the benefits of fertilizers and agro chemicals and source it primarily from their savings.The mean or average usage of fertilizer per hectare is 36.65 kg and the average cost per hectare is N5148.07. The average use of herbicide and insecticide per hectare by the respondents is 6.28 litres and 2.58 litres respectively. While the average cost of usage is N1261.64 and N316.49 respectively.

Findings imply that a higher proportion of respondents (88%) are willing to pay higher prices if fertilizers and agrochemicals can be made available to them at their vicinity. This decision is however premised on certain socio-economic variables of which education, income from farming, knowledge of mode of action of agrochemical, farming experience and cassava yield were all significant in explaining this willingness decision.

Based on the findings of the study, the following recommendations were made:

1- The problem of bureaucracy should be eliminated by the government in distributing agrochemicals, so that the rural farmers can have equal access to it both at farm centres and Agricultural Development agencies.

2- Private business owners can take advantage of this willingness and situate marketing outlets at strategic locations across rural areas.

REFERENCES

1- Ayoola, G.B. (1990). The marketing of agricultural pesticides in Nigeria workshop on the pesticides Industry in Nigeria, University of Ibadan, 6-8 February 1990.

2- Banfi, S., Farsi, M., Filippini, M., & Jakob, M. (2008). Willingness to pay for energy-saving measures in residential buildings, Energy Economics 30: 503-506.

3- Boyle, K.J. (2003). Contingent valuation in practice. In A primer on nonmarket valuation. Edited by Champ, P.A., Boyle, K.J., Brown, T.C. Kluwer Academic Publishers, Dordrecht.

4- Carson, R.T., & Hanemann, W.M. (2005). Contingent valuation. In Handbook on Environmental Economics, Vol 2. Edited by Mäler, K.G., Vincent, J.R. North-Holland, Amsterdam.

5- Diener, A., O'Brien, B., & Gafni, A. (1998). Health care contingent valuation studies: a review and classification of the literature. Health Economics, 7, 313–326.

6- Irene, S. E. (2012). Assessing the factor of adoption of agrochemicals by plantain farmers in Ghana using the ASTI Analytical framework. Kpmg: The growing importance of agrochemicals.

7- Krupnick, A., Alberini, A., Cropper, M., Simon, N., O'Brien, B., Goeree, R., & Heintzelman, M. (2002). Age, health and the willingness to pay for mortality risk reductions: a contingent valuation study of Ontario residents, Journal of Risk and Uncertainty, 24(2),161–186.

8- Lagat, J.K., Wangia, S.M., Njehia, B.K., & Ithinji, G.K. (2007). Environmental hazards African Agriculture: Factor influencing Application of agrochemicals in Nakuru district, Kenya. In Advances in Integrated Soil Fertility Management in Sub-SaharanAfrica: Challenges and Opportunities. Ed. A. Bationo. The Netherlands: Springer, 795- 804.

9- Lipscomb, C. (2011). Using contingent valuation to measure property value impacts. Journal of Property Investment and Finance, 29, 448-459.

10- Lusk, J.L., & Hudson, D. (2004). Willingness-to-Pay estimates and their relevance to agribusiness decision making. Review of Agricultural Economics, 26(2), 152–169.

11- Morris, M., Kelly, V.A., Kopicki, R.J., & Byerlee, D. (2007). Fertilizer use in Nigerian context.

12- Moser, C.M., & Barrett, C.B. (2003). The disappointing adoption dynamics of a yield- increasing low external input technology: The Case of SRI in Madagascar. Journal of Agricultural Systems, 76(3), 1085-1100.

13- Owusu-Bennoah, E., Anno-Nyako, F.O., Egyir, I.S., & Banful, B. (2007): Methodological framework: Analysing the agricultural science technology and innovation (ASTI) Systems in ACP Countries. The Ghana Case Study on Plantains. Wageningen, the Netherlands: Technical Centre for Agricultural and Rural Cooperation (CTA).

14- Thompson, E., Berger, M., Blomquist, G., & Allen, S. (2002). Valuing the arts: a contingent valuation approach. Journal of Cultural Economics, 26, 87–113.

15- World Resources Institute. (2010). Agriculture and food searchable database. <http:// earthtrends.wri.org/searchable_db/results.php?theme=8&years=& variable_ID=872&cID=&ccID=&years_rev=1>. Accessed May 10, 2010.

16- Zapata, S.D., Benavides, H.M., Carpio, C.E., & Willis, D.B. (2012). The economic value of basin protection to improve the quality and reliability of potable water supply: The Case of Loja, Ecuador. Water Policy, 14, 1-13.

Investigating Market Integration and Price Transmission of Different Rice Qualities in Iran

Amir Hossein Chizari [1*], Masoud Fehresti Sani [2] and Mohammad Kavoosi Kalashami [3]

Abstract

Rice production in most of Asian countries has been increased more rapidly than population and this has been led to increase in supply and proportionately decrease in the real price of rice in world and domestic markets. Furthermore, together with growth in production and national gross income of the country per-capita income has been increased and also demand for rice at national and international level quality has been increased. In this case studying the market conditions of different qualities of rice including marketing margins, causative relations among the prices, market integrations in long term and finally price transferring and market integration in short term is the important consequence that can help policymakers and planners in their decision makings on research, production, distribution and marketing of rice strategic product. So, using the statistics from Jihad Agriculture Organization of Guilan Province in case of the price of rice qualities (items) including Sadri momtaz (S1), Sadri darge yek (S2), Sadri mamooli (S3) and Khazar (K1) during 1999-2009 market conditions of different qualities of rice was studied. Results show that impulses in wholesale prices in Khazar rice rapidly influence on-farm prices, however, in case of other rice qualities the rate and speed of this influence is low. But in wholesale-retail market for Sadri quality rice impulses influence strongly in wholesale price and this shows intense integration of these two rice markets in Iran. It is suggested that according to the different quality of rice verities, support policy design and decision making process assigned separately.

Keywords:
Market integration, Rice quality, Wholesale price, Retail price, On-farm price

[1] *Assistant Professor, Department of Agricultural Economics, Faculty of Agricultural Economics and Development, Tehran University.*
[2] *Ph.D Student of Department of Agricultural Economics, Faculty of Agricultural Economics and Development, Tehran University.*
[3] *Assistant Professor, Department of Agricultural Economics, Faculty of Agriculture, Guilan University.*
* *Corresponding author's email: Chizari8000@ut.ac.ir*

INTRODUCTION

Rice is from millet family and it is one the main seeds to be used by human beings and is the staple food in Iran, with the quality of cooked rice outweighing all other considerations for Iranian consumers. This product needs more water under cultivation level of rice has been centered in northern provinces of the country. More than 615 thousand hectares of irrigated lands in Iran have been dedicated to rice cultivation the total area under rice is more than 615000 ha and rice is grown in 15 provinces and considering the yield of 2400 kg white rice, more than 1.4 million tons of white rice is produced in Iran every year. The remainder domestic need is compensated by importing. Under cultivation level of Iran's paddy rice during 1986-2005 had been changing from 471000 to 628100 hectares. The most central reasons for fluctuations in under cultivation level during the mentioned years are drought and shortage of water resources needed for rice cultivation. Average yearly production of paddy rice during the first development program (1990-1994) and the second development program (1995-1999) had been 2.25 and 2.35 million tons, respectively, that has had 3.93% growth compared to the first program. Average yearly paddy rice production during the third development program (2000-2004) had been 2.54 million tons that compared to the end of the second program has had 8.26% growth (rice self-sufficiency plan, 2006). However, more than 80 percent of rice area is distributed in two Northern provinces of Guilan and Mazandaran. It is estimated that 265000 ha those in Mazandaran (including areas in Gorgan province) and 230000 ha in Guilan are under rice cultivation. The monthly temperatures and rainfall of Guilan – which are similar to those Mazandaran – during the rice growing season vary from 19° to 25°C and 25 to 125 mm, respectively. From 1.8 million tons in the late1980s, rice production in Iran increased to 2.36 million tons 1993, with the average yield being 3780 kg/ha (rough rice). The per caput consumption of rice is around 28 kg per caput/year. As the demand and supply of rice in Iran are still not evenly balanced, the country imports around 400000 to 500000 tons of rice for domestic consumption. (Agronomic report of different rice varieties cultivation in Gulan province, (1996-2006))

Rice production in most of Asian countries (consequence of using different modern varieties, new irrigation systems, using fertilizer and so on) has been increased more rapidly than population and this has been led to increase in supply and proportionately decrease in the real price of rice in world and domestic markets. Furthermore, together with growth in production and national gross income of the countries per capita income has been increased and also demand for quality rice at national and international level has been increased. Studying the market conditions which include marketing margins, causative relations, and market integrations in short term, is the important consequence that can guide and help policymakers and planners in their research, production, distribution and marketing. Almost all rice is grown under irrigated conditions in normal soils (pH 7.0 – 7.5) and yields are high, at 3 to 3.5 tones/ha for local and 5 to 7 tones / ha for improved varieties. Normally one crop of rice is taken from April/May to August/September with 100- to 130- day varieties, with the appropriate duration being 110 to 125 days. Present study examines the rice marketing systems that facilitate the market integration at farm-wholesale-retail level for different rice qualities in Guilan province.

Varietal status

Despite the low yields of local varieties (averaging 2.5 to 3.5 tones/ha), because of their excellent quality traits, more than 80 percent of the total rice area in Iran is still under these varieties, which are similar to basmati types and are characterized by tall stature (125 to 135 cm), a weak Culm and droopy leaves. They have a long slender grain and a head rice recovery (HRR) of 60 to 63 percent, intermediate Amylase Content (AC), aroma and elongation qualities. They are prone to lodging and are also susceptible to blast and stem borer. The most popularly grown local varieties are *Sadri Momtaz* (S1), *Sadri Darge* Yek (S2), *Sadri Mamooli* (S3) and Khazar (K1).

Market integration

Spatial price behavior in regional rice markets is an important indicator of overall market performance. Markets that are not integrated may convey inaccurate price information distorting the marketing decisions of rice producers and contributing to inefficient product movements. Therefore, an important part of market performance analysis focuses on regional price analysis and rice market integration between different market places.

Analysis of market integration has been the concerned affair of most researchers during the recent years. According to Barrette and Lee (2002) market integration is often defined as commerce capability among different markets. This definition includes (place equivalence process) transparency in market and depending on the state of demand, supply and cost transfer in different markets it determines prices and commercial flow and also impulse transfers in prices from one market to other ones.

Barrette (2008) defines that commerce capability shows the fact that goods is exchanged between two economies or two markets and one of the markets is exporter and the other is importer. Signals of commerce capability are transfer of demand surplus from one market to another one that may take place potentially or actually. Most of market integration techniques have been formed on the basis of One Price Low (OPL). These techniques assume that if markets are integrated, prices will differ only due to exchange costs in each one of the markets. One to one changes in prices in a market will simultaneously be transferred to another market (short term integration) and/or together with some pauses (long term integration). Of course some adjustments took place concerning long term integration.

Sanogo *et al.* (2010) applying a threshold autoregressive model about Coarse rice market integration between Nepal and India analyzed. These results show that adjustments to negative price deviations from long-run stable equilibrium are faster than adjustments to the positive ones given a null threshold

MATERIALS AND METHODS

Engle - Granger's Co-integration method

One of co-integration tests is Engle-Granger's (1987) test. If a time series variable becomes stationary a times differencing, this integrated variable will be of a order or $I(a)$. If both time series variables of P_{1t} and P_{2t} are , then any linear combination of them will also be $I(a)$. And now, if there are fixed numbers of α and β, then residual related to P_{1t} and P_{2t} or mentioned linear combination of the two time series will be as follows:

$$Ut = P_{1t} - \alpha - \beta P_{2t} \tag{1}$$

Steps of this test are as follows:

At first the stationary state of the two variables are studied and if the two stationary variables are of the first order, regression 2 is estimated:

$$P_{it} = \varphi + \omega P_{jt} + e_t \tag{2}$$

where P_{it} and P_{jt} are price in market i and price in market j within the time t, respectively. φ and ω are parameters of the equation and ω is the error item. At the next step stationary state of residuals is studied with the help of the following equation.

$$\Delta e_t = \lambda e_{t-1} + \sum_{k=2}^{n} \theta_k \Delta e_{t-1} + \mu_t \tag{3}$$

If residual items are stationary, then the two integrated market will be of the same order that is they contain long-term integration.(Engle *et al.*, 1987).

Engle-Granger causality

According to this test, two variables will be causes for each other if they can be predicted using the past amounts. Following equations are estimated for this test:

$$y_t = \sum_{i=1}^{p} \beta_i y_{t-i} + \sum_{j=1}^{q} \delta_j x_{t-j} + \varepsilon_t \tag{4}$$

$$x_t = \sum_{j=1}^{r} \alpha_j x_{t-j} + \sum_{i=1}^{s} \gamma_i y_{t-i} + e_t \tag{5}$$

In above equations, p, q, r and s are the length of lags in the model. In order for reliable estimation and preventing from error in the numbers of optimal lags, has offered a systematic method for determining the lengths of lags (granger, 1969). This method combines causality of Granger and final prediction of error (FPE) for determining optimized length of the lag for each variable of the combination. To do so, at first, any variable is fitted to their lags and FPE amounts are calculated according to the following formula:

$$FPE_{(m)} = \frac{T+m+1}{T-m-1} ESS_{(m)}/T \tag{6}$$

Where, T is the sample size and m is lag length. Now, the regression with the least amount for FPE, will determine the length of optimized lag. Then lags of other variable are entered in the regressions that, at first step, had the least amount for FPE. The model with the least amount for FPE, will show the length of optimized lag that is obtained for the following relation:

$$FPE_{(m^*,n)} = \frac{T+m^*+n+1}{T-m^*-n-1} \tag{7}$$

m* is the length of optimized variable that has been fitted on its lags and n is the lag length of the second variable.

Price transferring between farm, wholesale and retail levels

If markets are efficient and policies are not an obstacle to their operation, changes in the one market price of rice should be similarly reflected in changes in often market prices know as price transmission (Rafeek, 2003). Model developed by Ravallion (1985) was used to study the price transfer and short term integration of farm, retail and wholesale rice market considering different qualities of rice. This model that formulates relations among the prices at different levels as simultaneous equations system is as follows:

$$P_{fit} = \phi_{i0} + \phi_{i1} P_{fi,t-1} + \phi_{i2} P_{wit} + \phi_{i3} P_{wi,t-1} + e_{ti} \tag{8}$$

$$P_{wit} = \psi_{i0} + \psi_{i1} P_{wi,t-1} + \psi_{i2} P_{rit} + \psi_{i3} P_{ri,t-1} + \varepsilon_{ti} \tag{9}$$

$$P_{rit} = \gamma_{i0} + \gamma_{i1} P_{ri,t-1} + \gamma_{i2} P_{wit} + \gamma_{i3} P_{wi,t-1} + \gamma_{i4} P_{fit} + \gamma_{i5} P_{fi,t-1} + v_{ti} \tag{10}$$

Where we have the followings:

P_{fit}: Farm price of rice product with *i* quality

$P_{fi,t-1}$: Farm price of rice product with *i* quality and a yearly pause

P_{wit}: Wholesale price of rice product with *i* quality

$P_{wi,t-1}$: wholesale price of rice product with *i* quality and a yearly pause

P_{rit} : Retail price of rice product with *i* quality

$P_{ri,t-1}$: Retail price of rice product with *i* quality and a yearly pause

e_{ti}, ε_{ti} and v_{ti}: residual items of equations

ϕ_{ij}, ψ_{ij} and γ_{ij}: parameters of regression equations

ϕ_{i2} and ψ_{i2} coefficients show the price transfer condition from the levels of farm to wholesale and retail and vice versa in different qualities of rice. In analyzing these coefficients it can be said that which quality of rice has operated more efficiently in transferring the price among retail and wholesale markets and farmers and in which case the short term integration has occurred (Rapsomanikis *et al.,* 2003).

DISCUSSION

Figures 1 to 4, present farm, wholesale and retail prices of selected rice qualities during 1999-2009. In all prices, the year 2008 has dedicated the highest price to itself during the investigated period. Wholesale and retail prices have taken the highest places in each year compared to other qualities as well as to farm price. And this has taken place more intensely in three final years of the investigated period. Among the qualities of S2, S3 and K1 the differences between prices have been small until 2007, however, they have been increased in 2008 and 2009.

Figures 5 to 8 show an overview of marketing

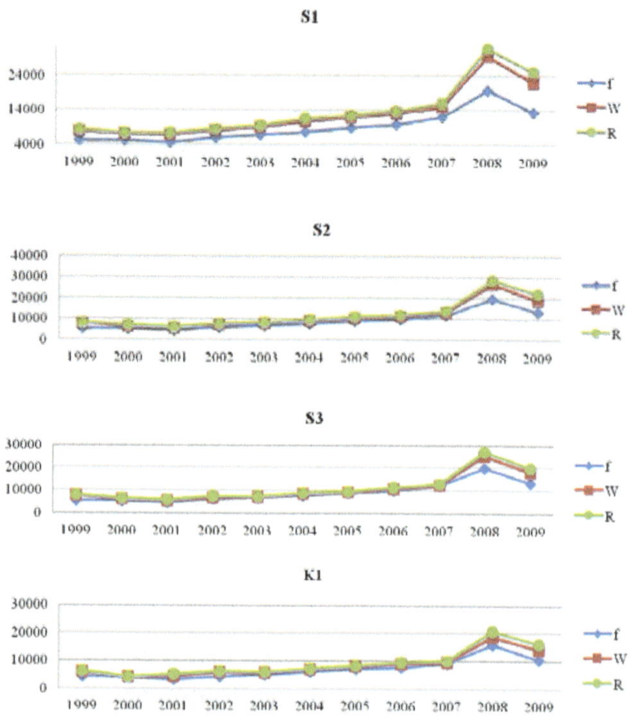

Figure: 1-4- Farm, Wholesale and Retail price of different rice qualities 1999-2009.

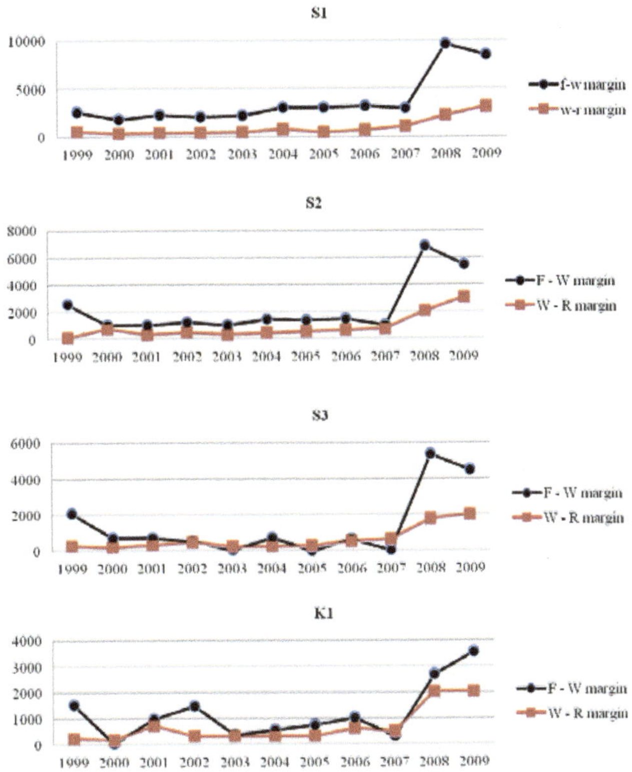

Figure: 5-8- Farm-Wholesale and Wholesale-Retail margins for different rice qualities.

margins in 2 levels of farm-wholesale and wholesale-retail for different qualities. For all qualities, until 2007 these margins have had fixed pattern. It is clear that in all figures fluctuations have occurred around an average and in other words, during 1999-2007 farm-wholesale and wholesale-retail margins have had stationary trend for all rice qualities. However, this has been changed in 2008 and 2009, and compared to wholesale-retail, price diffrences between farm–wholesale has intensely been increased.

ADF[1] stationary test results showed that all time series were I(1) and they become stationary after a differencing. Table 1 shows the results for cointegration tests of farm, wholesale and retail prices for different rice qualities. Except for K1 (wholesale-retail price) in other cases integration exist between different prices. In fact, it can be said that in all qualities of rice, market integration is present in long term and, markets have joined together so that created impulses in a market in long term are transferred to other markets.

Table 2 shows the results of Engle- Granger causality test for farm, wholesale and retail prices. In K1 and S2 there is bilateral relation in case of all prices. Table 2 support cointegration tests results and shows market integration in farm, wholesale and retail level for all rice qualities. So, applying system equations for investigating price transfer relations among marketing elements of different rice qualities has been considered.

Table 3 shows results from simultaneous equations system of price transfer model in different rice qualities. ϕ_{i2} and ψ_{i2} coefficients show the way the price transfers from farm level to wholesale and retail levels and vice versa. Estimation of ϕ_{i2} in different qualities shows that price transfer of wholesale and farm prices in short term, in other words, market integration of these two short terms in K1 rice is more than those of other qualities. Estimation of ψ_{i2} also shows that price transfer of wholesale and retail prices in short term, in other words, market integration in short term in case of K1 rice is less than those of other qualities. These results show that farm prices in K1 rice are quickly affected by shocks in its prices and in case of other rice qualities this influence takes place with a slow rate. Also, compared to other qualities, decisions made by farmer concerning changes in prices

Table 1: Cointegration tests of farm, wholesale and retail price for different rice varieties.

Tests	Null hypothesis	ADF statistic	P-value
1	Farm price is not cointegrated with wholesale price in S1	-2.02	0.04
2	Wholesale price is not cointegrated with retail price in S1	-3.75	0.02
3	Farm price is not cointegrated with wholesale price in S2	-2.17	0.03
4	Wholesale price is not cointegrated with retail price in S2	-3.14	0.005
5	Farm price is not cointegrated with wholesale price in S3	-2.04	0.04
6	Wholesale price is not cointegrated with retail price in S3	-2.2	0.03
7	Farm price is not cointegrated with wholesale price in K1	-2.56	0.01
8	Wholesale price is not cointegrated with retail price in K1	-1.88	0.06

Source: Research findings.

[1] *Augmented Dickey–Fuller*

Table 2: Engle-Granger causality test results for different rice qualities.

Tests	Null hypothesis	ADF statistic	P-value
1	For S1 retail price is not the causality of farm price	5.58	0.06
2	For S1 farm price is not the causality of retail price	3.13	0.15
3	For S1 wholesale price is not the causality of retail price	2.25	0.22
4	For S1 retail price is not the causality of wholesale price	2.08	0.23
5	For S1 wholesale price is not the causality of farm price	9.32	0.03
6	For S1 farm price is not the causality of wholesale price	5.32	0.07
7	For S2 retail price is not the causality of farm price	6.83	0.05
8	For S2 farm price is not the causality of retail price	3.53	0.13
9	For S2 wholesale price is not the causality of retail price	0.48	0.64
10	For S2 retail price is not the causality of wholesale price	0.85	0.45
11	For S2 wholesale price is not the causality of farm price	9.54	0.03
12	For S2 farm price is not the causality of wholesale price	5.4	0.07
13	For S3 retail price is not the causality of farm price	1.6	0.3
14	For S3 farm price is not the causality of retail price	0.85	0.85
15	For S3 wholesale price is not the causality of retail price	0.64	0.57
16	For S3 retail price is not the causality of wholesale price	0.86	0.48
17	For S3 wholesale price is not the causality of farm price	2.28	0.22
18	For S3 farm price is not the causality of wholesale price	1.61	0.3
19	For K1 retail price is not the causality of farm price	0.57	0.6
20	For K1 farm price is not the causality of retail price	0.14	0.86
21	For K1 wholesale price is not the causality of retail price	0.56	0.6
22	For K1 retail price is not the causality of wholesale price	0.83	0.49
23	For K1 wholesale price is not the causality of farm price	0.79	0.82
24	For K1 farm price is not the causality of wholesale price	0.38	0.7

Source: Research findings.

Table 3: Price transmission simultaneous equations system for different rice qualities.

Coefficients	S1	S2	S3	K1
ϕ_{i0}	1088.5 (0.97)	1096.9 (1.33)	1159.5 (0.96)	396.2 (0.7)
ϕ_{i1}	1.58 (0.012)	-0.49 (0.009)	0.56 (0.003)	0.19 (0.0007)
ϕ_{i2}	0.53 (0.001)	0.71 (0.0012)	0.65 (0.0005)	0.81 (0.00014)
ϕ_{i3}	-1.04 (0.008)	0.38 (0.0069)	-0.36 (0.0026)	-0.18 (0.00059)
ψ_{i0}	320.56 (0.69)	593.18 (0.45)	278.19 (0.65)	403.7 (0.64)
ψ_{i1}	1.61 (0.011)	-0.2 (0.0096)	0.77 (0.005)	0.58 (0.0069)
ψ_{i2}	0.96 (0.00002)	0.95 (0.00004)	0.93 (0.000005)	0.86 (0.00001)
ψ_{i3}	-1.59 (0.011)	0.11 (0.009)	-0.74 (0.005)	-0.53 (0.0063)
γ_{i0}	-601.16 (0.71)	-516.83 (0.47)	-222.49 (0.7)	-433.8 (0.71)
γ_{i1}	1.65 (0.011)	-0.12 (0.009)	0.79 (0.005)	0.61 (0.0073)
γ_{i2}	0.9 (0.0002)	1.12 (0.0001)	1.12 (0.00003)	1.2 (0.00003)
γ_{i3}	-1.41 (0.012)	0.25 (0.01)	-0.85 (0.005)	-0.68 (0.008)
γ_{i4}	0.25 (0.00001)	-0.99 (0.00001)	-0.066 (0.000002)	-0.08 (0.00002)
γ_{i5}	-0.39 (0.003)	-0.49 (0.001)	0.037 (0.00023)	0.015 (0.00005)
R^2	0.98	0.94	0.97	0.96

Source: Research findings.

are more affected by shakings on wholesale price of K1 high yielding rice. And in wholesale-retail market for Sadri quality rice, wholesale price is intensely affected by shocks on retail price and this shows intense integration of these two markets in rice product of Iran.

Since price transfer from wholesale to farm in high quality rice takes place rarely, price in wholesale level in case of such qualities is exclusive. In fact, increase in price at retail level rapidly transfers to wholesale level and consequently, this transfer takes place more from retail level to farm level in case of high yielding rice (lower qualities) compared to high quality rice transfer. It seems that, bargaining power of union of rice farmers concerning rich product quality that is also supported by executive authorities is considerable.

The point to be considered is that according to above said contents in introduction part of the present research, being careful about the quality for the purpose of attending international markets is an inevitable reality and that, supporting the high yielding qualities and paying less attention to the quality and more attention to supply rice market with large quantities will cause reduction in production of good quality rice in future and irreparable harm to rice economy of the country. So, it is proposed that union of rice farmers prioritize bargaining about price determination for rice with high qualities and concerned executive powers also change their directions towards the policymaking and planning for high qualities. It is suggested that according to the different quality of rice verities, support policy design and decision making process assigned separately.

REFERENCES

1- Agronomic report of different rice varieties cultivation in Gulan province (1996-2006). Iran's Rice research center.

2- Barrett, C.B. (2008). Spatial Market Integration. The New Palgrave Dictionary of Economics, second ed. London: Palgrave Macmillan.

3- Barrett, C.B., Li, J. (2002). Distinguishing between equilibrium and integration in spatial price analysis. American Journal of Agricultural Economics 84, 292–307.

4- Engle, R. & Granger, C.W.J. (1987), Co-integration and error correction: representation, estimation and testing, Econometrica, (2)55, 251-276.

5- Granger, C. (1969). Investigating causal relations by econometric models and cross-spectral methods, Econometrica, 37, 424-438.

6- Rafeek, M. (2003). Rice Marketing System: Implication For Rice Quality Improvement And Issue Of Affordability, socio economics and planning center. Peradeniya.

7- Rapsomanikis, G. & Hallam, D. (2003). Market integration and price transmission in selected food and cash crop markets of developing countries: review and applications, online: http://www.fao.org/DOCREP/006/Y5117E/y5117e06.htm.

8- Ravallion, M. (1986). Testing market integration, American Journal of Agricultural Economics, 68(1), 102-109.

9- Sanogo, I., & Maliki, M. (2010). Rice market integration and food security in Nepal: The role of cross-border trade with India, Food Policy, 35, 312–322.

Drought Risk Vulnerability Parameters among Wheat Farmers in Mashhad County, Iran

*Mojtaba Sookhtanlo, Hesamedin Gholami * and Seyyed Reza Es'haghi*

Abstract

Identification and analysis of farmers' vulnerability associated with their risk aversion degree is one of the necessary requirements for planning and reducing impacts of drought in Iran. So, this study was investigated three risk vulnerability parameters (economic, social and technical) among wheat farmers categorized in accordance with their risk aversion degree in the Mashhad County (Iran) between drought years of 2009-2011. Vulnerability parameters were determined by Delphi technique. For measuring vulnerability and risk aversion degree, formula of Me-Bar and Valdes and method of Safety First Rule were applied respectively. Findings revealed that in social vulnerability indicators; education level, collaboratively farming activities and dependency on government and in technical vulnerability; irrigation method, cultivation method and type of cultivation; risk averse farmers have had the highest vulnerability level under drought conditions. While respecting economic vulnerability, risk neutral farmers (in insuring for crops, sale prices of crops and the type of land ownership), have had the highest vulnerability level.

Keywords:
Drought, Wheat farmers, Vulnerability, Risk aversion degree

Department of Agricultural Extension and Education, University of Tehran, Karaj, Iran.
** Corresponding author's email: hagholami@ut.ac.ir*

INTRODUCTION

Drought is a slow-onset disaster that has economic, social, and environmental consequences. In Iran, drought is a re-current phenomenon and current drought management strategies in Iran are based on crisis management. For example, when drought occurs in different parts of the country, a state of emergency is declared and thus all resources are mobilized in that particular region.

However, this type of drought management strategy is proved to be ineffective. It is considerable that the year 2011 was the 13th continuous year that drought had been occurred in Mashhad County, northeast of Iran, and this has impacted most of the socioeconomic and technical dimensions of agricultural and rural sectors (The Agriculture Organization of Khorasan-e-Razavi Province, 2011). Therefore, this study is an abiding interest in how farmers cope with and overcome agricultural crises such as drought or natural disasters and provides a new and realistic vision for identifying of risk vulnerability indicators in drought. Drought risk is best defined as a combination of a location's exposure to drought and its vulnerability to drought (Ajijola *et al.,* 2011) and vulnerability is identified as the exposure to uninsured risk leading to an unacceptable level of well-being among farmers (Hoddinott and Quisumbing 2003, Hoogeveen *et al.,* 2005). Many studies (Gwimbi 2009, Deressa, 2010, Wilhelmi *et al.*, 2002, Kapoor and Ojha, 2006, Barbier *et al.*, 2008, Mongi *et al.,* 2010, Keshavarz *et al.*, 2011) highlight geographical situations and rainfall level as key factors on farmers' vulnerability. However, people within a locality and same area are not evenly vulnerable to drought (Slegers 2008). So, there is a growing appreciation that other factors such as farmers' characteristics including levels of their risk aversion (internal risk factors) have directly influenced drought vulnerability parameters but this still has been considered by rare studies (Hoogeveen *et al.*, 2005, Franke *et al.*, 2005, Brondizio and Moran, 2008, Ajijola *et al.*, 2011). Farmers' capacities to cope with drought, depending on ownership or access to a wide variety of resources such as land ownership, farmers' incomes, farming lands size, education level, access to governmental and bank credits (loans), crops insurance, technical assistance and information, social networking, and public support programs (Scoones 1998; Ellis 2000; St. Cyr 2006) are categorized in this study in three parameters of social, economic and technical. Eakin *et al.* (2006), Deressa (2010), Ajijola *et al.* (2011), Keshavarz *et al.* (2011) and Sharafi and Zarafshani (2010 and 2011) examined the impact of risk attitudes (level of risk aversion) on poverty and vulnerability level among rural farmers. The variety of information on household human resources and income sources, production and losses to climate hazards and pests, crop and livestock management practices, commercialization practices, input and machinery use, farmers' risk mitigation practices, landholding size, and farm profit, loans, selling of crop outputs, low income level, credits, irrigation method, household extension packages and farmers' access to resources and use of them considered importance of adaptation (technology, technical assistance, credit and insurance) on farmers' capacities to respond to stress and uncertainty conditions (drought). So, the main purpose of this study was to identify the most vulnerable farmers regarding their risk aversion degree in the Mashhad County (Iran). Particular interests are as follows:

- To identify wheat farmers based on economic, social and technical vulnerability indicators.
- To calculate wheat farmers' vulnerability.
- To calculate wheat farmers' risk aversion degree and categorize them according to it.
- To determine wheat farmers' vulnerability in each category of risk aversion degree.

MATERIALS AND METHODS

This study was conducted in the Mashhad County (rural areas) located in Khorasan-e-Razavi Province, Iran. The capital of Mashhad County is the Mashhad City. This County that is located in North East of Iran and is the most populous county in *Khorasan-e-Razavi* Province. This county is 992–1184 meters above sea level. The area of this county is 1490 km^2. This county consists of 591 villages. The cultivated land of the county is 56615 km^2. This county with an arid- semi-arid and arid climate receives an average rainfall of 256 mm (The agricultural statistics and information office of *Khorasan-e-Razavi* Province, 2009). Wheat is

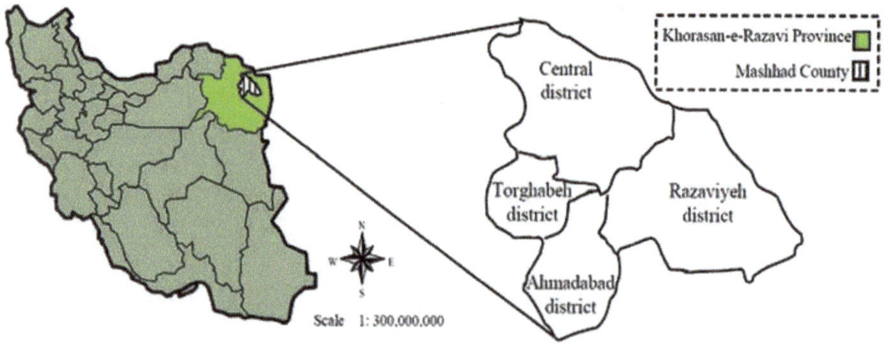

Figure 1: Area of study (Mashhad County, Khorasan-e-Razavi Province, Iran)
Reference: Statistical Centre of Iran (2012).

Table 1: Sample size in each district (Mashhad County)

Districts (Bakhsh)	Statistical population	Sample size
Central	2574	125
Razaviyeh	1320	65
Ahmadabad	1086	53
Torghabeh district	960	50
Total	**5940**	**293**

the dominant crop in the region, so the statistical sample of this study consisted of wheat farmers who live in *Mashhad* County. The selected region was severely affected by drought during the year 2009-2011. Mashhad County is divided into four districts (Bakhsh), with their capitals (Figure 1): Ahmadabad (capital: Malekabad), Central (capital: Mashhad), Razaviyeh (capital: Razaviyeh) and Torghabeh (capital: Torghabeh).

A proportional stratified random sampling was applied to access the respondents and using Cochran's test the size of sample was determined 293 wheat farmers (Table 1).

Two questionnaires and methods of interview were designed and used to gathering data. The first questionnaire included open questions to determine the most important socioeconomic and technical vulnerability indicators in the Mashhad County by Delphi technique. The second questionnaire was consisted of three parts. The first part was to collect data about farmers' personal and professional characteristics. The second part consisted of risk aversion indicators (according to formula of Safety First Rule). The third part consisted of vulnerability indicators was obtained through the first stage (Delphi technique) to calculate vulnerability level of farmers. This study is conducted in two main stages.

First stage (Delphi technique): This stage includes usage of the Delphi technique to identify and weigh major indicators of vulnerability in the study region as used in many previous studies (Kaly and Pratt 2000; Dercon 2004; Deressa 2010). Snowball method was used to determine experts related to the study objectives. In other words, we asked the experts who were known in the research process to introduce other experts to us. Finally 45 experts were chosen among which, 31 experts resend the questionnaires and their data was used. They were people who had field research about drought or extension experts who were directly engaged in programs or activities related to drought in the area of study. A primary questionnaire including open-ended questions (i.e. determine the most important socioeconomic and technical vulnerability indicators at Mashhad County) were distributed among experts. In the next step, first questionnaire data were used to determine and categorize common major social, economic and technical indicators with the most frequency. Acquired data were used to design another questionnaire including the primary indicators which were edited to send again to the experts to be confirmed by them. In the third step, the questionnaire was consisted of final confirmed economic, social and technical indicators and also a section for determining the weight (relative importance) (W_i=1… n) of each indicator in farmers' vulnerability by experts. They could weigh the indicators from 0 (the lowest importance) to 10 (the highest importance). It was emphasized in the questionnaire that, weighing must express the relative importance of indicators, so the indicators could not be weighted the same. These indicators used to design the next

Table 2: Status of the respondents, by the risk- aversion degree.

Risk- aversion coefficient	Status of wheat farmers	Frequency	percent
$0.1 \le Rj \le 1$	Risk taker	52	18
$-0.1 \le Rj \le 0.1$	Risk neutral	63	21
$-1 \le Rj \le -0.1$	Risk averse	178	61
Total	-	**293**	**100**

stage questionnaire.

Second stage (determining farmers' risk aversion degree and vulnerability level): In this stage another questionnaire was used. Formula of Safety First Rule was used to calculate the farmers' risk aversion degree and also categorizes them in three groups (namely risk adverse, risk neutral and risk taker). Furthermore, farmers' social, economic and technical vulnerability amount was determined by method of Me-Bar and Valdez (2005).

To calculate risk aversion degree Safety First Rule formula was used. Randhir (1991), Parikh and Bernard (1988), Sekar and Ramasamy (2001), Ajetomobi and Binuomote (2006) and Ajijola *et al.* (2011) used this method in their studies in order to determine the risk-aversion degree of farmers. Because of lack of access to accurate data needed for other common methods and lack of valid and categorized databases in the studied region, the mentioned formula was applied in this study. In this formula:

$$Rj = [E^*j - Ej] / [Sj], \quad j = 1, 2..., n$$

*(R j: Risk-aversion degree of farmer (j), E*j: Critical income level of farmer (j), E j: Expected income of farmer (j), S j: The standard deviation of the farmer (j)'s annual income (in the past three years of agricultural and non-agricultural activities))*

$$E^* = 7955936\,(FAM - CHI / 2) + DPT - (UAR + UAR')$$

$$E = VP\,(1 + DMG) - TC$$

The weighted crop damage variable was defined as:

$$DMG = (\Sigma k_i DMG_i) / (\Sigma k_i)$$

The parts of the above formulas are as the following:

-7955936: The per capita cost of supplying the least calorie supply in one year in Rial (the standard rate in Iran).

-FAM: The household's farm size (Hectare).

-CHI: Number of children (active members of the family in working of agriculture).

-DPT: Farmer's debt amount to formal and informal institutions (IRR).

-UAR: The farmers' annual income from activities other than wheat cultivation (IRR).

-UAR': The beneficiaries' annual income from non-agricultural activities (IRR).

- Total value of wheat production (IRR).

-DMG: The proportion of farmer's damage due to losses and abnormal incidents as a weighted average.

-TC: Total wheat production cost in the same year (IRR).

Among vulnerability assessment methods, a formula suggested by Me-Bar and Valdez (2005) was considered to be appropriate for assessment of socio-economic and technical vulnerability parameters in this study. Me-Bar and Valdez (2005) stated that vulnerability is a qualitative concept for which comparing societies should be measured quantitatively. Mentioned formula based on subjective assessment of factors is affecting drought vulnerability. Considering the lack of reliable resources of data and

Information which is a prerequisite for applying other common methods in the studied region and its successful application in other regions of Iran in previous studies (for example the studies of Sharafi and Zarafshani (2011) in Kermanshah Province and Keshavarz *et al.* (2011) in Fars Province) the applicability and efficiency of this method for the country condition was proved. So, this formula was applied for vulnerability assessment.

$$V = 1 / C_0 \sum (P_i W_i)$$

(V= each farmer vulnerability amount, C_0= sum of total vulnerability weight, Pi= each parameter amount, Wi= each parameter weight)

Also, in this formula:

$C_0 = \sum W_i$, $\sum W_i = (W_{max} \times n) / 2$, $C_0 = (W_{max} \times n) / 2$, $C_0 < W_{max} \times n$

(W_{max}: The maximum weight that can belong to each parameter (10), n: The number of each factor parameters)

RESULTS

Personal and professional characteristics

Among farmers, 84.5 % were men and 15.5 % were women. The most frequency of the wheat farmers education level was secondary education level which constituted 33 % of the sample and also 21 % of wheat farmers were illiterate and only 9 % of statistical population had a degree higher than diploma. The most experiences of wheat cultivation among the respondents were between 21 to 30 years. Also, by looking to the extent of farmlands, the highest frequency was related to the farmers who had 4 to 7 hectares. The average area of each farmer farmlands was 1.14 hectares and the most experiences of farming among the respondents were 31 to 40 years.

Risk-Aversion degree of respondents

In the table (2), risk-aversion coefficient (Rj) was calculated according to Safety First Rule formula. Based on the findings, 61 % of the respondents were risk averse, 23 % were risk neutral and 18 % were risk taker.

Parameters of vulnerability

Findings related to economic, social and technical parameters of vulnerability are shown in the tables (3, 4 and 5). First, for measuring indicators of any parameter, total vulnerability weight ($\sum W_i$) was calculated.

$$\sum W_i = (W_{max} \times n) / 2 = (10 \times 9) / 2 = 45$$

Indicators weight of parameters:

Findings showed that experts believed that economic parameter indicators (insuring crops (W_i=6.12), regional extension experts with economic advices (W_i=5.46), and access to governmental and bank credits (loans) (W_i=5.41)), social parameter indicators (farming collaborative activities (W_i=6.06), attending in extension

Table 3: The amount and weight of economic parameter indicators in three farmers' groups

Economic parameter indicators	Indicators weight (W_i)	Indicators amount in farmers groups (P_i)		
		Risk taker farmers (P_{i1})	Risk neutral farmers (P_{i2})	Risk averse farmers (P_{i3})
Insuring crops	6.12	1.98	3.49	2.87
Extension agents' economic advices	5.46	2.08	1.78	1.59
Farmers' incomes	4.95	2.12	1.54	2.17
Amount of liquidity	4.65	2.65	2.03	2.47
Pre-sale crops to middlemen	3.75	2.67	1.71	1.63
Sale price of crops	5.24	1.77	2.89	1.80
Land ownership type	4.82	1.54	2.75	3.26
Farming lands Size	4.60	1.63	1.57	2.04
Access to governmental and bank credits (loans)	5.41	3.25	2.68	2.74
Total	**45**	-	-	-

Table 4: Amount and weight of social parameter indicators in farmers groups

Social parameter indicators	Weight of indicators (W_i)	Indicators amount in farmers groups (P_i)		
		Risk taker farmers (P_{i1})	Risk neutral farmers (P_{i2})	Risk averse farmers (P_{i3})
Social esteem	4.93	2.63	2.63	2.56
Membership in rural associations / organizations	5.10	1.98	2.40	1.84
Dependency to government	4.77	3.27	2.65	2.62
Attending in extension education programs	5.52	1.65	3.15	1.60
Education level	5.35	2.29	1.94	3.43
Farming collaborative activities	6.06	2.21	1.67	2.76
Family members collaboration	4.81	2.50	2.15	1.97
The level of related to farming religious believe	3.82	1.85	2.13	2.11
Participation in rural development programs	4.64	1.90	2.25	2.51
Total	**45**	-	-	-

Table 5: The amount and weight of technical parameter indicators in groups of wheat farmers .

Technical parameter indicators	Indicators weight (W_i)	Indicators amount in farmers groups (P_i)		
		Risk taker farmers (P_{i1})	Risk neutral farmers (P_{i2})	Risk averse farmers (P_{i3})
Cultivation type (rain-fed/ watery)	6.06	3.03	2.37	3.00
Cultivation pattern (spring / autumn)	4.84	1.90	2.13	2.46
Cultivation method (traditional/ mechanized)	5.12	2.96	2.98	3.02
Use of drought resistant varieties	4.94	2.51	2.59	2.73
Irrigation method	5.65	2.46	3.11	3.42
Planting, saving and harvesting times	4.71	3.05	2.90	2.87
Use of chemical fertilizers	4.55	2.19	2.97	2.88
Weeds, pests and diseases control	5.29	2.64	3.19	2.90
Tillage implements	3.84	2.36	3.22	2.63
Total	**45**	-	-	-

education programs (W_i=5.52) and education level (W_i=5.35)) and technical parameter indicators (cultivation type (rain-fed / watery) (W_i= 6.06), irrigation method (W_{ii}=5.65) and weeds, pests and diseases control (W_i=5.29)), respectively were the most important indicators in order to explain parameters of vulnerability in target regions.

Indicators amount in parameters of vulnerability:

Considering the findings among risk taker farmers, the economic parameter indicators (access to governmental and bank credits (loans) (3.25), pre-sale crops to middlemen (2.67) and amount of liquidity (2.65)), the social parameter indicators (dependency to government (P_{i1}=3.27), social esteem (P_{i1}=2.63) and collaboration of family members (P_{i1}=2.50)) and the technical parameter indicators (planting, saving and harvesting times (P_{i1}=3.05), cultivation type (rain-fed /watery) (P_{i1}=3.03) and cultivation method (traditional / mechanized) (P_{i1}=2.96)), respectively were three indicators which had highest scores.

Among risk neutral farmers, the economic parameter indicators (insuring crops (P_{i2}=3.49), sale price of crops (P_{i2}=2.89) and land ownership type (P_{i2}=2.75)), the social parameter indicators (participation in rural development programs (P_{i2}=3.15), dependency to government (P_{i2}=2.65) and social esteem (P_{i2}=2.63)) and the technical parameter indicators (irrigation method (P_{i3}=3.42), cultivation method (traditional/ mechanized) (P_{i3}=3.02) and cultivation type (rain-fed / watery) (P_{i3}=3.00)), respectively had the highest intensity during drought. Also among risk averse farmers, the economic parameter indicators (Land ownership type (P_{i3}=3.26), insuring crops (P_{i3}=2.87) and access to governmental and bank credits (loans) (P_{i3}=2.74)), the social parameter indicators (education level (P_{i3}=3.43), farming collaborative activities (P_i3=2.76) and dependency to government (P_{i3}=2.62)) and the technical parameter indicators (Irrigation method (P_{i3}=3.42), cultivation method (traditional / mechanized) (P_{i3}=3.02) and cultivation type (rain-fed / watery) (P_{i3}=3.00)), respectively had the highest rank and means that during drought, these farmers have had the highest vulnerability in these indicators.

Total 6: total vulnerability amounts in farmers groups.

Vulnerability amount	farmers groups		
	Risk taker farmers	Risk neutral farmers	Risk averse farmers
Economic vulnerability	2.13	2.34	2.30
Social vulnerability	2.25	2.32	2.38
Technical vulnerability	2.59	2.83	2.90
Total vulnerability	6.97	7.49	7.58

Table 7: A summary of indicators priority in farmers groups.

Farmers groups	farmers groups		
	Economic vulnerability	Social vulnerability	Technical vulnerability
Risk taker farmers	Access to governmental and bank credits (loans) Pre-sale crops to middlemen Amount of liquidity	Dependency to government Social esteem Family members collaboration	Planting, saving and harvesting times Cultivation type (rain-fed / watery) Cultivation method (traditional/ mechanized)
Risk neutral farmers	Insuring crops Sale price of crops Land ownership type	Attending in extension education programs Dependency to government Social esteem	Tillage implements Weeds, pests and disease control Irrigation method
Risk averse farmers	Land ownership type Insuring crops Access to governmental and bank credits (loans)	Education level Farming collaborative activities Dependency to government	Irrigation method Cultivation method (traditional /mechanized) Cultivation type (rain-fed / watery)

Total vulnerability

Formula of Me-Bar and Valdez (2005) was applied to calculate total vulnerability amount. For example, economic vulnerability in risk averse farmers is calculated:

$V = 1 / C_0\sum (P_iW_i) = (6.12 \times 1.98) + (5.45 \times 2.08) + (4.95 \times 2.12) + \ldots + (5.42 \times 3.25) = 95.63/45 = 2.13$

According to table 6, the highest economic vulnerability was among risk neutral farmers and the lowest was among risk taker farmers. With respect to social vulnerability, the highest vulnerability was among risk averse farmers and the lowest vulnerability was among risk taker farmers. Also in technical vulnerability, risk averse farmers were the most vulnerable groups and risk taker farmers were the least vulnerable groups.

DISCUSSION

This paper describes an investigation of drought vulnerability in Mashhad County (Iran). The length of drought in studied region implies that it is a harsh reality of Iran agriculture and mitigation to the severe continuous impacts of that is critical. As the results pointed out, most of the farmers are vulnerable. Therefore, farmers are being extremely stressed to find alternative appropriate mechanisms to reduce their vulnerability. Although most of the farmers are risk averse, they hardly adopt the new advises with potential risks. This means that policy makers should significantly act different from what they currently do. Findings imply that the kind and amount of vulnerability among farmers with various risk aversion degree is different, so when we categorized them in three groups named risk taker, risk neutral and risk averse, they would completely had different and general vulnerability parameters and thus unspecific supports from these groups would be inefficient.

The interesting conclusion which could be made is that there is a relationship between farmers' risk aversion degree and their vulnerability level. In other words, various effects of drought on different farmers' groups have not been considered by policy makers and managers in the studied region and thus managerial, educational and support programs have not been appropriate for these groups. In sum, findings revealed that risk taker farmers had the least vulnerability in all three vulnerability parameters named economic, social and technical parameters, while risk neutral farmers were the only most vulnerable group in economic parameter. Risk averse farmers were the most vulnerable group, because they were the most vulnerable group in two parameters named social and technical parameters. So, it can be said that they are the most vulnerable farmers. Some other parts of findings are shown in table 7.

Considering the common indicators in all three farmers groups, it can be said that as emphasized by Vásquez-León *et al.* (2003) and Nelson and Escalante (2004), in order to manage and reduce negative impacts of drought economic vulnerability, mechanisms such as granting gratuitous or low interest loans based on farmers livelihood level, establishing small rural banks, more governmental attention to crops insurance fund (Hazell, 2004), and developing and enriching local credit funds should be regarded as high priority actions.

With respect to social vulnerability, findings revealed that among risk taker farmers, dependency to government (consistent with Sharafi and Zarafshani (2011)), social esteem and family members' collaboration have had the most effect on social vulnerability. Among risk neutral farmers, attending in extension education programs, dependency to government and social esteem and education level among risk averse farmers which is consistent with Vásquez-León *et al.* (2003), Sengestam (2009) and Deressa (2010), farming collaborative activities (consistent with Iglesias *et al.* (2009)) and dependency to government have had the most effect on economic vulnerability.

With respect to technical vulnerability indicators for risk taker farmers, planting, saving and harvesting times, cultivation type (rain-fed/watery) and cultivation method (traditional/ mechanized) have had the most effect on their technical vulnerability. Among risk neutral farmers, weeds, pests and diseases control and irrigation method and among risk averse farmers, irrigation method, cultivation method (traditional/ mechanized) and cultivation type (rain-fed/watery) have had the most effects on their technical vulnerability. Hence, identification and promotion of varieties and species which are suitable for each group of farmers and also compatible with continental conditions as substitutions for crops with high water requirements, providing infrastructures for sustainable development of water resources such as draining, pressured irrigation systems and helping farmers to control pests and common diseases during drought is recommendable. The results of this study can imply that drought relief programs should be based on the rate of socio-economic and technical vulnerability among farmers' groups in term of their risk aversion. Furthermore, an up-to-date vulnerability assessment helps extension agents to plan more effective content for their educational programs.

REFERENCES

1- Ajetomobi, J.O., & Binuomote, S.O. (2006). Risk aversion among poultry egg producers in southwestern Nigeria. International Journal of Poultry Science, 5 (6), 562-565.

2- Ajijola, S., Egbetokun, O.A., & Ogunbayo, I.E. (2011). Impact of risk attitudes on poverty level among rural farmers in Ogun State. Journal of Development and Agricultural Economics, 3(12), 581-587.

3- Barbier, B., Yacouba, H., Karambiri, H., Zorome, M., & Some, B. (2008). Human vulnerability to climate variability in the sahel: Farmers' adaptation strategies in northern Burkina Faso. Environmental Management, 43,790-803.

4- Brondizio, E. S., & Moran, E.F. (2008). Human dimensions of climate change: the vulnerability of small farmers in the Amazon. Philosophical Transactions of the Royal Society B: Biological Sciences, 363(1498), 1803-1809.

5- Dercon, S. (2004). Growth and Shocks: Evidence from rural ethiopia. Journal of Development Economics. 74(2), 309-329.

6- Deressa, T.T. (2010). Assessment of the vulnerability of Ethiopian agriculture to climate change and farmers' adaptation strategies. Ph.D thesis in environmental economics, Department of Agricultural Economics, Extension and Rural Development. University of Pretoria, Pretoria, available from: http: //upetd.up.ac.za/thesis/available/etd-10232010-165854/unrestricted/thesis.pdf.

7- Eakin, H., Webhe, M., Ávila, C., Torres, G.S., & Bojórquez-Tapia, L. (2006). A comparison of the social vulnerability of grain farmers in Mexico and Argentina. AIACC Working Paper 29, 47-55, Assessment of Impacts and Adaptation to Climate Change in Multiple Regions and Sectors Program, Washington, DC. USA.

8- Franke, G., Richard, C.S., & Subrahmanyam, M.G. (2005). Incremental risk vulnerability. University of Konstanz. Journal of Economic Literature, 05-08: 1-19, available from:http://cofe.uni-konstanz.de/ Papers/ dp05 _08. pdf.

9- Gwimbi, P. (2009). Cotton farmers' vulnerability

to climate change in Gokwe District (Zimbabwe): impact and influencing factors. Journal of Disaster Risk Studies, 2(2), 81-92.

10- Hazell, P. (2004). Climate change and management of catastrophic risk. Environment and Policy, 39, 385-395.

11- Hoddinott, J., & Quisumbing, A. (2003). Methods for micro econometric risk and vulnerability assessments. Social Protection Discussion Paper Series, 0324,1-59. Social Protection Unit, Human Development Network. The World Bank. Washington D.C. USA.

12- Hoogeveen, J., Tesliuc, E., Vakis, R., & Dercon, S. (2005). A guide to the analysis of risk, vulnerability and vulnerable groups. Social Protection Unit, Human Development Network. The World Bank, Washington, DC. USA.

13- Iglesias, A., Garrote, L., Cancelliere, A., Cubillo, F., & Wilhite, D. A. (2009). Coping with drought risk in agriculture and water supply systems: Drought management and policy development in the Mediterranean. Springer Netherlands. Advances in Natural and Technological Hazards Research, 26 (2), 153-159.

14- Kaly, U., & Pratt, C. (2000). Environmental vulnerability index: development and provisional indices and profiles for Fiji, Samoa, Tuvalu and Vanuatu. Phase II. SOPAC Technical Report 306, 88-89.

15- Kapoor, S., & Raj Kumar, O. (2006). Vulnerability in rural areas: potential demand for microinsurance. Sage Publications. International Journal of Rural Management, 2(1), 67-83.

16- Keshavarz, M., Karami, E., & Zamani, G. (2011). Drought vulnerability of farm households: a case study. Iranian Agricultural Extension and Education Journal.

17- Me-Bar, Y., & Valdez, F. (2005). On the vulnerability of the ancient Maya society to natural threats. Journal of Archaeological Science, 32, 813-825.

18- Mongi, H., Majule, A.E., & Lyimo, J.G. (2010). Vulnerability and adaptation of rain fed agriculture to climate change and variability in semi-arid Tanzania. African Journal of Environmental Science and Technology, 4(6), 371-381.

19- Nelson, C.H., & Escalante, C.L. (2004). Toward exploring the Location-Scale condition: A constant relative risk aversion Location-Scale objective function. European Review of Agricultural Economics, 31(3), 273-287.

20- Parikh, A., & Bernard, A. (1988). Impact of risk on HYV adoption on Bangladesh. Agricultural Economics, 2 (2), 167–78.

21- Randhir, T.O. (1991). Influence of risk on input decisions in Tankfed farms of South India. Indian Journal of Agricultural Economics, 46 (1), 57-63.

22- Scoones, I. (1998). Sustainable rural livelihoods: A framework for analysis, Institute of Development studies, IDS Working paper. University of Sussex. Brighton. England, 72, 1-18.

23- Sekar, I., Ramasamy, C. (2001). Risk and resource analysis of rainfed tanks in South India. Journal of Social and Economic Development, 3 (2), 208-215.

24- Sengestam, L. (2009). Division of capitals- What role does it play for gender-differentiated vulnerability to drought in Nicaragua?. Community Development, 40, 154-176.

25- Sharafi, L., & Zarafshani, K. (2011). Economic and social vulnerability assessment among farmers towards drought (Case of study: Wheat farmers in Kermanshah, Sahne, and Ravansar townships). Journal of rural research, 1(4), 129-154. In Farsi.

26- Sharafi, L., & Zarafshani, K. (2010). The assessment of vulnerability in respect of drought mitigation and management for farmers whom cultivating wheat (case study: sarpol-e-zahab, islamabad, and javanrood). Arid regions geographic studies, 1(1), 106-120. (In Farsi).

27- Slegers, M.F.W. (2008). If only it would rain: Farmers' perceptions of rainfall and drought in semi-arid central Tanzania. Journal of Arid Environments, 72, 2106- 2123.

28- Statistical Centre of Iran. (2012). Population and housing census, administrative units of razavi khorasan and their populations.

29- St. Cyr, J. F. (2006). At risk: natural hazards, people's vulnerability, and disasters. Journal of Homeland Security and Emergency Management. Published by The Berkeley Electronic Press, 2(2),1-5. Available from: http:// www.geo.mtu.edu/volcanoes/06upgrade/SocialKateG/Attachments%20Used /AtRiskReview.pdf

30- The Agricultural statistics and information office of Khorasan-e-Razavi Province. (2009). Agricultural statistic annals in 2009, Department of Planning and Economic Affairs. Pp, 1-164.

31- The Agriculture organization of Khorasan-e-Razavi Province. (2011). The eighth leaf. Magazine of news and Information Science, Public relations of Agricultural Jihad Organization of Khorasan-e-Razavi Province, 10, 1-12.

32- Vásquez-León, M., West, C.T., & Finan, T.J. (2003). A comparative assessment of climate vulnerability: agriculture and ranching on both sides of the US-Mexico border. Global Environmental Change, 13, 159-173.
33- Wilhelmi, O.V., & Wilhite, D.A. (2002). Assessing vulnerability to agricultural drought: a Nebraska case study. Natural Hazards, 25, 37-58.

Livestock Farming Systems and Cattle Production Orientation in Eastern High Plains of Algeria, Cattle Farming System in Algerian Semi Arid Region

Lounis Semara [1], Charefeddine Mouffok [2] and Toufik Madani [3]*

Keywords:
Livestock, typology, Farming system, Cattle, Management

Abstract

This study was an attempt to devise productive orientations of cattle herds in eastern high plains of Algeria. In this regard, 165 farms randomly identified were investigated. The selection of breeders was based to existence of cattle on the farm, and the farmer proposed to investigation must have at least two cows. The approach taken was to identify all systems adopted by farmers in a region through the analysis of the relationship between the maintenance of different types of cattle and preferred marketing policies. The model has been emerged as a result of functional typology established using the procedure categorical principal components analysis (CATPCA) of optimal coding in SPSS [19. 2010]. Following this approach, five types of cattle productive orientation have been identified, the balanced mixed system (dairy-beef), beef mixed system, dairy mixed system, dairy system and beef system. These results showed that the breeders were oriented towards specialization (dairy or beef) in less than 20% of situations. Farmers in our context prefer mixed systems when beef mixed system was the model type frequently encountered in the region (over than 50% of farms).

[1] Ph.D Student of Animal Production, Department of Agriculture and Animal Science, Setif 1 University, Algeria.
[2] Assistant Professor, Department of Agriculture and Animal Science, Setif 1 University, Algeria.
[3] Professor, Department of Agriculture and Animal Science, Setif 1 University, Algeria.
** Corresponding author's email: mouffokcharefeddine@yahoo.fr*

INTRODUCTION

Livestock and their products provide direct cash income and the animals are living the assets for many farmers (FAO/ILRI, 1995). Breeding cattle assume also the roles of job creation and income very important for social stability (Srairi *et al.*, 2009). In Algeria, animal production especially dairy cattle were always at the center of occupation of public authorities, well as several policies and actions have been applied. However, the dairy industry operates almost with imported powdered milk where 60% of milk requirements are imported. Several researchers have attempted to explain the poor performance of the cattle sector in the Algerian context by a constraint which opposes the development of a strong dairy activity, in particular, problems of adaptation of exotic breeds in different agro-ecological zones of countries (Madani and Mouffok, 2008) and the lack of fodder production required for intensive dairy farm.

Researches on livestock have always been guided by the search for efficiency improvement activity (Dedieu, 2009). Madani and Mouffok (2008), provide that the deficiency of milk production in Algerian's farms requires changes in technical choices and especially the type of animals and livestock systems implanted. Systemic vision on cattle farms prospection is therefore essential to understand better the factors influencing the elaboration of performances. Many authors suggest two conceptual approaches, one focused on the analysis of farmers' practices, it comes to technical, economic and social farmers practical (Chapman *et al.,* 2008; Dufumier, 1996) and the other attempts to understand how farmers make their decisions (Shalloo *et al.,* 2004).

This research can be considered as a contribution to characterization diversity of cattle farming systems. Its aims through the adjustment of some technical and economic practices in of Algerian Eastern high plains farms to analyze the organization of cattle production systems and identify pathways to explain management and planning strategies adopted.

MATERIALS AND METHODS

Methodological approach

Investigation in a single passage was conducted among farmers and herders of cattle. Selection of farmers' was based on existence of cattle breeding activities and farm proposed for investigation must have at least two cows. In study region, these categories of farms correspond to plus than 90% of all breeders. In this regard, a sample of 165 farms were randomly chosen and visited. Livestock farms that are the subject of our investigation were located in two provinces of the eastern high plains of Algeria, Setif and Bordj Bouarraridj departments (Figure 1).

We also selected scale of aridity gradient which increases from north to south. The investigations have been developed in a questionnaire consisting of three components (socio economics of farmers family, structure and resources of farms and functionning practices of cattle herd) with more than 150 questions. The objective of this survey was to collect among those surveyed as much information about the livestock, but

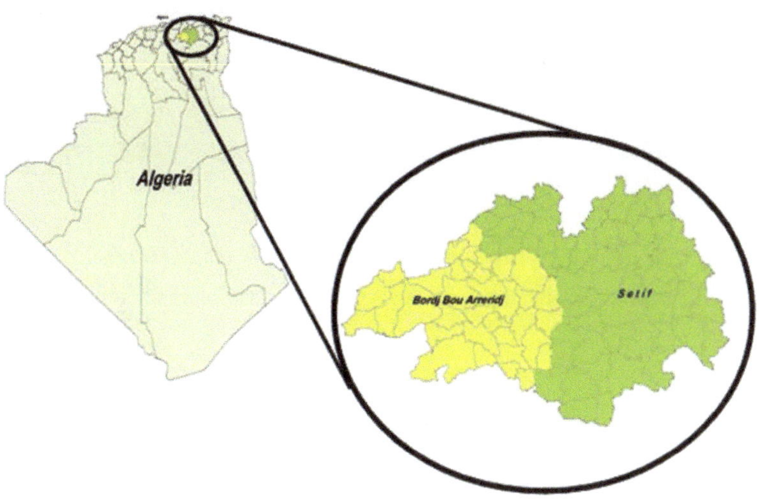

Figure 1: Localisation of studied area

Table 1: Data of cattle categories number in all farms'

	Cattle (LU)	Cows	Heifer	Beef	Male Calf	Femelle Calf
Mean	12.63	7.69	2.53	1.53	2.02	2.13
Standard error of mean	0.78	0.46	0.22	0.32	0.20	0.17
Standard deviation	10.00	5.49	2.80	4.01	2.55	2.10
Minimum	02.00	2.00	0.00	0.00	0.00	0.00
Maximum	71.45	45.00	14.00	40.00	18.00	10.00

LU : Livestock Unit

also variables related to the production environment and the diversity of functionning practices of cattle and strategies developed.

Diagnostics tools

A graphics typology was established using Categorical Principal Components Analysis (CATPCA) optimal coding procedure of SPSS (19.2010) software. This procedure was most appropriate to research aims to analysis the relationship between quantitative variables describing the structure of cattle herd (dairy cows effective, beef cattle effective, heifers effective, effective male and female calves) and the different modalities of qualitative variables describing practice-policies adopted by farmers such as type de breeding choice and cattle product commercialization (amount of sold milk and calves sale age). This categorization was reinforced by two-step cluster automatic classification procedure. All variables was presented by means and standard deviation.

RESULTS

Overall characteristics of farms

Effective and structure of cattle herd

The descriptive analysis of cattle herd size by farm was summarized in table 1. Results show that all farms exploit average herd of 12.6±10.0 LU. The number of cows was 7.6 ± 5.4 by farm and this category represents more than 60% of the total cattle population. Those farms mark the permanent presence of 2.5±2.80 heifers' and 1.5±4.0 young beef. A large standard deviation recorded reflects a high divergence in the compositional structure of cattle herd between farms, which was the first indicator of the diversity of cattle production policies.

Description of economic practices

Theses farms were mostly cattle farms alone (46.7%) or cattle-sheep (41.8%). The association of cattle-sheep-goat was observed in less than 5% of cases and about 6.7% producers have developed a new trend to the association

Table 2: Farming system and economic practices of farms

Variable	Modality	Percentage (%)
Breeding species	C.S.G	4.8
	C.S	41.8
	C.P	6.7
	C	46.7
Milk soled	Total	45.4
	Part of	49.7
	Never	4.9
Age of calf sale	Pre weaning	12.5
	After weaning	13.5
	Old age	64.4
	As needed	10.0

C : Cattle ; S : Sheep ; G : Gaot ; P : Poultry.

Table 3: Model parameters' of CATPCA

Dimension	Alpha of Cronbach	Proper Value	Explain variance
1	0.691	2.52	31.55
2	0.409	1.55	19.42
Total	0.809 [a]	3.41[b]	42.67 [b]

a. The total value of Alpha of Cronbach is based on the total proper value
b. Due to the presence of multiple nominal variables. the proper value and the total percentage of explained variance does not correspond to the sum on the dimensions

of cattle with intensive poultry production.

The analysis of economic practice (Table 2) shows that 45.4% and 49.7% of farmers commercialize respectively all or part of milk produced in the local and regional market and only 4.9% of producers refuse the sale of milk. Therefore, a large part of farmers surveyed (64.4%) announce that calves were sold at later age (more than one year of age). Therefore, about 12.5% of farms visited declare that the sale of male calves was programmed early before weaning and 13.5% of cases a marketing of male calves were done shortly after their weaning. In addition, 10% of producers using calves as saving money to mobilize when their economic needs (selling as needed).

Multivariate analysis

Statistical model presentation

Categorical Principal Components Analysis (CATPCA) was defined two axes with 43% of total variance. First axis represented about 32% of total variation. It was interpreted as an axis of dairy orientation. It was highly correlated to variables related to dairy activities such as number of cows' and heifers. The second axis explains 19% of total variation and was positively correlated to beef number and negatively correlated to dairy parameters' (Table 3).

Cattle farming systems identified

The approach adopted enhanced with two step classification has demonstrated five types of cattle system according to productive orientation of cattle herd (Figure 2 and table 4). The term "dairy farms" includes different levels of orientation, development and integration of milk production.

Type 1. Dairy system

Cattle breeders in this system (about 15% of all farms visited) prefer to exploit the potential of animals in dairy production. The principal concern of these farmers was the commercialization of all milk produced on their farms in order to ensure highest possible income. Male calves born on the farms were for this category of breeders a co-product that gets rid rapidly before their weaning. Livestock is mostly specialized (cattle alone) or associated with intensified production of poultry. Animal material is formed by a lower size of herd (10±8.5 LU/farm) characterized by a large dominance of cattle (over 90%). The milk production was ensured by the presence of 7±6 dairy cows.

Type 2. Dairy mixed system (Dairy-beef cattle oriented milk)

This type covers about 20% of cattle farms in the region. In this livestock system, farmers adopt strategies of mixed cattle production but producing and sales milk was their essential income. The fattening of theirs calves was an unplanned act used to cope with economic uncertainties (sale of calves according to the economic needs). The livestock exploited was in order of 17±12 LU. Cattle alone breeding mark this type of farms in 65 % of situation and cattle herd represents more than 80% of all exploited ruminants. This system promotes the highest number of dairy cows (10±8) and a low number of beef (1.2 ±1.3) per farms' due to sold of calves at an early age.

Type 3. Balanced mixed system (Dairy and beef cattle)

It was recorded in only 4% of total farms surveyed. In this first model, the cattle complete two different functions, complementary and reasonably balanced, milk and beef production. This system represents a small sample which are generally a large farms distinguished by prac-

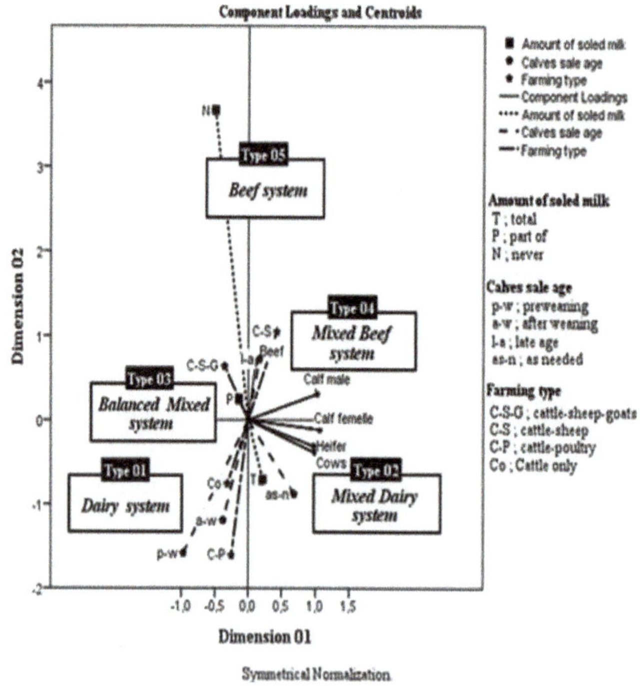

Figure 2: Graphical presentation of obtained model

tices of the partial marketing of milk and early sale of male calves. These farms do not give special attention to the type of production compared to the other due to large diversification of crops and livestock offered a multiple returns. In this regard, livestock herd was important (18.5 LU/farm in average) marks the breeding of fifty ewes and about ten goats near the cattle herd. Cattle represents about 50 % of all animals exploited characterized by the presence of 6.14±3.02 cows and 1.57±0.98 beef per farm.

Type 4. Beef Mixed System (Dairy-beef cattle oriented beef)

This system dominates the study region and farms shown here make up the model frequently found in the context of the eastern high plains of Algerian (more than 56% of farms). These farmers adopt policies of mixed cattle system but more directed to the beef production. Suckling calves was a priority in farming practices as far as farmers reasoning was based on the earnings of beef compared to milk. A ruminant livestock contains 18±13 LU per farm. Over 60% of breeders who belonging to this group combine the cattle with sheep have subsequently more than thirty sheep, the rest of farmers exploits mostly cattle alone. Approximately 70 % of total livestock exploited per farm was formed by the cattle herd defined by the presence of 7.42±5.21 cows and 2.44 ±2.78 beef.

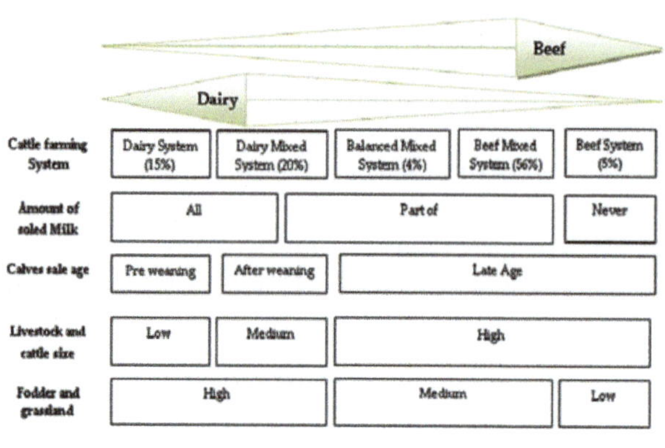

Figure 3: Characteristics of cattle farming systems'

Table 4: Characteristics of cattle farming system

Variable	Modality	Cattle Farming System				
		Type 1 Dairy system (14.5%)	Type 2 Dairy Mixed System (20.0%)	Type 3 Balanced Mixed System (4.2%)	Type 4 Beef Mixed System (56.4%)	Type 5 Beef System (4.8%)
Land	Arable Land	24.7 ±23.9	23.6 ±29.9	41.5 ±71.9	23.9 ±33.2	11.9 ±9.9
Fodder production	Fodder land	2.2 ±4.7	2.5 ±4.6	2.0 ±2.6	2.5 ±4.3	1.0 ±1.4
	Grass land	1.5 ±2.5	1.1 ±1.9	1.6 ±1.8	0.9 ±1.6	0.4 ±0.7
Livestock	Livestock	10.4 ±8.6	17.2 ±12.1	18.3 ±6.0	18.0 ±13.2	19.4 ±23.7
	Unit	7.9 ±22.6	18.9 ±36.5	45.3 ±22.8	35.4 ±47.4	12.7 ±15.6
	Ewes (head)	0.0 ±0.0	0.0±0.0	13.6±7.8	0.3 ±1.9	0.0 ±0.0
	Goats (head)					
Cattle	LU Cattle	9.2 ±7.5	14.3 ±10.2	10.2 ±3.7	12.7 ±8.9	17.5 ±23.2
	Cows (head)	6.9 ±5.6	9.9 ±7.9	6.1 ±3.0	7.4 ±5.2	6.4 ±6.4
	Beefs (head)	0.8 ±1.2	1.2 ±1.3	1.9 ±1.4	3.0 ±2.6	5.7 ±13.9
	Heifers (head)	1.9 ±1.6	3.2 ±2.9	1.9 ±1.5	1.7 ±3.0	0.0 ±0.0
	Calf Male	0.8 ±0.9	1.6 ±1.7	1.6 ±1.0	2.4 ±2.8	2.9 ±4.9
	Calf Femele	1.4 ±1.7	2.1 ±1.9	2.0 ±1.0	2.5 ±2.3	1.0 ±2.1

LU: Livestock unit

Type 5. Beef system

This category of farms encountered with a low frequency (less than 5% of farmers) is similar to the model of suckle cattle system in temperate regions. If reproduction and fattening calves are the center of interest of policy makers on these farms, the milk can not be sold. It was valued in suckling of future beef. In these situations, ¾ breeders exploit in parallel sheep if were not specialized farmers (cattle only). It has a biggest animal herd (more than 19 LU). The cattle herd separately form more than 90% of the overall livestock distinguished operated by breeding 6.8 ± 6.4 cows and 5.7 ±13.9 beefs.

DISCUSSION

The study showed that more than 80% of farmers in the context of Algerian semi arid area adopt mixed cattle farming systems (Dairy-Beef). These producers operate in an unfavorable agricultural environment characterized by several economic and technical problems (instability of farm product price in the internal market and lack of technical backstopping). In this particular environment where milk production is low and unfavorable to these constraints, the profitability of cattle livestock specializes in the production and marketing of milk is unsecured. Only the profits generated by the fattening of calves born on the farm can encourage these farmers to continue their activities. On these farms, lack areas of grassland and feeble forage production for various reason preventing farmers to achieve a satisfactory of intra-farm level of forage autonomy (Figure 3). In several situations, forest grazing or cereals residues are sources of food for herds which housed under these livestock systems. Such obstacles are pushing farmers logically to avoid the attachment of the productivity of theirs farms to a single product (milk or beef). In this condition maximizing production is a secondary goal after the survival of the farm (Abbas, 2004). However, in temperate countries with high predisposition to the specialization activities Chatellier and Jacquerie (2004) reported that 25% and 20% of farms respectively in Belgium and Austria are mixed (Dairy-Beef) due to various reasons. In Tunisia, according to Jaoad (2004), a mixed dairy-beef systems can be observed in medium-sized farms. The reason for this is that beef production in combination with milk can be carried out with fewer animals than in beef production systems. Milk and beef production systems are closely connected and changes in milk production systems will cause alterations in beef production systems (Christel and Magnus, 2003).

The mixed system (Dairy-Beef) but more oriented towards beef production is the system model dominant cattle breeding in the study area (over 56%). Logical approach which these farmers have melted their policies are still economic profitability of beef production against milk and the efficiency of work organization in farms. Partial sale of milk guaranteed the coverage of life family expenses and daily cost of livestock activities, while the selling of beef and calves promotes the creation of funds using for new investment and modernization of farm. A difficulty of integration on milk collection networks and lack of milk conservation instruments in farms have also contributed in these policies. In France for example and according to a report established by Livestock Institute, meat produced comes in 35% from dairy cattle, and mixed and dairy farms supply 50% of young cattle for fattening.

The mixed system (Dairy-Beef) but more oriented towards the production of milk corresponds perfectly to farms ''cattle alone''. These farms sold all milk produced to the public or private dairies to benefit a subvention for milk production and other advantages. An important number of these farms were in mixed system (Dairy-Beef) oriented beef evolved gradually to Diary-Beef system oriented milk in search of stability and consistency of income provided by the sale of milk at a price substantially improved over the last years. Jaouad (2004) report that in Tunisia mixed systems can be found in small-scale irrigated farming which is predominantly oriented to dairy production.

The dairy system can be encountered only in farms with cattle-sheep breeding dominated by cattle herds or in cattle-poultry farms. These specialized farms are rarely only cattle farms. The early sale of young males born on the farm offers more facility in the sale of all milk produced by dairy cows. Farms structure factor is not to call into question but rather the search for stable sources of income that have guided this policy. In Morocco, the specialized diary system was observed in large farms (Srairi and Kessab, 1998) or in irrigated perimeter smallholders' (Srairi *et al.*, 2003) that 100% of arable land was used in fodder production. In this region only farms' directly committed to the way of specialization arrive at high economic performance.

In beef system, it was absolutely normal to accept that the sale of milk is never done on these farms for technical reasons relating to the valorization of milk producing in suckling of calves following the example suckling systems in temperate regions. However, cows' of local or cross breed in this environment are conducted in extensive on limited areas or without forage resources. It is reasonable also to assume that these practices are largely inflicted by traditional and socio cultural reasons. In Maghreb, the breeding of calves or beef fattening, are based on very limited areas (less than 5 ha) that much of the feed is purchased (Jemai and Saadani, 2000; Srairi *et al.*, 2003).

CONCLUSION

Clearly, breeders in conditions of Algerian semi arid area prefer mixed systems. In cattle production strategies, it was the interaction of several factors that oriented breeders to favorite such system compared to the other. So it is logical to assimilate that the maximization of profits by reducing costs and optimizing production potential of herd, were the objectives of the breeder whatever manner with which it is organized. However, maximization of production per speculation was a secondary goal after the survival of the farm.

REFERENCES

1- Abbas, K. (2004). La Jachère Pâturée dans les zones céréalières Semi-arides: Pour une approche de développement durable. Cahiers options méditerannienes, 62, 169-173.

2- Chapman, D.F., Kenny, S.N., Beca, D., & Johnson, I.R. (2008). Pasture and forage crop systems for non-irrigated dairy farms in southern Australia. 1. Physical production and economic performance.

Agricultural Systems, 97(3), 108-125.
3- Chatellier, V., & Jacquerie, V. (2004). La diversité des exploitations laitières européennes et les effets différenciés de la réforme de la PAC de juin 2003. INRA Prod. Anim. 17 (4), 315-333.
4- Dedieu, B. (2009). Adaptation des systèmes d'élevage et incertitudes sur l'avenir. CRA-W & FUSAGx - Carrefour Productions animales 2009.
5- Dufumier, M. (2006). Diversité des exploitations agricoles et pluriactivité des agriculteurs dans le tiers monde. Cahiers d'Agriculture, 15 (6), 584-588.
6- FAO/ILRI. (1995). Livestock development strategy for low income countries. Proceedings of a Roundtable held at ILRI, Addes Ababa, Ethiopia, 27 Feb. to 02 March 1995.
7- Frensh Livestock Institute. (2005). La production de viande bovine en france. Le dossier economie de l'élevage.
8- Jaoad, M. (2004). Dynamique des cheptels bovins en Tunisie et contraintes alimentaires et fourragères. Cahiers Options Méditeranniennes, 62, 421-424.
9- Jemai, A., & Saadani, Y. (2000). Evolution des systèmes d'élevage dans les zones montagneuses du nord ouest de la Tunisie. Options méditerranéennes. 39, 39-56.
10- Madani, T., & Mouffok, C. (2008). Production laitière et perfomances de reproduction des vaches Mmontbéliardes en région semi aride algérienne. Revue Elev. Méd. Vét. Pays trop, 61 (2), 97-107.
11- Shalloo, L., Dillon, P., Rath, M., & Wallace, M. (2004). Description and validation of the moorepark dairy system model. Journal of Dairy Science. 87, 1945-1959.
12- SPSS, Inc. (2010). SPSS GLM 18.0. Chicago: M. J. Norusis.
13- Sraïri, M.T., Kiade, N., Lyoubi, R., Messad, S., & Faye, B. (2009). A comparison of dairy cattle systems in an irrigated perimeter and in a suburban region: Case study from morocco. Trop Anim Health Prod, 41 (2), 835–843.
14- Sraïri, M. T., & Kessab, B. (1998). Performances et modalités de production laitière dans six étables spécialisées au Maroc. Productions Animales-Paris-Institut National De La Recherche Agronomique-, 11, 321-326.
13- Sraïri, M.T., Leblond, J.M., & Bourbouze, A. (2003). Production de Lait et/ou de Viande: Diversité des Stratégies des eleveurs de bovins dans le périmètre irrigué du gharb au maroc. Revue d'élevage et de médecine vétérinaire des pays tropicaux, 56 (3-4), 177-186.

An Investigation into Credit Receipt and Enterprise Performance among Small Scale Agro Based Enterprises in the Niger Delta Region of Nigeria

Ubon Asuquo Essien, Chukwuemeka John Arene and Noble Jackson Nweze

Abstract

The study was designed to analyze credit receipt and enterprise performance by small scale agro based enterprises in the Niger Delta region of Nigeria. A multistage sampling technique was adopted in selecting 264 agro based enterprises and 96 agro based enterprises that accessed informal and formal credit respectively. The Heckman model was used to examine the factors affecting amount of informal and formal credit received by the enterprises. Financial ratios such as the current ratio and return on capital employed ratio were used in addition to the t-test to examine the performance of enterprises that borrowed from informal and formal credit markets in the area. Analyses of informal credit amount received reveal that gender, age and social capital are significant for the first hurdle, whereas gender, size, income, guarantor and social capital are significant for the second hurdle. Similarly, gender, education, age, size, and collateral are significant for the first hurdle for formal credit, while the second hurdle reported significant results with age, size, income, collateral and social capital. Formal credit was less accessible than informal credit but enhanced greater performance. Formal credit should be made to be easily accessible and efficiently utilized. This will go a long way in complementing the amnesty programme of the federal government of Nigeria in the region.

Keywords:
Credit amount received, Credit Access, Enterprise performance, Agro-based enterprises, Niger delta region, Nigeria

Department of Agricultural Economics, University of Nigeria, Nsukka.
** Corresponding author's email: essienua@gmail.com*

INTRODUCTION

A little over four decades, the issues confronting the Niger Delta region of Nigeria have caused increasing national and international concern. The region produces immense oil wealth and has become the engine of Nigeria's economy, but it also portrays a paradox as the vast oil revenues barely touch Niger Delta own pervasive poverty, hence giving birth to formidable challenges to sustainable human development in the region (UNDP, 2006). People are more volatile, resulting in youth restiveness, conflicts between youths and community leaders, youths and government agencies, youths and multinational companies (UNDP, 2006). These propagated negative nominal and real shocks in every sector of the economy, including small business sector, with the economy operating under atmosphere of politically unstable environment, eroded productivity and declined private investments (Ministry of Niger Delta Affair, 2011).

Studies on developing economies have considered financial development vital for economic growth and poverty reduction. Strong financial systems have helped delivered rapid growth as well as direct and indirect benefits, across income distributions (Honohan and Beck, 2007). Beck and Demirguc-Kunt, (2005) indicate that financial development reduces inequality by disproportionately boosting the income growth of the poor. Hence across Africa, access to finance is rightly seen as a key to unlocking the income growth for poor families, as much as for expanding trade (Honohan and Beck, 2007).

In this regard, policy makers have held the conception that micro and small scale firms in developing countries lack access to adequate financial services for efficient inter-temporal transfers of resources and risk coping (Besley, 1995). Without well-functioning financial markets, small scale firms may lack much prospects for increasing their productivity in many significant and sustainable ways (Nwaru, 2004). Based on these reasons, and the fact that traditional commercial banks typically have minimum interest in lending to small firms due to their lack of viable collateral and high transaction costs associated with the small loans that suit them, most developing country governments, have set up credit programs aimed at improving access to credit (Arene, 1993; CBN, 2010).

Efforts targeted at small businesses are based on the premises that they are the engine of economic development, but market and institutional failures impede their growth, thus justifying government interventions (Gomez, 2008). However, the failure of government supported financial institutions is a convincing evidence of the need for a better understanding of how these firms in the Niger Delta, often operating in highly risky environment insure against risk and conduct their inter-temporal trade in the absence of well functioning financial markets (Ministry of Niger Delta Affairs, 2011). In response to these failures and in recognition of the critical role that credit can play in alleviating poverty in a sustainable way, innovative credit systems are being developed and promoted in Nigeria as a more efficient mechanism of improving micro and small scale firms' access to credit (CBN, 2010). This inefficient nature of the credit market presupposes the lack of adequate information relating to empirical issues on credit receipt by small scale agro-based enterprises and performance in post conflict Niger Delta.

Evidently, small scale enterprises have performed at very abysmal level (Hassan and Olaniran, 2011). This low performance has exacerbated poverty, hunger, unemployment and low standard of living of people in a country whose economics is ailing (Hassan and Olaniran, 2011). Considering the emergence of many formal and informal financial institutions in the Niger Delta, there is hope for small Agro-based enterprises, but to what extent has credit advanced to these enterprises influence performance? Assessment of the influence of financing is popular, but lacking among small agro based enterprises in a post-conflict context. Therefore, attempt to formulate credit policies without substantial information on how Agro-based firms respond to the credit market and factors militating against their response in a region such as the Niger Delta may be deficient, since it is not backed by empirical evidence. The study therefore sets out to investigate credit receipt and enterprise performance in the Niger Delta region of Nigeria.

MATERIALS AND METHODS

The study area was the Niger Delta Region of Nigeria. It lies between latitudes 4°2" and 6°2" north of the equator and longitudes 5°1" and 7°2" east of the Greenwich meridian (Tawan, 2006). Nine of Nigeria's constituent states make up the region, namely; Abia, Akwa Ibom, Bayelsa, Cross River, Delta, Edo, Ondo, Imo and Rivers states, with an area of 112,000 sq. km, a population of 27 million people, 185 LGAs, about 13,329 settlements; 94% of which have populations of less than 5,000 (Ojameruaye, 2008).

According to the Ministry of Niger Delta Affairs (2011), the climate of the Niger Delta Region varies from the hot equatorial forest type in the southern lowlands to the humid tropical in the northern highlands and the cool montane type in the Obudu plateau area. Further, the wet season is relatively long, lasting between seven and eight months of the year, from the months of March to October.

The region has huge oil reserves and ranks sixth exporter of crude oil and third as world's largest producer of palm oil after Malaysia and Indonesia (Omafonmwan and Odia, 2009). Further, the Delta leads in the production of timber, pineapple and fish, also; cocoa, cashew, rice, yam and orange are produced in large quantities in the area (Omafonmwan and Odia, 2009). While cassava resources can stimulate the

Table 1: Distribution of small scale Agro-based enterprises by their Socio-economic characteristics.

	Informal Credit Borrower Enterprises		Formal Credit Borrower Enterprises	
Variables	Frequency	Percentage	Frequency	Percent-
Age				age
1-4	146	55.30	41	
5-8	68	25.76	26	42.71
9-12	38	14.39	13	27.08
13-16	6	2.27	12	13.54
17-20	4	1.52	2	12.50
21-24	2	0.76	2	2.08
Total	264	100	96	2.08
Mean	5.35		6.92	100
Gender				
Male	184	69.70	61	
Female	80	30.30	35	63.54
Total	264	100	96	36.45
Accessibility of Credit Market				100
Informal Credit	264	60.13		
Formal Credit	96	21.86		
No Access	79	17.99		
Total	439	100		
Years of Borrowing Experience				
1-3	155	58.71	53	
4-6	76	28.79	28	55.21
7-9	21	7.95	8	29.17
10-12	8	3.03	5	8.33
13-15	4	1.52	2	5.21
Total	264		96	2.08
Mean	3.79	~~NN~~	4.09	1.49
Level of Formal Education				
No Formal Education	9	3.14	1	
Primary	83	31.44	18	1.04
Secondary	102	38.64	36	18.87
Tertiary	70	26.52	41	37
Total	264		96	42.7

Source: Field Survey, 2012

growth of local processing industries for fufu, garri, chips, flour, glucose, starch and pellets; massive furniture, building and craft industries can be built on the regions huge bamboo resources.

The major occupation of the people is fishing and agriculture but activities of oil companies have impacted on the environment with poor access to water, transport, telecommunication, power and fuel, housing, poor waste management, and poor educational structure (Igbuzor, 2006); this lead to conflict in the region some years back. Traditional industries in the area include canoe carving, pottery, cloth weaving, mat-making, thatch making (roofing materials), palm oil processing, food processing (garri, fufu and starch from cassava), local gin distillation etc. Small and Medium scale enterprises are found almost everywhere in the region. The main characteristics of these industries found in varying proportions throughout the region, are that they are based on manual artisanal technologies, local inputs and skills transferred chiefly through family upbringing and not via formal training or education (Ministry of Niger Delta Affairs, 2011). Further, the major lending credit institutions in the region are formal and informal credit institutions.

A multistage sampling technique was used in this study. Of the 9 Niger Delta States of Abia, Akwa Ibom, Bayelsa, Cross river, Delta, Edo, Rivers, Imo and Ondo states, three states were purposively selected based on high concentration of economic activities which are Agro-based The States were Bayelsa, Delta and River States. Further, three local Government Areas each were purposively selected from each of the three states, from which one each was randomly selected for the study. The Local Government Areas were Brass, Warri North, and Phalga This was possible with the help of staff of the Ministry of Economic Development/trade, the Small and Medium Scale Enterprise Associations resident in each state and by oral interview.

In the third stage, a list of Small Scale Agro-based enterprises was obtained from the Small and Medium Scale Enterprises Associations and the Local Government Business registration office. This list was stratified into three sectors namely manufacturing, services and trading sectors, out of which two enterprise types were randomly selected from each of the three sectors, making it six. The enterprises selected were Bakery and Capentry/furniture- Manufacturing; Restaurants and Cold Room Services- Services; Poultry Feeds and drugs- Trading. Twenty of each of the selected enterprises from each sector was randomly selected for study. One hundred and twenty enterprises were selected from the three sectors in each local Government Areas of each state. In all, three hundred and sixty enterprises were selected from the three states. Furthermore, the 360 enterprises were stratified along credit source lines. On the whole, two hundred and sixty four enterprises that accessed informal credit and 96 enterprises that accessed formal credit were used for detailed study.

Data from the study were obtained from primary sources through the use of structured questionnaire and oral interview.

Before undertaking the actual data collection, research assistants were briefed on the use of the data collection instruments. This was implemented during the pilot study where the same personnel were used for pre-testing of the questionnaires. This was to ensure clear understanding of the instrument to avoid inconsistency and incomplete response. Changes were however considered on the questionnaire and problem statement after the pilot testing. Following the actual data collection, examination of the questionnaires was made in other to determine and drop questionnaires with inconsistent as well as incomplete answers. Though meticulously implemented, notwithstanding, however missing firms or non-respondents were encountered. The missing data was dealt with by matched sample from the frame. That is, those firms not in existence were matched by another firm in frame because of random sampling.

Data analysis

Data were analyzed by the use of descriptive statistics such as frequency, means, percentages, etc. The Heckman model, selected financial ratios and the t-test were also employed in the study. The Heckman model is illustrated by the following equations:

(a) *Index Equation* $d_i^* = X'_{Ii}\beta_1 + U_i$, $U_i \sim N(0,1)$... ...

Threshold index equation=$\{1$ *if* $d_i^* > 0$*, and is 0*

if $d_i^* \leq 0$

(b) Amount of Credit received: $t^* = X'_{2i}\beta_2 + V_i$, $V \sim N(0, \delta^2)$

Threshold equation: $t_i = \{t_i^*$ *if* $d_i = 1$. 0 *if* $d_i = 0$

Where d_i = Probability of access to credit

t^* = amount of credit received

t_i = amount of credit received if respondent i has access to credit, 0 otherwise

Where d_i = Probability of access to credit

t^* = amount of credit received

t_i = amount of credit received if respondent i has access to credit, 0 otherwise

Other variables in the model were defined below:

GEN (Gender of the entrepreneur. Defined as dummy, takes the value of 1 for male and 0 for female.

EDU=Entrepreneur's Education. (This is the level of formal education attained by the owner/manager of firm. Measured by the total number of years the entrepreneur spent in receiving formal education).

AGE= Enterprise Age (Defines the total number of years that the business has been in existence. Measured in years).

SIZ= Enterprise Size (This describes the worth of the enterprise; total assets of the enterprise valued in Naira)

INT=Interest Amount; this is the total amount the borrower pays as interest charges on money borrowed.

INC= Income of firm (Receipts of the enterprises from sales in the last one year (Measured in Naira)

COL= Collateral (Defined as any valuable asset that eases the approval of formal credit (Measured as Dummy: 1 if firm provided collateral to access credit, 0 otherwise)

GUA= A person who pledges that a debt will be paid. (Binary; 1 if guarantor was available and 0 if not)

SOC=Social Capital (For informal credit; it describes borrowers acquaintance with lender. Measured as dummy, 1 if borrower is acquainted with lender, 0 otherwise. For formal credit, it describes membership of cooperative society, hence the number of people in the cooperative.

Further, financial performance of enterprise was analysed using the:

(1) Return on Capital employed (ROCE)

(2) Current Ratio

1-Returns on Capital Employed Ratio = Net Profit after Tax/Capital employed

These ratios indicated how small scale agro based enterprises in the study area have used capital employed. Higher ratios implied greater efficiency.

2-Current Ratio = CA/CL

Where CA = Current Asset of enterprise valued in Naira (₦)

CL = Current liabilities of enterprise valued in Naira (₦)

Current Assets included cash and assets converted into cash within the last year, such as marketable securities, inventories, and prepaid expenses. Obligations that matured within the year were included in current liabilities and they included creditors, bills payable, accrued expenses, income tax liability and long term debts maturing in the current year. Ratios greater than one meant that small scale agro based enterprises in the study area has more current assets than current liabilities (or claims) against it.

RESULTS AND DISCUSSION

Socio-economic characteristics of respondents

The distribution of sampled small scale agro-based enterprises according to age of enterprise as shown in table 1 below reveals that 95.45% of informal credit enterprise borrower and 83.33% of formal credit enterprise borrower are under 12years of age. The mean ages are 5.35 and 6.92 for the informal and formal credit borrower enterprises respectively. This implies that most of the small scale agro-based enterprises that borrowed from the formal credit market are older than their informal credit borrower counterpart. The most common age fell within the range of 1-4years.

Gender of respondents show that 69.70% of the male entrepreneurs borrowed from the informal credit market whereas 63.54% borrowed from the formal credit market. Further, 30.30% and 36.46% of informal and formal credit borrower entrepreneurs are females. This implies that most male entrepreneurs tend to borrow from the informal credit market than the formal credit market. Doan *et al.* (2010) explain that gender does not really matter in credit participation but plays a role in explaining loan size.

While 60.13% of agro-based enterprises have access to the informal credit market; only 21.86% have access to formal credit market. Further, 17.99% of the enterprises do not have access to formal or informal credit market. Access to external resources is needed to ensure flexibility in resource allocation and reduce the impact of cash flow problems (Bigsten *et al.*, 2003). Firms with access to funding are able to build up inventories to avoid stocking out during crises while availability of credit increases the growth potential of the surviving firms during periods of macro-economic instability (Atieno, 2009). In appraising financial constraints to small scale farming in Etsako Local Government Area of Edo State, findings show that only 7% of small scale farmers have access to basic loan while 93% access loan from other sources like co-operative societies, personal savings and relations.

Further, more than 80% of agro-based enterprises who access informal and formal credit have been borrowing for more than 3 years. Further, average borrowing age for formal credit enterprise borrower is 1.5 years whereas informal credit enterprise borrowers have been borrowing for about 4 years.

Also, 96.6% of informal credit borrower entrepreneurs and 98.96% of formal credit borrower entrepreneur had one form of primary to tertiary education. This is significantly high and consistent with MNDA (2004) which indicates that the adult literacy status of the Niger Delta states is about 78%, slightly higher than the national average of 54%, although marked differences exist among the states. Most entrepreneurs who borrowed from the informal credit market however had achieved secondary education level supplemented with training compared to formal credit borrower entrepreneurs. This may imply that most people with this level of education failed to find employment in the formal sector and thus resort to small scale enterprise activities.

The table 2 below presents the maximum likelihood estimates of the first part of the Heckmann model. The estimated probit regression model gave the Mc Fadden R-Squared of about 0.52 which implies that all the explanatory variables included in the model were able to explain 52% of the probability of the decision of small scale agro-based enterprises to access informal credit.

Informal credit access by small scale agro based enterprises (First Hurdle)

The table 2 presents the maximum likelihood estimates of the first part of the Heckmann model. The estimated probit regression model gave the Mc Fadden R-Squared of about 0.52 which implies that all the explanatory variables included in the model were able to explain 52% of the probability of the decision of small scale agro-based enterprises to access informal credit.

The coefficient of the first hurdle indicates

Table 2: Estimated determinants of informal credit access (First Hurdle).

	Coefficient	Std. Error	Z	Slope*	p-value	
Const	0.32658	0.572032	0.5709	-	0.56806	
GEN	-1.35722	0.426915	-3.1791	-0.10347	0.00148	***
EDU	-0.0435472	0.0310248	-1.4036	-0.0045343	0.16043	
AGE	0.0694754	0.0376283	1.8464	0.00723403	0.06484	*
SIZ	8.46569e-08	1.06843e-07	0.7923	8.81479e-09	0.42816	
INC	-6.2595e-09	7.12192e-08	-0.0879	-6.51762e-010	0.92996	
INT	-8.58228e-09	1.68174e-07	-0.0510	-8.93619e-010	0.95930	
SOC	2.58999	0.33906	7.6387	0.676166	0.00001	***

McFadden R-squared	**0.526944**	**Adjusted R-squared**	**0.415245**
Log-likelihood	**-49.25958**	**Akaike criterion**	**114.5192**

Source: Estimated From Field Survey Data, 2012
*** P < 0.01, ** P < 0.05 , * P < 0.10

how a given variable affects the likelihood of access to credit. The result of the first hurdle (probit model) indicates that the co-efficient of enterprise Age (AGE at 10%) and social capital (SOC at 1%) are positive and statistically significant with respect to the decision or probability to access informal credit by small scale agro-based enterprises in the study area. The result implies that as Enterprise Age and relationship with lender increases for informal credit borrower enterprise, the chance to have access to credit increases too. Result is in line with a priori expectations as increase in enterprise age implies increase in experience and growth. Also, increase in the social capital, will enhance relationship and hence will bring about confidence and trust in business between the lender and the entrepreneur. The result for Enterprise Age as reported corroborates the research findings of Atieno (2001) and Mwangi and Oumar (2012) in Kenya. The result for Social Capital is in consonance with findings of Togba (2009) in Cote d'voir. On the other hand, coefficient of Gender (GEN at 1%) is negatively signed and statistically significant with respect to decision to access credit by the small scale agro-based enterprises in the study area. The result reveals that the probability of accessing informal credit decreases among male small scale agro-based entrepreneurs in the study area. This result could be attributed to the fact that men engage in large scale entrepreneurial activities compared to women hence informal credit may not be adequate enough for investment.

The marginal effect of the Probit model show changes in the probability of access to credit for additional unit increase in the decision variables. The probability of access increases by 0.7% and by 67% for every unit increase in Enterprise Age and Social Capital, while 10% reduction in the chance to access informal credit occurs for every unit increase in male respondents. However, based on the magnitude of the slope co-efficient in the estimated model, Social Capital and Gender appear to be the most important policy variables that impact on the decision of small scale agro-based enterprises to have access to informal credit in the study area.

Informal credit amount received by small scale agro based enterprises (Second Hurdle)

The second hurdle indicates how a decision variable influences informal credit amount received by small scale agro based enterprises.

The maximum likelihood estimates of the Truncated Tobit model are presented in table 3. The estimated truncated Tobit regression model reveals a normal distributed regression residual

The result of the Tobit model reveals that coefficient of Gender is significant at 5% level and negatively related to the amount of Informal credit received by small scale agro-based enterprises in the study area. The result implies that increase in number of male entrepreneurs will

Table 3: Estimated determinants of informal credit amount received (Second Hurdle).

	Coefficient	Std. Error	Z	p-value
Const	70627.9	189919	0.3719	0.70998
GEN	-175070	85407.5	-2.0498	0.04038**
EDU	-848.123	8435.06	-0.1005	0.91991
AGE	-5078.49	9499.96	-0.5346	0.59294
SIZ	0.0195456	0.006454	3.0283	0.00039***
INC	-0.000692011	0.0001709	-4.0476	0.000202***
INT	-0.0544053	0.0607931	-0.8949	0.37083
GUA	62183.5	24359.89	2.5527	0.004045**
SOC	422993	117559	3.5981	0.00032***
Chi-square(8)	**19.43569**	**p-value**		**0.012695**
Log-likelihood	**-3325.222**	**Akaike criterion**		**6670.444**

Source: Estimated From Field Survey Data, 2012
*** P < 0.01, ** P < 0.05 , * P < 0.10

lead to a decrease in the amount of credit received from informal credit sources in the area. This result could be attributed to the fact that women borrow smaller amounts for businesses unlike men, hence they tend to patronize the informal credit markets more. This result is substantiated by the findings of Kimuyu and Omiti (2000) that a greater proportion of female entrepreneurs borrow from NGOs and Non-bank financial Institutions hence credit –source-related differences partly account for the gender disparities in the amounts borrowed.

Further, the co-efficient of enterprise size is positively signed and significant at 1%. This is in consonance with a priori expectation. The result implies that increase in size of enterprise will lead to increase in amount of informal credit received by small scale agro-based enterprises in the study area. Informal credit suppliers will disburse funds based on information on value of assets owned by the respondents' assets as a form of security for their loans. Informal credit suppliers have informational advantage over any other lending source type as they are found among the people.

Furthermore, the result of the Truncated Torbit model reveals that the coefficients of income of the firm is significant at 1% level and negatively related to the amount of informal credit supplied to small scale agro-based enterprises in the study area. The result implies that there is an indirect relationship between income and amount of informal credit received; therefore, as firm's income increases informal credit amount received decreases. Firms who have increased income, will tend to borrow less because the increase will be reinvested into the business. The result reflects the pecking order theory which postulates that firms first prefer internal financing, and then debt, lastly raising equity as a "last resort. Findings agree with the empirical research reports of Kedir *et al.* (2009) in Uruguay, and Nwaru *et al.* (2004) in Nigeria.

The Social Capital co-efficient is positively signed and significant at 1% level. This is in line with a priori expectation. A direct relationship, implying that increase in social capital will lead to increase in informal credit amount supplied. Informal credit suppliers are usually within and around the neighborhood and the borrowers acquaintance with the lender goes a long way to enabling him have access to credit. The result of this study is substantiated by the findings of Mwangi and Ouma (2012) in Kenya.

Access to formal credit by small scale agro based enterprises (First Hurdle)

The table 4 presents the maximum likelihood estimates of the first part of the Heckman model. Again, the result is similar to that obtained for formal credit access in the first objective. The estimated Probit regression model gave the Mac Fadden R-Squared of about 0.87 which implies that all the explanatory variables included in the model where able to explain 87% of the probability of the decision of small scale agro-based enterprises to access credit from formal credit institutions.

The coefficient of the first hurdle shows how a given variable affects the likelihood to access formal credit. Those in the second hurdle indicate how a decision variable influences the amount of formal credit received by the respondent entrepreneur. The result of the first hurdle (Probit model) indicates that coefficient of Gender (GEN at 10%), Education (EDU at 5%), Enterprise age (Age at 5%), Enterprise Size (SIZ at 10%), Collateral (COL at 5%) are all positive and statistically significant with respect to the decision or probability to access formal credit by small scale agro-based enterprises in the study area. The implication of the result is that, as Gender, Education, Enterprise Age, Enterprise Size, and value of Collateral increases, the greater the chances to access formal credit. Further, there is greater chance for male entrepreneurs to access credit than female entrepreneurs. This result is in line with a priori expectations because increase in Enterprise Age implies Experience in business; Firm size is a concession to stimulate greater investment, hence greater access to investible funds. Further, the pecking order theory best explains the positive relationship between increased income and credit access; a firm's debt ratio will therefore reflect its cumulative requirements for external financing. The result for firm size, firm income and age of firm corroborates the research findings of Lawal *et al.* (2002) in Osun State, Nigeria and Fatoski while that of Gender is consistent with Ajagbe *et al.*

Table 4: Estimated determinants of access to formal credit by small scale Agro-based enterprises (First Hurdle).

	Coefficient	Std. Error	Z	Slope*	p-value
Const	-24.6353	10.8426	-2.2721		
GEN	1.16775	0.841978	1.3869	0.136854	0.16547*
EDU	0.174152	0.0921485	1.8899	0.0159772	0.05877**
AGE	1.17937	0.545003	2.1640	0.108199	0.03047**
SIZ	8.42606e-07	3.26089e-07	2.5840	7.7302e-08	0.00977*
INC	5.63213e-08	2.97144e-07	0.1895	5.1670e-09	0.84967
INT	2.67719e-07	5.06448e-07	0.5286	2.4561e-08	0.59707
COL	16.4717	7.81895	2.1066	1	0.03515**
SOC	0.0594197	0.0490305	1.2119	0.0054513	0.2255
McFadden R-squared	**0.875574**		**Adjusted R-squared**		**0.732665**
Log-likelihood	**-7.835945**		**Akaike criterion**		**33.67189**

Source: Estimated From Field Survey Data, 2012
*** P < 0.01, ** P < 0.05, * P < 0.10

(2012) in Osun.

The table 5 presents the maximum likelihood estimates of the second part of the Heckman model. The estimated Tobit regression model gave the Mac Fadden R-Squared of about 0.89 which implies that all the explanatory variables included in the model where able to explain 89% of Credit amount received from formal credit institutions.

The coefficient of age of enterprise is positive and significant at 10% level. This is a direct relationship with formal credit amount received by the enterprises in the region. It implies that enterprises in the region tend to receive more credit as their age increases. Number of years in business is usually perceived as an incentive to loan access. It is a form of security for the lender for the loan amount he is giving out as trust increases with years of business dealings.

Further, the co-efficient of enterprise size is positively signed and significant at 5%. This is in consonance with a priori expectation. The result implies that increase in size of enterprise will lead to increase in amount of formal credit received by small scale agro-based enterprises in the study area. Formal credit suppliers will disburse funds based on information on value of assets owned by the respondents. This is because as size of the enterprises increases, the stock of inventory will increase, hence increase in assets of the enterpises. If these assets are liquidated, they can be used to repay loans incased of any eventuality. Therefore enterprise size will attract more credit amount all things being equal.

Further, the result of the truncated Torbit model reveals that the coefficients of income of the firm is significant at 1% level and negatively related to the amount of formal credit received by small scale agro-based enterprises in the study area. The result implies that there is an indirect relationship between income and amount of formal credit received; therefore, as firm's income increases formal credit amount received decreases. Firms who have increased income, will tend to borrow less because the increase will be reinvested into the business. Also, start-up businesses may not necessarily go for credit abnitio, irrespective of increased income. The result reflects the pecking order theory which postulates that firms first prefer internal financing, and then debt, lastly raising equity as a "last resort. Findings agree with the empirical research reports of Kedir *et al.* (2009) in Uruguay, and Nwaru *et al.* (2004) in Nigeria.

Collateral is a prerequisite for credits in the formal sector. It enhances easy access to funds. The coefficient of collateral is positive at 1% level. This implies that the more the availability of adequate security for loan, the more the amount of formal credit the enterprise will be able to access. Collateral therefore is a great incentive to formal credit amount supplied. This is consistent , de-

Table 5: Estimated determinants of formal credit amount received

	Coefficient	Std. Error	Z	p-value
Const	1.97263e+06	666081	-2.9615	0.00306***
GEN	205785	333579	0.6169	0.53730
EDU	3295.75	31285.2	0.1053	0.91610
AGE	57680.5	30835.5	1.8706	0.06140*
SIZ	0.095229	0.039657	2.4013	0.03671**
INC	-0.0179366	0.004246	-4.2235	0.00313***
INT	-0.017621	0.131784	-0.1337	0.89363
GUA	1.95356e+06	494901	3.9474	0.00008***
SOC	30044.9	15976.2	1.8806	0.06003*
Chi-square(8)	38.03476		p-value	7.42e-06
Log-likelihood	-1009.351		Akaike criterion	2038.702
McFadden R-squared	0.89543			

Source: Estimated From Field Survey Data, 2012
*** P < 0.01, ** P < 0.05 , * P < 0.10

sirable and in line with a priori expectation.

The Social Capital co-efficient is positively signed and significant at 10% level. This is in consonant with a priori expectation. It is desirable and consistent. Formal credit institutions would usually want to loan money to groups rather than individuals for business. Therefore as more small scale enterprises form themselves into cooperative groups, they will be able to access more funds from the formal credit market. The result of this study is substantiated by the findings of Mwangi and Ouma (2012) in Kenya.

Performance of enterprises in the region

The ability of small scale agro based enterprises to cope with turbulence, and to provide an entrepreneurial engine of job creation and innovation in the region is heavily dependent upon their financial position. Instability in the Niger Delta made clear that the financial position of firms in the region, particularly small firms, was less adequate than it might have been. Ratio analysis is commonly used to interpret the adequacy of financial performance. The current ratio gives current assets relative to current liabilities. A ratio of less than 1.00 indicates that current liabilities exceed current assets, and thus the liquidity of the firm is poor.

Table 6 and 7 therefore represent the current ratio for small scale agro-based enterprises that received credit from formal and informal credit sources in the region.

The table reveals that 29.55% of enterprises that received credit from the informal credit market and 25% of the enterprises that received from the formal credit market in the study area, had a current ratio of less than one. This implies that these groups of enterprises cannot meet up their current obligations. More than 60% of the enterprises in the region can meet up their current obligations. From these, a greater number (51.04%) of the enterprises that received credit from formal credit sources with current ratio > 5.99 are well able to meet up current obligations compared to the smaller percentage (45.45%) of enterprises who received from the informal credit sources with similar current ratio. Further, the implication of this is that majority of enterprises who received credit from formal credit sources perform better than those small scale enterprises who received credit from the informal credit sources. This result is expected, desirable and in line with a priori expectation. This is because formal credit is always larger than informal credit and useful for meaningful production. If well employed, large size credit amount which is characteristic of formal loans should enhance performance through economies of scale occasioned by larger credit amount, ceteris paribus the result of this work corroborates the research findings of Majumder and Rahman (2011) in Bangladesh.

Table 6: Current ratio for formal credit borrower enterprise

Category	Frequency	Percentage
0.00-0.99	24	25
1.00-1.99	4	4.16
2.00-2.99	9	9.37
3.00-3.99	5	5.20
4.00-4.99	2	2.08
5.00-5.99	3	3.12
>5.99	49	51.04
Total	**96**	

Source: Estimated From Field Survey Data, 2012.

Table 7: Current ratio for informal credit borrower enterprises

Category	Frequency	Percentage
0.00-0.99	78	29.545
1.00-1.99	41	15.530
2.00-2.99	6	2.272
3.00-3.99	5	1.893
4.00-4.99	8	3.636
5.00-5.99	6	2.27
>5.99	120	45.454
Total	**264**	

Source: Estimated From Field Survey Data, 2012.

Difference in means of current ratio

There is no significant difference between mean current ratio by the two groups of enterprises, that is, enterprises that borrowed from the informal credit market and enterprises that borrowed from the formal credit market. The t-cal is insignificant (.579), implying that there is no difference in performance between enterprises that borrowed from formal credit market and enterprises that accessed funds from the informal credit market. That is, even though majority of enterprises that borrowed from formal credit market were able to meet up current obligations compared to a lesser percentage of enterprises that borrowed from the informal credit market, their performance do not actually vary.

Return on capital employed by small agro based enterprises

Further, tables 8 and 9 represent the return on capital employed for small scale agro-based enterprises that received credit from informal and formal credit sources in the study area. The most independent ratio for assessment of profitability is the return on capital employed. Lower ratios suggest that management is not efficient in the use of funds. It reflects the overall efficiency with which capital is used.

Table reveals that 19.69% of informal credit borrower enterprise and 12.5% of formal credit borrower enterprise have a return on capital ratio of 0.10 and below while 60% of the enterprises that received from the informal credit source and 72% of enterprises that received from the formal

Table 8: Return on capital employed for Informal credit borrower enterprise.

Category	Frequency	Percentage
0.00-0.10	52	19.696
0.11-0.20	17	6.439
0.21-0.30	2	0.75
0.31-0.40	11	6.875
0.41-0.50	2	6.489
0.51-0.60	3	1.136
0.61-0.70	9	3.409
0.71-0.80	3	1.136
0.81-0.90	3	1.136
0.91-1.00	2	0.75
>1.00	160	60.606
Total	**264**	

Source: Estimated From Field Survey Data, 2012.

Table 9: Return on capital employed for formal credit borrower enterprise.

Category	Frequency	Percentage
0.00-0.10	12	12.5
0.11-0.20	3	3.125
0.21-0.30	2	2.083
0.31-0.40	3	3.125
0.41-0.50	-	-
0.51-0.60	-	-
0.61-0.70	-	-
0.71-0.80	3	3.125
0.81-0.90	2	2.083
0.91-1.00	1	1.041
>1.00	70	72.916
Total	**96**	

Source: Estimated From Field Survey Data, 2012.

credit source have a return on capital of 0.100 and above. In theory, the return on capital employed (ROCE) should be above borrowing rate. The current commercial bank borrowing rate in the country is fixed at about 12.5%, however, this is not obtainable in the banks as bank lending rate are observed to be as high as 25%. Against this backdrop, the result implies that majority of the enterprises from both group have an ROCE below the lending rate for the formal credit borrower, hence there is inefficient use of resources among small scale agro based enterprises in the study area. More enterprises that received credit from the formal credit market however are more efficient in use of capital than those that received from the informal market. Again the result of this work corroborates that of Majumder and Rahman (2011) in Bangladesh.

Difference between means of return on capital employed

There is a significant difference in the mean return on capital employed by the two groups of enterprises, that is, enterprises that borrowed from the formal credit market and enterprises that borrowed from the informal credit market. The T-cal is -3.57. This is significant at 1% level at 1% level, implying that enterprises that depended on formal credit sources performed more efficiently than enterprises that accessed funds from the informal credit sources. This result may be due to consistent and efficient monitoring of loan-use by financial institutions in the area. The result corroborates that of Majumder and Rahman (2011) in Bangladesh.

CONCLUSION

The study was conducted to identify factors influencing small scale agro based enterprises access to credit, actual credit amount accessed and the performance of the enterprises in Niger Delta Nigeria. The study uses the Heckman model to analyze the two stage decision of credit access and acquisition by small scale agro based enterprises in the study area. The Probit model regression analysis reveals that enterprise age, social capital and gender are statistically significant decision variables influencing the probability of accessing Informal credit by small scale agro based enterprises in the study area, whereas enterprise size, income of the enterprise, guarantor and social capital significantly influenced the actual informal credit amount received by these enterprises.

Similarly, gender, age, enterprise size, income and social capital significantly influences formal credit access by those groups of enterprises that had access to formal credit sources, whereas age, gender, enterprise size, income and collateral were variables that significantly influenced actual formal credit amount received by the enterprises in the region. It was however observed that apart from the age of the enterprise which influence formal credit amount accessed, similar factors affect credit amount obtained from formal and informal credit sources. It was therefore recommended that operators of formal credit institutions should endeavour to review their lending policies in other to favour start-up businesses; this will enhance performance in the sector.

1- To improve small scale agro based enterprises access to credit, the study recommended that entrepreneurs in the study area should form cooperative societies as this will ensure appropriate information sharing, risk reduction and increase awareness on matters relating to credit.

2- Operators of credit institutions should endeavor to locate some of the lending institutions or outfits nearer to these entrepreneurs.

3- Adult education programe should be implemented for agro based entrepreneurs as this would affect their access to credit positively.

4- To increase credit amount received by entrepreneurs in the study area, the study advocated for the re-assessment of the collateral needs of the lending agents and the duration of credit to the entrepreneurs.

AKNOWLEDGEMENT

This paper forms part of the first author's ongoing Ph.D thesis. The co-authors are the thesis supervisors. The authors are grateful to members of the departmental postgraduate committee for their constructive criticisms of an earlier draft.

REFERENCES

1-Ajagbe, F. A., Oyelere, B. A., & Ajetomobi, J. O. (2012). Determinants of small-scale enterprise credit demand: evidence from Oyo state, Nigeria. Am. J.

Soc. Mgmt. Sci, 3(1), 45-48.
2-Arene, C. J. (1993). An analysis of loan repayment potentials of small holder soya bean group farmers in Nigeria. Quaterly Journal of International Agriculture, 32(2),160-169.
3-Atieno, R. (2001). Formal and informal institutions lending policies and access to credit by small scale enterprises in kenya: an empirical assessment. AERC Research Paper 111, African Economic Research Consortium, Nairobi. 1-46.
4-Atieno, R. (2009). Linkages, access to finance and performance of small scale enterprises in Kenya. Research Paper No. 2009/06. United Nations University, World Institute for Development Economics Research. 15.
5-Beck, T. & Demirguc-Kunt, A. (2005). Small and Medium-Size Enterprises: access to finance as growth constraints, Journal of Banking and Finance, 30(11), 2931-2943.
6-Besley, A. (1995). Non-Market institutions for credit and risk sharing in low income countries: Journal of Economic Perspectives, 9(3), 115-27.
7-Bigsten, A. Collier, P., Dercon, S., Gauthier, B., Fafchamps, M., Gunning, J., & Zeufack, A. (2003). Credit constraints in manufacturing enterprises in Africa. Journal of African Economies, 12(1), 104-25.
8-Central Bank of Nigeria (CBN) (2010). Monetary, Credit, Foreign trade and exchange policy guidelines for fiscal years 2010/2011 (Monetary Policy Circular), 38, 1-78.
9-Doan, T., Gibson, J., & Holmes, M. (2010). What determines credit participation and credit constraints of the poor in Peri-Urban areas, Vietnam? Munich Personal RePEc Archive 36p. http://mpra.ub.uni-muenchen.de/27509/.
10-Fatoki, O., & Asah, F. (2011). The impact of firm and entrepreneurial characteristics on access to debt finance by SMEs in King Williams town, South Africa. International Journal of Business and Management, 6 (8), 107-179.
11-Gomez, G.M. (2008). Do Micro-enterprises promote equality and growth? Institute of social science. Woord Daad. 1-28.
12-Hassan, M.A., & Olaniran, S.O. (2011). Developing small business entrepreneurs through assistance institutions: The role of industrial development centre, Osogbo, Nigeria International Journal of Business and Management, 6(2). Retrieved from: www. ccsenet.org/ijbmwww.ccsenet.org/ijbm.
13-Honohan, P., & Beck, T. (2007). Making finance work for Africa, World Bank, Washington, DC, USA.
14- Igbuzor, O. (2000). Niger delta human development report. United Nations Development Programme (UNDP). Abuja, Nigeria, 218.
15- Johnson, S. (2006). Gender and microfinance; guideline for good practice. University of Bath, Bath.
16-Kedir, A., Ibrahim, G., & Torres, S. (2009). Household level credit constraints in urban ethiopia, university of Leicester, Nottingham Trent University and Catholic University of Uruguay 34.
17- Kimuyu, P., & Omiti, J. (2000). Institutional impediments to access to credit by micro and small scale enterprises in Kenya (Vol. 26). Institute of Policy Analysis and Research.
18- Lawal, J.O., Omonona, B.T., Ajani, O.I.Y., & Oni, O.A. (2009). Effects of social capital on credit access among cocoa farming households in Osun State, Nigeria. Agricultural Journal, 4(4), 184-191.
19- Majumder, T.H., & Rahman M.M. (2011). Financial analysis of selected pharmaceutical companies in Bangladesh. European Journal of Business and Management. ISSN2222-1905(Paper) ISSN 2222-2839, 1-22.
20 Ministry of Niger Delta Affairs (2011). Federal Republic of Nigeria. Retrieved from: http:// www. MNDA.gov/ng/resources/download-resources.
21- Mwangi, I., & Shem, O. (2012). Social capital and access to credit in Kenya. American Journal of Social and Management Sciences, 3(1), 8-16.
22- Nwaru, J.C. (2004). Rural credit markets and resource use in arable crop production in Imo State of Nigeria. Ph.D Dissertation, Michael Okpara University of Agriculture, Umudike, Nigeria.
23-Nweze, N.J. (2002). Rural development in Nigeria: Past approaches, Emerging Issues and Strategies for the Future. Nigerian Journal of Cooperative Studies, 2(1), 41-45.
24- Ojameruaye, E. (2008). An action plan for Poverty reduction, peace and sustainable development in the Niger delta region. A memorandum submitted to the Niger Delta Technical Committee, September 25, 2008. International Foundation for Education & Self-Help (IFESH), USA.
25-Omafonmwan, S.I., & Odia, L.O. (2009). Oil exploitation and conflict in the Niger-delta region of Nigeria. Journal of Human Ecology, 26(1), 25-30.
26- Oni, O.A., Oladele, O.I., & Oyewole, I.K. (2005). Analysis of factors influencing loan default among poultry farmers in Ogun state, Nigeria. Journal of Central European Agriculture, 6 (4), 619-624.

27-Tawari, C.C. (2006). Effectiveness of agricultural agencies in fisheries management and production in the Niger Delta. Ph.D Thesis, Rivers State University of Science and Technology (RSUST) Port Harcourt, Nigeria(unpublished) 180p.

28-Togba, E.L. (2009). Microfinance, social capital and household access to credit; Evidence from Cote d' voire. Paper presented at the 7th international conference on Inclusivee growth, Inovation and Technological Change; Education, Social Capital and Sustainable Development. GLOBELICS, 6-8 October, 2009, Dakar, Senegal, 1-30.

29- United Nations Capital Development Fund UNCDF (2005). UNCF Business Plan 2005-2007. Investing in the LDCs to Achieve the Millennium Development Goals. Annual Session, 13-24 June, New York. DPP 2005, 1-20.

Investigation of the Potential Market and Estimation of WTP for Insurance of Pistachio Tree Trunk (Case Study Rafsanjan-Iran)

Mostafa Baniasadi [1]*, Saeed Yazdani [2] and Habib Allah Salami [2]

Abstract

Capacity of garden productions in Iran is such that is accounted as a country that produces thirteen garden products in the world but despite excellent condition in Iran for producing garden products, natural disasters damage production of fruits in the country therefore farmers incur a loss. Pistachio tree has been in danger of destruction and dryness. Thus, in order to reduce loss incurred on trees, it is necessary to insure the tree. This study is aimed to investigate factors affecting willingness towards insurance of pistachio tree and to estimate willingness to pay premium for pistachio tree in Rafsanjan located in Kerman province. For this purpose, methods of contingent valuation and double bounded dichotomous have been used. Research data were obtained by field method and interview with 184 pistachio gardeners in 2012. Results suggest that willingness to pay premium of pistachio tree in central part, Anar and Kashkuieh has been estimated by 1953, 3255.8 and 1183.3 IRR per tree respectively. Considering results and high risk destruction of pistachio trees, it is suggested that premium of pistachio tree is offered to reduce risk and loss of pistachio gardeners. In order to determine premium in Rafsanjan, WTP calculated in this study can be used.

Keywords:
Pistachio tree, Contingent valuation, WTP, Logit model, Rafsanjan

[1] *MSc Graduated of Agricultural Economics, Department of Agricultural Economics, University of Tehran.*
[2] *Professors of Agricultural Economics, Department of Agricultural Economics, University of Tehran*
* *Corresponding author's email: Baniasadi.m65@gmail.com*

INTRODUCTION

As one of poles in production of garden products, Iran has a special potential. Capacity and climatic varieties have provided production and development of numerous garden products. According current statistics, Iran has been ranked among first to seventh countries in the world for producing 14 garden products. 17% of agricultural lands are planted for gardens. 17.5% of total agricultural production, 13% of employment and 80% of total agricultural experts are allocated to garden subsector (Agricultural Ministry 2009). Pistachio production in Iran and Rafsanjan is famous throughout the world. Rafsanjan with planting area of 110000 hectares is the main center of pistachio production in the world, Iran and Kerman province such that contribution of this city to fertile planting area for this product has been 34 and 60% in the world and Iran respectively (Mirzaee and Heidari 2007). In recent years, production of pistachio is prone to various risks due to production, weather, technological and market uncertainties in Rafsanjan county. Also the pistachio trees face always with risk of drying and ruin. Among economical activities, agricultural activity faces the highest risk. Agricultural production is different from other commercial and productive activities. the most important difference is that this sector relies in the nature highly and faces a wide range of natural events and dangers such as flood, hail, cooling and warming, pests and plant diseases that changes activity in this sector into a high risk one (Anderson 2003). All these risks affect the income stability and welfare of their gardeners' households. Risk management plays a really important role in the development of agriculture. Considering risky condition and uncertainty, one of the ways to cope with this phenomenon in agricultural productions is insurance of agricultural products (Kiani Rad 2004). Public support to agricultural insurance is necessary for its development especially in the incipient stage which can encourage farmers to take an active role in risk management and participate in insurance systems. Insurance of agricultural products is a strategy for participation in risk taking so that by cooperation with producers in risky condition, losses incurred on producers will be prevented (Nelson and Loehman 1987). Unfortunately, tree trunk insurance has not been performed in Iran and also, no serious researches have been done yet. Pistachio tree is one of those that have always been in danger. Much money has been spent for years to have a fruitful tree. If this tree is destroyed by natural disasters after paying money for years, loss will be incurred on farmers. In addition to direct costs, opportunity cost and time consumed for this tree make the damage double. Assuring fruitful tree and other productive factors is accounted as one of goals of fourth economical development program (Baniasadi 2011). Therefore, high risk condition in the country, legal obligation and in order to invest on producing garden products, it seems necessary to design a wide insurance system for fruit trees in the country. Codification of such pattern for fruit trees of the country due to new subject of tree trunk insurance needs to many studies in order that this new service will start and implement in the country as a scientific base. The purpose of the study is investigation of potential market and effective factors on willingness to adoption of gardeners for the new insurance. Then, the willingness to pay of gardeners for tree insurance (including; determining price and premium) will be estimated. Establishing a price for a product is not always as straightforward as finding the intersection point of the supply and demand curves as taught in Microeconomics. One may run into particular difficulties when attempting to price a product which is a public or non-market good. Numerous methods have attempted to solve this problem e.g. hedonic pricing, cost-benefit analysis, travel cost and cost-effectiveness to name a few (Asfaw and von Braun 2005). Much of the current WTP literature uses CVM method which elicits directly what individuals would be willing to pay for a particular product or good (Wright *et al.*, 2009). In topic of tree insurance (destruction of tree trunk), Willingness to pay, analysis of the potential market for new insurance and effective factors on Willingness to adoption of this insurance, there is not any research background. In this paper, the CVM method is used to study the demand for new tree trunk insurance in Rafsanjan county of Iran.

Ahsan *et al.* (1989) found that record of confrontation with the danger is as one of the significant factors on adoption of the agricultural insurance. He believed that insurance of agricultural products is one of the main ways to reduce the fluctuation of profit and income changes. Shaik and Atwood (2003) examined effective factors on demand of insurance for cotton products by using Logit model and the results showed that producers of more efficiency and bigger farm have more willingness to insure their products. Ogursov and Marcel (2006) analyzed explanatory factors of buying insurance in Indian dairy section. The results showed that there is a direct relationship among number of cows, level of income, and extent of the farm with buying the insurance. Wang (2010) analyzes characteristics of farmer behavior in agricultural insurance and the factors affecting their behavior. He with using Von. Norman-Morgenstern Utility Model to analyze the risk preferences of individual farmer. The research result is that under the present stage, agricultural insurance behavior is influenced by many factors. In the voluntary insurance and nor a certain amount of subsidy, the vast majority of farmers would not choose insurance and the demand of agricultural insurance can only be regarded as a potential demand rather than effective demand.

Moreover, there are many studies about scrutiny of potential market and estimation of willingness to pay for species of insurances. For example in circle of health insurance, Wright *et al.*, (2009) analyze the willingness to pay for health insurance and hence the potential market for new low-cost health insurance product in Namibia, using the double bounded contingent valuation (DBCV) method. The findings of this study show that 87 percent of the uninsured respondents are willing to join the proposed health insurance scheme and on average are willing to insure 3.2 individuals (around 90 percent of the average family size). On average respondents are willing to pay NAD 48 per capita per month and respondents in the poorest income quintile are willing to pay up to 11.4 percent of their income. However, there is no study about insurance market and the willingness to pay for fruit tree trunk insurance. But, in other countries some studies have been done in this subject, that majority of studies are around agricultural insurances (except insurance of fruit tree). For example Fengli Xiu *et al.* (2012) analyze farmers' willingness to pay (WTP) for cow insurance and factors influencing farmers' participation in and WTP. The results of this study show that the current premium is in the range that famers can accept. The acceptance of premium and farmers' knowledge of allowance significantly influenced both of their participation and WTP.

Considering newness of tree insurance, formulating such model for trees requires numerous researches in order that this new initiative is performed on a scientific basis. In first step to perform tree insurance design, this study is aimed to investigate factors affecting willingness of farmers for accepting this new insurance. Since most of farmers resist against new ideas, in first step, factors affecting acceptance of tree insurance should be studied. Then willingness of farmers to pay for tree insurance is estimated. Information required for this research has been obtained by questionnaire in 2010-2011 in a field method from pistachio gardeners of Rafsanjan.

MATERIALS AND METHODS

Due to lack of time series data to measure risk of trees destruction' in order to determine a rate for insurance premium in Iran, one of the best methods for determine temporary of insurance premium is estimation of willingness to pay of gardeners for pistachio trees insurance. Estimating willingness to pay for insurance is a kind of economic valuation. Contingent valuation method (CVM) is one of the typical non-market valuation methods (Turner *et al.*, 2001; Batmane *et al.*, 2003). Because insurance of tree trunk is not supply and there is not any market for this insurance in Iran, contingent valuation method is used to determine the value of this kind of insurance. Value of goods or service (WTP) is gained in CVM by choice technique, which is the main factor in contingent valuation method (Mitchell and Carson 1989). In this study, double bounded dichotomous choice (DBDC) method was used. This method used for the first time by Hanemann *et al.* (1991). Double bounded method is used when data are normal. Double bounded dichotomous is statistically more efficient than single bounded technique (Hanemann *et al.*, 1991). To determine number of necessary sample from society studied in

CVM, researcher must determine of the society which is influenced by existence or non- existence new good or service (such as; the insurance of pistachio tree as a service that has not been supplied). After choice of society (that are pistachio gardeners of Rafsanjan county), number of sample is determined. In CV method, first some questionnaire is completed as a pre-test and then sample volume is calculated by using the coefficient of variation of WTP in following equation (Mitchell and Carson 1989):

$$n = \left[\frac{t \times \hat{V}}{d}\right]^2 \quad (1)$$

Where, n is the sample volume, t is t-student statistic (that in level of 5% is approximately equal to 1.96), $\hat{V}$ is the coefficient of variation of WTP, d is the error percentage of the difference between WTP calculated and real WTP. Acceptable amount of d is 0.3-0.5 in contingent valuation studies (Mitchell and Carson 1989). Then, with using equation (2) the number of samples based-on percent of under-sown level of each tree in every district of the county is determined:

$$n_l = n\frac{h}{H} \quad (2)$$

Where, n_l is the number of necessary sample in 1st district, h is the amount of under-sown pistachio level in 1st district, and H is the whole level of under-sown pistachio in Rafsanjan county. In double-bounded choice method, it is assumed that gardener bear utility functions. Each gardener is ready to pay some amount of his agricultural income to tree trunk insurance titled as proposed amount (B) and this usage from new insurance causes utility to be created for him. The amount of created utility due to the usage of tree trunk insurance is more than the case in which he doesn't use insurance and relation (3) shows it (Amirnejhad *et al.*, 2006; Judge *et al.*, 1998; Pattanayak and Evan Mercer 1998):

$$U(1, Y\text{-}B;S)+\varepsilon_1 \geq U(0, Y; S)+\varepsilon_0 \quad (3)$$

Where U is indirect utility function, Y is individual's income; S is a vector of other eco-social factors of individuals. and ε_1 and ε_0 are random variables with average 0 that have been distributed randomly independent of each other. Created difference in utility (ΔU) due to the effect of using tree trunk insurance is calculated from relation (4) (Amirnejhad *et al.*, 2006):

$$\Delta U = U(1, Y\text{-}B;S) - U(0,Y;S)+(\varepsilon_1\text{-}\varepsilon_0) \quad (4)$$

Double-bounded questionnaire structure in studying the WTP of individuals has a dependent variable with dual selection. Hence, logit model for studying the effect of different descriptive variables on the amount of WTP of gardener was used to determine the premium. According to the logit model, probability of acceptance of the proposed amount by individual is expressed as relation (5) (Amirnejhad *et al.*, 2006; Hanemann *et al.*, 1991; Bishop and Heberlein, 1979):

$$P_i = F_\eta(\Delta U) = \frac{1}{1+\exp(-(\Delta U)} = \frac{1}{1+\exp(-(\alpha+\beta B+\theta S))} \quad (5)$$

Where F_η (ΔU) is accumulative distribution function with a standard logistic difference and in this paper it includes some eco-social variables. β and θ are coefficients that can be estimated and it is expected that $\beta \leq 0$ and $\theta > 0$. In order to calculate WTP a method known as truncated mean WTP is used, because this method protects the stability and compatibility of limitations with theory, statistical effectiveness and aggregation. The expected amount of WTP in this method is calculated from relation 6 by numerical integration within the range of 0 to maximum bid (B) (Amirnejhad *et al.*, 2006; Batmane *et al.*, 1995):

$$E(WTP) = \int_0^{B_{max}} F_\eta(\Delta U)dB = \int_0^{B_{max}}\left[\frac{1}{1+\exp(-(A+\beta B))}\right]dB \quad (6)$$

Where B is bid amount variable and A is calculated from relation (7):

$$A=\alpha+\Sigma_i\beta_i M_i=\alpha+\beta_a M_a+\beta_{dL}M_{dL}+\beta_{dI}M_{dI}+\beta_y M_y + \beta_{nt}M_{nt}+\beta_{ce}M_{ce}+\beta_{ei}M_{ei}+\beta_{tl}M_{tl} \quad (7)$$

Where α is intercept of model and β_a, β_{dL}, β_{dI}, β_y, β_{nt}, β_{ce}, β_{ei}, β_{tl} are age of gardener, dummy variable of literacy, dummy variable of agricultural income, yield of pistachio, number of pistachio tree, dummy variable of crop insurance contract extension, qualitative variable of new insurance effect on damage reduction (Likert scale) and number of destroyed trees in the previous year, respectively and also M_a, M_{dL}, M_{dI}, M_y, M_{nt}, M_{ce},

M_{ei}, M_{tl} are their averages, respectively. Logit model could be estimated in linear or logarithmic form. The interpretation of two parameter is important in logit model results include elasticity and marginal effect. The elasticity of k explanatory variable (X_k) is as relation (8) (Hayati *et al.*, 2011; Kavoosi Kalashami *et al.*, 2012):

$$E = \frac{\partial(B'X_k)}{\partial X_k} \cdot \frac{X_k}{B'X_k} = \frac{e^{B'X_k}}{(1+e^{B'X_k})^2} B_K \frac{X_k}{B'X_k} \quad (8)$$

Elasticity of an explanatory variable explained the percentage change in the probability of bid acceptance for tree trunk insurance by individual when X_k amount changed by one percentage. Also, the marginal effect showed the percentage change in the probability of bid acceptance for green chicken buying by individual when Xk amount changed by one unit. The marginal effect of k explanatory variable (X_k) is as relation (9) (Kavoosi Kalashami *et al.*, 2012):

$$ME = \frac{\partial P_i}{\partial X_k} = \frac{\exp(B'X)}{(1+\exp(B'X))^2} . B_k \quad (9)$$

In above relation, the extent of change in the probability of bid acceptance depends on the initial probability and initial value of independent variables and their coefficients.

RESULTS AND DISCUSSION

Collecting data and researches process was done in Rafsanjan county where includes Central district and suburbs, Anar, Nogh and Koshkoo'iyeh. Before the completing of questionnaire, with regard to the study that for the first time in Iran has been performed; CVM questionnaire is provided by analyzing conditions of each area, other questionnaires related to the insurance and CVM, and also corresponding with experts and professors. First, 29 questionnaires is accomplished as a pre-test to determine all of the questionnaires required in Rafsanjan county. In this pre-test, the necessary samples of Rafsanjan County are determined 170 individuals and then 184 questionnaires are accomplished.

Table 1 shows some of the statistics about the variables of age, level of education, agricultural income, under-sown level, yield and number of pistachio trees of the under-question individuals.

To Investigation of willingness to pay for pistachio tree insurance, first we should see whether there is any destruction risk of pistachio tree and if so, whether there is any willingness to adoption for insurance of this risk or not. Table 2, showed some reports from responses of questions related to the subject.

As can be seen in table 2, destruction risk of pis-

Table 1: Statistical characteristics (economic- social) of quantitative variable.

Variables	Average	Min	Max	Standard deviation	Coefficient of variation
Age (year)	50.4	24	80	13.1	0.26
Level of education (Years of education)	7.9	0	18	5.8	0.73
Agricultural income (10 million rials)	23.7	0.6	300	37.9	1.61
under-sown level (hectare)	4.8	0.12	70	7.4	1.55
Yield (tone per hectare)	1.3	0.06	3	0.7	0.56
Number of pistachio tree in 2010	4361	100	60000	8429.2	1.93

Source: research findings

Table 2: Willingness to adoption of pistachio tree insurance

Question	Whether have been destroyed your trees because of existing risks?		Whether you desire to insure your trees? (willingness to insurance)	
Answer	Yes	No	Yes	No
Frequency	172	12	132	52
Percentage	93.5	6.5	71.7	28.3

Source: research findings

tachio tree is reported in a high level of the sample as 93.5% of gardeners were faced with destruction risk of their trees. Therefore, it is expected that gardeners will accept the new kind of insurance. In according to table 3, among 184 gardeners asked in this sample, 132 of them (approximately 72%) have willingness to insure their trees and demand this new kind of insurance.

The main question of questionnaire is the WTP amount of gardeners to insure pistachio tree that is suggested which close-ended and double bounded dichotomous choice method. The first bid is based-on mean of maximum WTP of gardeners with regard to the pre-test (Batemen *et al.,* 1995). Generally in pretest, median of willingness to pay is considered as the first bid amount. If the first bid amount is accepted, double of first bid amount is considered as the second bid. If the first bid is not accepted, the second bid amount will be half of the first bid amount (Batemen *et al.*, 1995). Table 3, showed bid amount and adoption or not adoption of gardeners for bid amount.

According to table 3, from 184 gardeners asked to interview, 67 of them (approximately 36.4%) accepted the first suggestion and 117 of them (approximately 63.6%) did not accept it. As generally, at least one of the 3 suggestions is accepted by 118 of owners of pistachio gardens (approximately 64.1%) and none of the suggestions is not accepted by 66 of owners of pistachio gardeners (approximately 35.9%).

Table 4 reports results of the model. According to the results of estimated model, variables

Table 3: The results of accepting of bid amounts by gardeners

Bid amounts (Rials)	Acceptatio		Not acceptation		Total	
	Frequency	Percentage	Frequency	Percentage	Frequency	Percentage
2000	67	36.4	117	63.6	184	100
1000	51	43.6	66	56.4	117	100
4000	28	41.8	39	58.2	67	100

Source: research findings

Table 4: Effective factors on probability of adoption of pistachio tree trunk insurance

Variable	Logit model			
	Coefficient	t-student statistic	Elasticity at mean	Marginal effect
Intercept	-1.95*	-2.38	-	-
Age	-0.016***	-1.60	-0.51	-0.004
DVL	0.68**	1.97	0.33	0.11
Agri-income	0.73*	2.47	0.18	0.16
Yield (tone per hectare)	0.44*	2.24	0.36	0.10
NPT	-0.00006*	-2.33	-0.17	-0.00001
DVCICE	0.93*	3.63	0.22	0.21
QVIEODR (Likert scale)	0.52*	4.31	0.95	0.12
Number of destroyed tree in 2010	0.00003	0.009	0.0004	0.000009
Bid amount (10 Rials)	-0.004*	-2.97	-0.46	-0.0008
Anar district	0.20	0.52	0.02	0.04
Koshkoo'iyeh district	0.53****	1.53	0.06	0.11
Nogh district	-0.80*	-2.18	-0.08	-0.13
Percentage of right predictions			70	
Likelihood ratio statistic			60.15	
P-value of Likelihood ratio			0.01	

Source: research findings. *** $P < 0.01$, ** $P < 0.05$, * $P < 0.10$

Table 5: The results of estimation of expected WTP

District	A	WTP for per tree (Rials)
Central section and Anar	0.42	2573
Koshkoo'iyeh	0.95	3548
Nogh	-0.37	1454

Source: research findings

of age in level of 10% and number of pistachio trees in level of 1% with a negative sign are significant. Also, dummy variable of literacy, dummy variable of agricultural income, yield of pistachio, number of pistachio trees, dummy variable of period extension of crop insurance contract and qualitative variable of new insurance effect on damage reduction with positive sign and orderly in level of 10, 1, 5, 5, 1 and 1 percentage are significant. Therefore, analyzing effects of change related to every descriptive variable on probability percentage of acceptable bid amount is done.

The percentage of correct prediction of model is 70%. Therefore, considered model is reliable for analysis of the next results. The amount of expected WTP is obtained from relation below:

$$WTP_R = \int_0^{50000} \left[\frac{1}{1+\exp(-(0.42+0.004B))} \right] dB = 2573.1$$

According to relation above, the amount of expected WTP of gardeners for pistachio trees insurance is calculated 2573 IRR for central district and its suburbs and Anar section. In regarding to sample studied, average of number of the pistachio trees for per hectare is 866 trees that with multiply this number at expected WTP of a tree, the premium for per hectare are achieved. Calculated insurance premium for per hectare of pistachio garden is 2228212 IRR in Central section and Anar.

To estimate WTP of gardeners in districts of Nogh and Koshkoo'iyeh, only coefficients of two these districts should have to put into an equation differently. Then, WTP is estimated. With regard to relation below, the amount of A is 0.95 and -0.37 for Koshkoo'iyeh and Nogh, respectively.

$A_{Kosh}=A+\beta_{Kosh}=0.95$

$A_{Nogh}=A+\beta_{Nogh}=-0.37$

Where, β_{Kosh} and β_{Nogh} are dummy variables coefficients of Koshkoo'iyeh and Nogh in model. According to relation below, WTP for pistachio tree insurance are calculated 3547.6 and 1453.9 rials in Koshkoo'iyeh and Nogh restrict, respectively.

$$WTP_{Kosh} = \int_0^{50000} \left[\frac{1}{1+\exp(-(0.95+0.004B))} \right] dB = 3547.6$$

$$WTP_{Nogh} = \int_0^{50000} \left[\frac{1}{1+\exp(-(-0.37+0.004B))} \right] dB = 1453.9$$

According to average of number of pistachio trees for per hectare, expected insurance premium for per hectare of pistachio garden are calculated 3072568 and 1259164 IRR in Koshkoo'iyeh and Nogh, respectively. The results of WTP estimated are shown in table 5.

CONCLUSION

Purpose of the study is examine potential market of insurance of pistachio tree trunk, estimate the maximum willingness to pay of gardeners for the insurance, also analyzing the effective factors on adoption of pistachio tree insurance in Rafsanjan county of Iran. Selecting pistachio tree is done in regard to high risk of destruction of this tree, under-sown level and the importance of pistachio product in economy of Rafsanjan county and Iran. The results show that pistachio tree encounter with high risk of tree destruction. To estimating of WTP, 4 district of central and its suburbs, Anar, Nogh, and Koshkoo'iyeh have been studied. According to surveys, Nogh and Koshkoo'iyeh have significant differences in WTP for the insurance rather than central section, whereas Anar has no significant difference. The expected WTP of gardeners is 2573 IRR in central district and its suburbs and Anar district, but it is 3547.6 and 1453.9 IRR in Koshkoo'iyeh and Nogh, respectively. Also, with regard to the average of number of pistachio trees per hectare (866 trees), the average of willingness to pay for insurance in three district of central section, Koshkoo'iyeh, and Nogh are estimated 2228218, 3072568, and 1259164, respectively. Therefore, there is a potential market to demand of pistachio tree trunk insurance in the county and agricultural insurance fund can use this potential and supply this new insurance. As above is said, it is concluded that the destruction risk of pistachio tree is high and every year gardener encounter with withering of number of pistachio tree that should cut their trees. The problem was severe for low-income gardeners (yeoman farmer) that own few pistachio trees and it is the only income source for them, as a result, it is required a supporting system, such as tree trunk insurance and this insurance is necessary. According to the results, due to inadequate information to measure risk and to determine of insurance premium by common methods, it is suggested to consider WTP of gardeners to insure their pistachio trees trunks as a temporary insurance premium until determining the fair insurance premium. To do experimental test of pistachio tree insurance and detection its problems, degree of efficacy and willingness of gardeners will be determined. According to considered elasticities, income of gardeners has significant effect on accepting the bid price to use new service for pestachio trees. Therefore, a policy advice to supply more efficiency of tree insurance is reinforcing the income levels by supporting production especially for low-income gardeners. To performing experimental of the tree insurance's project, it is begun from wealthy gardeners (great owners or esquire) because they have more income, so they will accept the insurance with higher probability. Positive attitude to new service of tree insurance influences on accepting of tree insurance and its bid prices. Therefore, before supplying this kind of insurance, with sending experts and promoter of the insurance in each area and introducing and making known this service and its requirements, agricultural insurance fund should make positive attitude towards tree insurance. Variable coefficient of crop insurance contract extension has significant and positive influence on accepting pistachio tree insurance. Extension of the period of of crop insurance contract considerably depends on economic logic, so client benefits from the contract and satisfies it. Therefore, it shows indirectly the influence of satisfied client from the general performance of insurance fund to attract gardeners' attention. The results of this study are usable for pistachio tree in Rafsanjan county, so it cannot be used for other trees and areas unless in areas which overlap the same characteristics. According to the characteristics of geographical, agricultural, and natural hazards each area, gardeners encounter with various destruction risks for fruit trees. So, it is suggested that the whole country will be categorized and estimated willingness to pay for every tree based on areas' features.

ACKNOWLEDGMENT

This study was supported by the Agricultural Insurance Fund thereby we appreciate from this institution.

REFERENCES

1- Agricultural ministry. (2009). Results of the sample program of garden products. Tehran: Office of statistical and information technology.

2- Ahsan, S.A., Ali A.G., & Kurian, N.G. (1987). Toward a theory of agricultural insurance. American Journal of Agricultural Economics, 69(3), 520-529.

3- Amirnejad, H., Khalilian, S., Assareh, M.H., & Ahmadian, M. (2006). Estimating the existence value of north forests of Iran by using a contingent valuation method. Ecological Economics, 58(4), 665-675.

4- Anderson, J.R. (1979). Impacts of climate variability in Australian agriculture. Review of Marketing and Agricultural Economics, 49(31), 147-177.

5- Asfaw, A., & von Braun, J. (2005). Innovations in health care financing: new evidence on the prospect of community health insurance schemes in the rural areas of Ethiopia. International journal of health care finance and economics, 5(3), 241-253.

6- Baniasadi, M. (2012). Effective factors on willingness to adoption for insurance of selected fruit trees in Kerman province. MSc thesis, University College of Agriculture and Natural Resources, Department of Agricultural Economics, University of Tehran.

7- Batmane, I.J., Lovett, A., & Brainard, J.S. (2003). Applied environmental economics, a GIS approach to cost-benefit analysis. Cambridge: Cambridge University press.

8- Batmane, I.J., Langford, I.H., Tuner, R.K., Willis, K.G., & Garrod, G.D. (1995). Elicitation and Truncation affects in contingent valuation studies. Ecological Economics, 12, 161-179.
9- Bishop, R., & Heberlein, T.A. (1979). Measuring values of extra-market goods: Are indirect measures baised?. American Journal of Agricultural Economics, 6, 926-930.
10- Fengli, X., Fengguang, X., & Siegfried, B. (2012). Farmers' willingness to pay for cow insurance in Shaanxi province, China. Procedia Economics and Finance, 1, 431-440.
11- Hanemann, M., Loomis, J., & Knninen, B. (1991). Statistical efficiency of double-bounded dichotomous choice contingent valuation. American Journal of Agricultural Economics, 73(4), 1255-1263.
12- Hanemann, W.M. (1992). Pricing in European environment. Oslo: Scandinavian University press.
13- Hanemann, W.M. (1984). Welfare evaluation in contingent valuation experiments with discrete responses. American Journal of Agricultural Economics, 66, 332-341.
14 Hayati, B.A., Ehsani, M., Ghahremanzadeh, M., Raheli, H., & Taghizadeh, M. (2010). Factors affecting the willingness to pay for Elgoli Tabriz park visitors. Iranian Journal of Economics and Agricultural Development (Agricultural Science and Technology), 24(1), 91-98.
15 Judge, G., Hill, R. C., Grifith, W. E., Lutkepphi, H., & Lee, T. C. (1988). Introduction to the theory and practice of Econometrics. 2nd Edition, New York: Wiley.
16 Kanninen, B.J. (1993). Design of sequential experiments for contingent valuation studies. Journal of Environmental Economics and Management, 25, 1-11.
17- Kavoosi Kalashami, M., Heydari, M., & Kazerani, H. (2012). Investigating consumers' willingness to pay for organic green ghicken in Iran (case study: Rasht city). International Journal of Agricultural Management and Development, 2(4), 235-241.
18- Kianirad, A. (2004). Determination of income insurance Pattern of selected agricultural products with emphasis on price risk importance. Ph.D thesis, University College of Agriculture and Natural Resources, Department of Agricultural Economics, University of Tehran.
19- Mirza'ee, H.R., & Chizari, A.H. (2008). Inputs management in pistachio production (case study Rafsanjan County). 6th Iranian conference of Agricultural Economics, Mashhad, Iran.
20- Mitchell, R.C., & Carson, R.T. (1989). Using surveys to value public goods: The contingent valuation method. Washington, D.C: Resources for the Futher.
21- Nelson, C.H., & Loehman, E.T. (1987). Further toward a theory of agricultural insurance. American Journal of Agricultural Economics, 69, 523-531.
22- Ogursov, A., & Marcel, V. (2006). Factors explaining farmer's insurance purchase in the Dutch dairy sector. 99th EAAE seminar.
23- Pattanayak, S., & Evan Mercer, D. (1998). Valuing soil conservation benefits of agroforestry: Contour hedgerows in the eastern Visayas Philippines. Agricultural Economics, 18, 31-46.
24- Shaik, S., & Atwood, J. (2003). Demand for optional units in crop insurance. American Agricultural Association Annual Meeting. Montreal, Canada, July 27-30.
25- Turner, R.K., Pearce, D., & Batmane, I.J. (2001). Environmental economics: An elementary introduction. New York: Harvester Wheatsheaf.
26- Wang, Q.S. (2010). The farmers behavior in agricultural insurance under the VonNeuman-Morgenstern utility model. International conference on agricultural risk and food security, 2010. Agriculture and Agricultural Science Procedia, 1, 226–229.
27- Wright, E.G., Asfaw, A., & Van der Gaag, J. (2009). Willingness to pay for health insurance: An analysis of the potential market for new low-cost health insurance products in Namibia. Social Science and Medicine, 69, 1351–1359.

The Economic and Welfare Effects of Different irrigation Water Pricing Methods, Case Study of Khomein Plain in Markazi Province of Iran

Gholamreza Zamanian [1], Mehdi Jafari[2] and Shahram Saeedian [3]*

Abstract

The scarcity of water resources and supply resources limitation, have caused an increasing gap between water supply and demand specially in recent decades in almost all regions of the globe. One of the best known solutions proposed by the economists is using the different water pricing approaches thereby obtaining the optimal allocation and social justice. To this purpose, this paper uses the positive Mathematical Programming (PMP) and Econometric Mathematical Programming (EMP) in a comparative analysis to study the economic and welfare impacts of alternative water pricing approaches in the agricultural sector during agricultural period 2011/2012 in Khomein plain of Markazi province in Iran. Results show that the EMP can be a better alternative approach instead of PMP to better analyze of agricultural policies. According to the final outcomes, it is suggested to apply the block tariff in place of volumetric pricing method to reach the optimal allocation and promoting the water efficiency in the price range of 198 to 853 Rials.

Keywords:
Economic and welfare effects, Water pricing, Mathematical programming, Khomein plain

[1] Assistant Professor in Economics, University of Sistan and Baluchestan, Iran.
[2] Ph.D Student of Agricultural Economics, University of Sistan and Baluchestan, Iran.
[3] Graduate student of Agricultural Economics, University of Sistan and Baluchestan, Iran.
* Corresponding author's email: m.jafari365@gmail.com

INTRODUCTION

During recent decades, as for to population growth and improving life standards, water demand faced a dazzling speed. To deal with this, primarily, the strategy of discovering and exploiting new water resources came up. Increasing healthy and reliable water resources, producing more food and electricity and rural economic development were some of the benefits of this policy. It has to be mentioned that, nowadays, more than 70 percent of irrigation water is provided from fresh water and the increasing growth of this demand can be easily predicted (Jafari 2013). Extending the exploitation of non-renewable water resources has been one of the main approaches to provide with this increscent water demand in Iran. However, continuing the expansionist policies which were followed previously, due to impairment losses of species, ecosystems and water resources pollution are not possible any more. Additionally, the difficulty of finding new water resources and also the externalities of constructing huge water projects has increased the marginal cost of water extracting. To solve these problems, water management pattern with brand new policies like as concentrating on productivity improvement, managing the water demand and reallocating water between consumers as more suitable solutions are changing. A substantial number of studies show that governments, in order to reach the optimum allocation and rising water productivity used some policies like decentralization of irrigation water management, pricing systems, water laws and commercial plans (see Dinar and Maria, 2005; Johansson *et al.*, 2002; Tiwari and Dinar, 2002; Tsure, 2004; Roe, 2005; Veettil, 2011 for surveys). What emerges from these studies is that the relationship between variables and different existent characteristics in agricultural environment as the special irrigation type, water laws, structural frameworks and alternative cropping systems can affect the results significantly. The results of Liao *et al.* (2007), Frija *et al.* (2008), Herrera *et al.* (2004), Speelman *et al.* (2010 and 2011) mentioned that farmers willingness to pay could be influenced by environmental conditions and when water laws are not defined properly, it leads to inefficiency of water pricing systems, non-optimum water allocation, increasing trade-off costs and expenditures and finally inappropriate evaluation of water resources (Fragoso and Marques, 2013). More than 60 percent of Iran, Markazy province and particularly Khomein region have a dry and semi-dry climate. Khomein as a flat and talented agricultural region as for 240 mm annual rainfall and also increasing population growth as well as extending agricultural activities, faces rising water demand and contrary to its shortage. With considering the substantial relationship between water resources stock and rainfall, surveys show that despite the moderate and extreme drought happened during 2008 to 2011 in khomein region, annual exploitation from underground water resources increased 13.1 million m3 on average (Mosayebi and Maleki, 2012) which led to completely drying of 164 deep and shallow wells, 172 Ghanats, 57 natural fountains, 38 rivers and 13 soiled-dams. And also 349 deep and shallow wells, 79 Ghanats, 21 natural fountains, five rivers and 5 soiled-dams have 1 to 10 liters in second water that are so likely to get dried in the near future (Agricultural organization of Markazi province, 2012). Water has been regarded as a free commodity in Iran, historically. The act of pricing this scarce input and increasing the current prices encounters many problems. Currently water pricing in Iranian agricultural sector is done on the basis of "Justly Distribution of Water" law and regarding the underlying crop. As in this system, pricing is not based on water consumption volume; there is not enough motivation to efficient and economic allocation of water and its marginal return is often higher than the price and providing and distributing costs. The extension limitation of water resources and weak management companies with huge water losses make the applying of water demand-side policies as complementary inputs taxes or product taxes unavoidable. These policies have been investigated by different researchers in Iran. Hossain zad (2004) and Asadi *et al.* (2007) showed that as for low elasticity of water demand in agricultural sector of Iran, increasing the price of this input decreases the water demand slightly. So, the water price has to be increased substantially or alternative policies are to be introduced. But, it is to be noted that efficiency improvement and water allocation without suitable economic policies and instruments is not conceivable. Moreover, the result of a policy or its impact depends highly on the farmers' reaction to the applied policies. The farmers' reaction is depended on farm condition, individual attitudes and characteristics. It is not possible to examine alternative policies in laboratory conditions and the policy maker is seeking to get a good intuition about policy implications in agricultural sector and farmers' reaction to the policies. To this end, this paper is going to study the different water pricing methods impacts on water demand, water allocation among irrigated agricultural crops, farmers' revenue, costs and other inputs demand in Khomein plain. The rest of the article is organized as follows: In part twowe discuss he analytical framework. Part three, elaborates the data and empirical models. Part four presents the empirical results and discussion and finally in part five concludes.

Analytical framework

The water demand and supply

Generally, the demand for Irrigation water is come from the market demand of agricultural products. Suppose a farm with n products and an input of water, the profit is defined as:

$$\pi = \sum_{j=1}^{n} \left[p_j f_j(q_j) - w q_j \right]$$

$j=1, 2, \ldots, n$ (1)

Where $fj(qj)=yj$ is an ascending and strictly concave function, j is production yield, pj indicates the market price of j-th product and water price is shown by w. Essential prerequisite to maximize the profit is as

$$\frac{\partial \pi}{\partial q_i} = 0 \text{ or } f'(q_j(w)) = \frac{w}{p} \Rightarrow q_j(w) = f_j'^{-1}(w/p) \quad (2)$$

Where $qj(w)$ shows the amount of entering water with price w. In other words, the water demand function of farmer is

$$q(w) = \sum_{j=1}^{n} q_j(w) = \sum_{j=1}^{n} f_j'^{-1}(w/p) \quad (3)$$

The individual water demand is specified by $qj(w)$ and the aggregative water demand for all farmers is the sum of individual demands presented as

$$q(w) = \sum_{i}^{k} q_i(w) = \sum_{i=1}^{k} \sum_{j=1}^{n} f_{ij}'^{-1}(w/p) \quad (4)$$

Water demand could be measured with considering it as a free commodity and also with supposing it limited in x liter. Here, the thing which seems important is that we have to know the farmers willingness to pay for Δ unit more water. When they use water at x level, their revenue is $p \times f(x)$. Expectedly, the additional income from using Δ unit more water is $p[f(x+\Delta)-f(x)]$. The additional income $pf(x)$ which is due to the little amount Δ, is indeed the maximum price that farmers are willing to pay for consuming additional units of irrigation water. This price is called Shadow price of water and its value is positive if the water constraint is binding. In other words, the problem of allocating the water between products can be solved by maximizing profit condition as:

$$\pi(p,x) = \max \sum_{j=1}^{n} p_j f_j(q_j)$$

$$s\,t \sum_{j=1}^{n} q_j \le x \quad (5)$$

Which its lagrangian form is as

$$\ell = p_j f_j(q_j) - \lambda \left[x - \sum_{j=1}^{n} q_j \right] \quad (6)$$

Where λ here is a coefficient of constraining factor of water, and shows the shadow price of it. This strategy can be applied for more inputs, variables, constraints, infinite puechased inputs and crops that use the water in their production process. In words, a combination of non-linear production functions with linear programming can be combined into a non-linear programming frame. Also, in both of these cases, the Irrigation water demand function could be obtained subject to maximizing the profit at different water levels which allows to obtain different allocative amounts of qj with shadow price of water λ. In the usual approach, the irrigation water demand can be extracted by regression analyzing of the observed information of water price and quantity. However, due to some problems like as unavailability of information and the variability of water price in small scales, this method causes imprecise estimations (Tsure, 2005).

Water pricing

The intersection between the non-decreasing marginal cost function and the descending slope derived from demand function determines the marginal cost of water. This happens In the great irrigation projects, when the average cost function is decreasing and the marginal cost curve placed under the average cost *(w* <AC (w*))*. Therefore, the real profit of supplier does not meet the fixed costs and in order to continue the activity in long-run, subsidies have got to be given to the suppliers. In the long-run, financing the suppliers' costs increases in order to cover their costs which often results in decreasing the average pricing costs. In this case, water price is set up following the exploited demand function and average cost. By the way, despite the farmer is able to return the overall water cost, but this policy is not efficient in the average pricing cost because it does not ensure the maximization of producer and farmers welfare. As it was mentioned by Tsure (2005), determining the water price by moving through the marginal cost curve toward the average cost could provide producers with positive profit and simultaneously decreases the farmers' profit. Since this decreasing profit is greater than that increasing profit, so the total welfare will experience a down fall by taking this method of pricing (Tsure, 2005). Hence, according to Tsure and Dinar (1997), water pricing on the basis of marginal cost can give the optimal water allocation but implementing this method requires some prohibitive operations like monitoring and management and collecting exact data. Thus, alternative water pricing methods are applied across the world including volumetric approach under which, the water costs are measured directly by estimating the water volume consumpted; input-output approach that irrigation water valuing is done based on products or inputs (except water) used in the production process ; regional method that water is priced on the basis of irrigation methods used in the region. Usually the differences in irrigation costs come as for the kind and amount of irrigation, irrigation method and irrigation season in a special region; blocking method in which variable volumetric tariffs are used proportional to an specified level of water consumption; two component tariff which usually comprises the pricing method based on marginal cost and annual fixed costs for water right (which has different values in each region depending on irrigation method); the last approach is the tax method in which water costs payments are considered according to the added value of the sown area which is caused by the irrigation water. Each of these water pricing methods, leads to different levels of welfare and net benefits and choosing one of them is based on the implementation costs which vary from one region to another as for the climatic issues, demographic, social structure, water rights, time and economic conditions. Thus, the pricing method is considered which has the most benefit. Without considering the implementation costs, one of the efficient approaches is the volumetric method. Tsure and Dinar (1997) compared the results of the volumetric and regional pricing methods. Results showed that if 7.5 percent of the outcome from water would used for operational expenditures, the regional pricing method has a much better return in comparison to the other approaches. Most notably, the supply and demand function and also the pricing methods based as the theoretical base of this paper are derived from Tsure (2000, 2005), Dinar (2000), Dinar and Maria (2005), Dinar and Mody (2004).

Positive mathematical programming (PMP) models

Recently, there appears an increasing interest to apply sort of generalized mathematical programming in agricultural sector. Heckely and Britz (2005) ratiocinate this interest by some reasons. First, an expanded range of political tools in addition to supportive policies based on pricing are come up. Also, as for to developing the multipurpose agriculture which is so important, it is more likely that with existing technical constraints, most of the old mathematical programming models give incoherent results. After presenting the positive mathematical programming by Howitt (1995) for calibration, it was applied in agriculture widely. Positive mathematical programming (PMP) was developed to overcome the difficulties of normative mathematical programming (Howitt, 1998). At most one concave profit function and the MC parameter as well in the non-linear variable cost function are used for PMP models. Therefore, this model is able to reproduce the observed situation and evaluate the policies and to suggest more reliable policies. The main model which was presented by Howitt (1995) had two components. The first component was a linear model with calibration constraints in order to build the dual value of resources and constraints and in the second component given obtained dual, calibration parameters (including MC as the coefficient of concave cost function(in the short-run as the coefficient of the non-linear profit function)) are estimated to maximize the model given the linear constraints. The main idea hidden in this method is using the dual values for non-linear calibration of the objective function in order to obtain the simple and exact basic situation observed data). Sabuhi *et al.* (2006) put it well"In order to specify the non-linear object function, each type of non-linear function which can set the marginal cost of preferential activities to their related prices in the level of observed activities, is possible to be utilized. So, in this study, for analyzing the policies we use the quadratic cost function which its general presentation is outlined in Heckely and Britz (2005), Henry *et al.* (2007) surveys. As it was shown in equations (1) and (11), these models show the maximum sum of farmers' surpluse.

$$\max \pi = gm'l \tag{7}$$

s.t

$Al \leq b[\lambda]$

$l \leq l^0 + \varepsilon[\rho]$

$l \geq 0$

Where π indicates the profit function in the short-run which is corresponding to the gross yield of farm in the short-run. *n* and *gm* present the vector of gross yield corresponding to each activity and the non-negative variable of sown area of each product, respectively; shows the technical coefficients matrix; *b* is the $m\times 1$ vector of available inputs (like land, water, labor and chemical fertilizer); indicates the $m\times 1$ vector of shadow prices for each input; ρ and l^0 present the $n\times 1$ vector of observed sown area for each product in the base year and the corresponding shadow price of it, respectively; ε is the littlest number as the calibration constraint which is used to prevent the linear dependency.

When ρj the the is determined then in the second step, using the PMP approach, the variables of non-linear cost function $C^v(l^0)$ are estimated in which the marginal cost of the activity $MC^v(l^0)$ has formed from two components: the known costs of activity *(c)* and the unknown marginal cost which are given below

$$MC^v = \frac{\partial c^v(l^0)}{\partial l} = d + Ql^0 = c + \rho \tag{8}$$

Where Q and d are a $n\times 1$ positive, determined and symmetric vector of linear coefficients and the quadratic matrix of variable. To simplify, the diagonal elements of matrix Q as for the standard estimation approach from $q_{jj}= \rho j/l_j^0$ for determining the quadratic function of costs are placed in the model below

$$\max \pi = gm'l - \frac{1}{2} l'Ql \qquad 0$$
$$s.t \quad Al \leq b[\lambda], \quad l \geq 0$$

The econometric mathematical programming (EMP) models

According to Heckely and Britz (2005), PMP approach faces some important limitations for instance the calibration constraints have to have the zero degree of freedom while this issue needs much of data or a so flexible functional form to cover all constraints. The other limitation is that different approaches to estimate the calibration parameters lead to considerable differences in simulation behavior. Buysse *et al.*, (2007) state that to obtain the more realistic simulation behavior, econometric programming models which can estimate the objective function and constraints given the external information are suitable alternatives for PMP models. The main axiom of this strategy using the lagrangian model is presented like below

$$\ell = gm'l - \frac{1}{2} l'Ql + \lambda(b - ul) \tag{10}$$

If the land is the only fixed resource, then $A=u$ and u is a $n\times 1$ vector of the sum of them. The first optimization condition is

$$\frac{\partial \ell}{\partial l} = gm - Ql - \lambda u = 0$$
$$\frac{\partial \ell}{\partial \lambda} = b - ul = 0 \qquad (11)$$

Thus, the unknown parameters λ and Q can be estimated using some of econometric measures. In the case that the observed data are less than the parameters that has to be estimated, we face the *III-posed* situation. In this case, using the generalized Maximum Entropy (GME) the aforementioned situation can be solved (Golan *et al.*, 1996). Similar to Heckely and Wolf (2003) combining information related to land demand elasticity extracted from the sample, we can have a better estimation. The simple structure of the GME model constrained to the optimization conditions used in the programming model is

$$\max_{w_t, w^e, Q, l, \lambda} H(w_t, w^e) = -\sum_{t=1}^{T} w_t' \ln(w_t) - w^{e'} \ln(w^e)$$
$$s\,t \qquad (12)$$
$$gm_t^0 - \lambda_t u - Q(l_t^0 - \varepsilon_t) = 0$$
$$u'(l_t^0 - \varepsilon_t) = b_t^0 \qquad (13)$$
$$\varepsilon_t = Vw_t = \sum_{s=1}^{2} \sigma_{jts} w_{jts} \qquad (14)$$
$$diag\left[E\right] = V^e w^e = diag\left[(Q^{-1} - Q^{-1}u (u'Q^{-1}u)^{-1} uQ^{-1})(\frac{\overline{gm^0}}{l^0})\right] \qquad (15)$$
$$Q = LL', \qquad L = 0 \ \forall \ i \geq j \qquad (16)$$
$$\sum_{s=1}^{2} w_{jts} = 1, \qquad \sum_{s=1}^{2} w_{js}^{e} = 1 \qquad (17)$$

Where H is the entropy variable, w_t and w^{ε} are the probabilities values as for to the error and the estimated elasticity E; gm_t^0 and l_t^0 are vector of production marginal yield and the production physical amount for each observation t, respectively; λ is the shadow price of fixed resources (like land); Q is the symmetric positive and determined matrix of production marginal cost coefficients; V and V^{ε} are the known matrix of errors and supportive values of elasticity. Equation (12) indicates the maximum entropy; (13) is the first optimization condition; (14) and (15) allow to calculate the error term (ε_t) and elasticity (E) as for to the second optimization condition that the variable cost function has to be non-descending. (16) is included to ensure the concavity of variable cost function and being positive and determined of Q; (17) makes sure that the sum of the probabilities and elasticity are equal to unity.

The stochastic errors of each observation (ε_t) have zero mean and a standard deviation of σ_{jts}. To apply the GME approach, it was necessary to carry out re-parameterization of the error term as expected values of a probability distribution (Vw_t). This is calculated based on known values of standard deviation, which are spread by two support points (the n×n×2 V matrix). Incorporation of out of sample information through the use of priors on elasticities allows us to obtain more accurate estimates for the Q matrix. In our case the elasticity estimates (E) are given by the product between the n×n Jacobian matrix of the land demand functions $\{Q^{-1}-Q^{-1}u(\acute{u}Q^{-1}u)^{-1}u^{-1}\}$ and the mean of observed gross profit divided by the mean of observed land allocation to crop $(\frac{\overline{gm^0}}{l^0})$. As for the error estimates, the elasticities (E) also have to be re-parameterized as the expected values of a probability distribution (we). In this case, for the central value of prior elasticities two support points were also considered and the values of standard deviations are bounded in the $n \times n \times 2V^e$ matrix (Fragoso and Marques, 2009).

After estimating the σ, Q, ε^t, w^e, wtvalues, the values are placed in the defined programming frame and it will be used to simulate the water pricing policies.

It is to be mentioned that in order to survey the different impacts of either of two EMP and PMP models on cropping pattern in this paper, after estimating the two models, we investigate the resulted diversity pattern. Genarally, for measuring the diversity of determined optimal cropping plans, regardless the different definitions that are presented for cropping diversity, we can measure it using two indexes including sown area and gross income. There exist Numerous indexes for calculating the diversity of a cropping plan. Shanon and bor, Simpson, Herfindal, Entropy and corrected concentration index are some of the most famouse indexes that are used to this end (Karbasi *et al.*, 2010). In this paper, we use Entropy Diversity index as it is used for large-scales like our case. This index is measured according to the following equation (Chang and Mishra, 2008):

$$EI = \sum_{i=1}^{i=n} \frac{X_i}{\sum_{i=1}^{i=n} X_i} \log\left(\frac{X_i}{\sum_{i=1}^{i=n} X_i}\right)^{-1} \quad (18)$$

Where X_i indicates the sown area of the activity. In this equation, if the *EI* is greater than zero, the cropping diversity is high and if it is equal to zero or less than it, there is no cropping diversity.

Data

In the current study, Khomein plain accommodated 7543 farmers, was used as the underlying statistical population and in order to collect the data related to the quantity of inputs consumption required for producing crops which are included as: water, labor, machinery, chemical fertilizer and maure and herbicide were gathered through a three-stage stratified sampling and given the Cochran-Orcut formulation according to number of farmers in rural districts and villages over 2011-2012 agricul-

Table 1: Descriptive statistics of variables

variable	mean	SD	Min	max
Yield				
Irrigated wheat	3113.7	1840.5	200	8000
Dry wheat	1157	909	125	4000
Irrigated barely	2615	1421	200	7000
Dry barely	1411	852	2000	8000
Dry pea	370	250	150	1350
Bean	2568	889	1000	5000
Potato	19470	12890	3333	32000
Onion	5000	0	5000	5000
Alfalfa	8696	5658	750	20000
Irrigated corn	40000	0	40000	40000
Inputs				
Labor(rial)	360163.8	453615.4	19845.24	6775431
Chemical Fertilizer phospat(kg/ha)	136.2	33.6	63.5	171.9
Chemical Fertilizer azot(kg/ha)	199	68.8	83.3	318.8
Animal fertilizer(kg/ha)	5527	6208	0	14428
Herbicide(kg/ha)	0.683	0.589	0	1.7
Machinery (rial)	63269.5	74904.9	4328.1	984926.1
Irrigated wheat	4933.3	56.6	4820	5010
Dry wheat	4773.6	169.41	4560	4890
Irrigated barely	5944	198.8	5370	5948
Dry barely	5944	0	5944	5944
Dry pea	5850	0	5850	5850
Bean	37100	2325	36400	42000
Potato	2100	0	2100	2100
Onion	6000	0	6000	6000
Alfalfa	6500	3250	3180	7000
Irrigated corn	1000	0	1000	1000
Labor(hour)	5945.9	1164.6	4926.1	7215
Chemical fertilizer(kg)	118.4	84.7	68.9	216.4
Animal fertilizer(kg)	9.8	1.03	8.7	10.8
Herbicide(kg)	5627.6	1708.2	4926	7575
Machinery (hour)	6566.5	651.8	5911.3	7215

Source: Research findings

tural year, 36, 41, 30, 32, 39, 45 and 27 questionnaires were distributed among Chahar cheshme, Khoram dasht, Ashena khor, Hamze loo, Rastagh, Salehan and Gale zan (Totally 250 questionnaires) rural districts, respectively. Farmers of each village were chosen by systematic sampling method so that on the basis of farmers number and sample size related to each rural district, the sample of each village was determined. Also, the information related to the sown area and the production quantity of the under study region was derived from the Markazi province's Jihad-e-Agriculture organization data bank The descriptive statistics are reported in the Table 1.

RESULTS

The results are presented in two sections. The first section compares the results from PMP and EMP approaches using observed data aimed to choose a model which is able to explain the farmers' behavior in the best way. The second section is related to survey the alternative water pricing policies on water consumption, irrigated land area, farm profit and total welfare.

Results of EMP and PMP

Results of measuring the Entropy index for the PMP and the EMP models are showing that changing the cropping pattern is on the basis of reproducing the PMP model and also decreasing the diversity of cropping pattern is based on the EMP model reproduction. In the next step, as for which model predicts the farmers' behavior, we compare the results exploited from the PMP and EMP models with observed data. Also in the second section of the results, evaluation of the impact of alternative water pricing policies on water consumption, the irrigated area, farm profit and total welfare is discussed.

Irrigation water demand

Given that Evaluating the irrigation water demand and irrigated land area is not only beneficial for choosing the best policy analysis model but it can be used for creating pricing scenario assumptions which are essential for simulation. In order to survey and compare the EMP and PMP models in Figures 1 and 2 respectively, we evaluate the water demand quantity and the percentage of irrigated land versus the shadow price in either of the models.

As shown in Figure 1, the PMP pattern has a more flexible curve than EMP which indicates that the water constraint has a more considerable effect in PMP model for simulation of products substitution. This result is more obvious in figure 2 where the sown area is stated as a function of shadow price. Although, the EMP curvature is more than PMP, but it is a more suitable model to predict farmers' behavior regarding policy changes. Evaluating the irrigation water

Table 2: Comparative results of PMP and EMP models in comparison to the base year.

Activity	Sown area in the base year(Hectare)	Models PMP	EMP	EMP-base
Irrigated wheat	8531	8531	8533	0.02
Dry wheat	9422	9422	9450	0.3
Irrigated barely	2708	2708	2700	-0.3
Dry barely	24	24	0	-100
Dry pea	151	151	157	0.03
Bean	3496	3496	3500	0.11
Potato	114	114	110	-3.51
Onion	266	266	293	10.5
Alfalfa	1494	1494	1498	0.27
Irrigated corn	35	35	0	-100
Total sown area	26241	26241	26241	0
Entropy Index	0.6576	0.6576	0.6572	-34
Water consumption('000 M3)	23558	23558	20106	-14.8
Dual value of land (Rial /hectare)	71685	71685	68385	-4.8

Source: Research findings

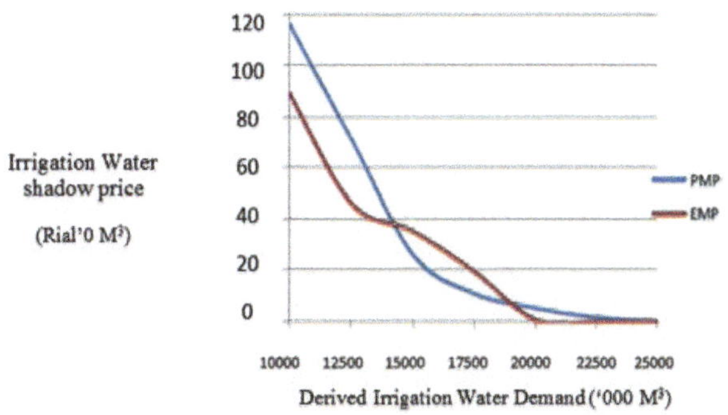

Figure 1: Derived Irrigation Water Demand from PMP and EMP models

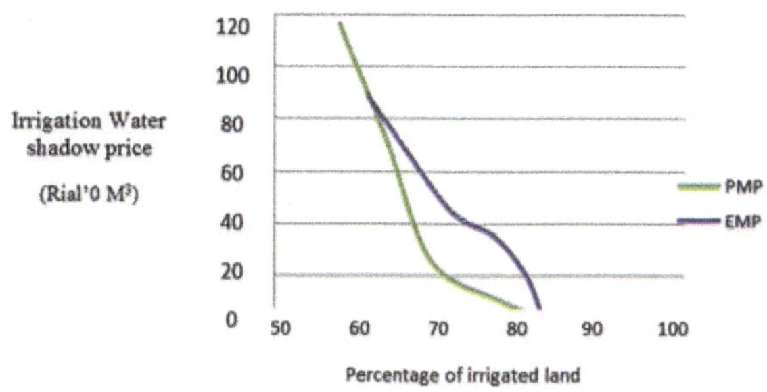

Figure 2: Percentage of irrigated land area from PMP and EMP models

demand and irrigated land area is not only beneficial to choose the best policy analysis model but it can be used for creating pricing scenario assumptions which are essential for simulation.

In the first part of demand curve i.e., 17500 to 25000 M3, The shadow price of water is in the range of 0 to 198 rials which have an elasticity of 0.008. In this price range, the changing percentage in water consumption is much less than the price changing percentage. In the second part of water demand curve i.e., 15000 to 17500 M3, price is in the range of 198.2 to 382 and the elasticity increases to 0.017 which means that more changing is expected from farmers regarding the price changes. In the third part and the availability range of 12500 to 15000 M3, price is between 382 and 420.3 and the elasticity comes out as 0.049. Interestingly, in the last part of demand curve, the elasticity increases to 0.108 which is the largest change in the consumption regarding to the price changes.

As for to the aforementioned results and the objective of this paper the simulation of the irrigation water pricing policies was done regarding the volumetric and the block tariff. In the volumetric tariff, simulation was done for 198, 382, 420.3 and 853.3 rials as optimal prices for each cubic meter of irrigation water. For the block method, the water costs were divided to three parts and in each part 50, 100 and 150 percent of water costs coverage was simulated.

Evaluation of the irrigation water pricing policies

In this section, the effect of water pricing policies on water consumption in all over the region,

Table 3: The water demand elasticity resulted from water demand for each shadow price.

Number	Water Demand	Water shadow price	Demand elasticity
1	17500-25000	0-198	0.008
2	15000-17500	198.2-382	0.017
3	12500-15000	382-420.3	0.049
4	10000-12500	420.3-853	0.108

Source: Research findings

Table 4:The economic effects of alternative irrigation water pricing policies using the EMP model

Policy	Welfare Changes	Sown area	Water consumption	Total Profit
Tariff 1	-24%	-16%	-8%	-1.5%
Tariff 2	-33%	-20%	-14%	-4%
Tariff 3	-42%	-38%	-19%	-25%
Tariff 4	-52.1%	-50%	-23%	-40%
Block Pricing	-30%	-17%	-21%	-26%

Source: Research findings

gross profit and total welfare of the society is tested. The final out comes are featured in the Table 4.

As featured in Table 4, total welfare (supplier plus consumer welfare) under the volumetric tariff of 198 rials, had the least reduction of 24% and in the second place, block tariff with 30 percent reduction proportional to the base year shows the least reduction in the total welfare. The distance between block and volumetric methods even exceeds 20 percent (for Tariff 4 case) which shows that generally, the block tariff provides a more satisfying total welfare level in comparison to the volumetric tariff. In the volumetric tariff of 198 rials, the water saving is 8% and in the block tariff, it reaches to 21%. As it is seen, the difference between these two policies in upper tariffs is negligible. The less reduction of farm profit is related to the tariffs of 198 and 382 rials for each cubic meter of volumetric method which shows 1.5% and 4% reduction, respectively. Considering the results, it can be said that in the lower tariffs, the block and volumetric tariffs effect on farm profit is almost identical but in the higher levels of pricing tariffs, the profit reduction in block tariff is less than volumetric one so that this difference in 420.3 and 853.3 rials for each cubic meter of water has 11% and 26% more reduction of profit. The sown area in the tariff of 198 rials for volumetric tariff is 84% and under block tariff is 83%. On the hand, attention has to be paid that increasing water price results in reducing the irrigated land areas thereby reducing the water consumption. Therefore, for this level of pricing tariff, block tariff allows to have a 3 percent water saving for 1 percent of reduction in the sown areas. Furthermore, placing 382, 420.3 and 853.3 tariffs will reduce the irrigated lands to 80, 62 and less than 50 percent, respectively.

CONCLUSION

In a brief summarizing, given the comparative results of the two models including Positive Mathematical Programming (PMP) and Econometric Mathematical programming (EMP) in reproducing the observed values and also the water demand and irrigated water amounts, it is understood that econometric mathematical programming model is more suitable and it is suggested to us this approach to better analyze the simulation of the effects of agricultural policies on farmers behavior. On the other hand, the simulation results show that pricing policies in irrigation sector are extremely affected by the local, structural and institutional situation. Also the pricing policies often are seeking objectives as economic efficiency, reducing costs, justice and resources conservation which are opposite to each other. The simulation analysis of alternative irrigation water pricing policies indicates that the block pricing policy is considerably capable to influence the allocation, efficiency improvement and water saving with taking into account of farmers' profit and the total welfare of suppliers and consumers. So, in order to sustaine the water resources and management and influencial reduction in irrigation water demand, it needs to incease the water price significantly but this plan will face serious reactions by beneficiaries of surface water and and also the policy makers and administrative authorities related to water issue. Given the affornantioned issues, there appears that for properly managing the water demand, they should go forward with accurate planning and scheduling the water price increasing (so that, the average water price approaches to long-run marginal cost companied with reforming the economic structure of the country) and consolidated which are likely to improve the irrigation water demand management. In this line, management and planning the water resources distribution and also correcting the water rules and presenting a suitable pattern of determining the rate of water price as block tarrif which is appropriate for the khomein region or other similar plains are the most influencial policies to reconstruct a progressive irrigation water managent system.

REFERENCES

1- Agricultural Static Yearbook, (2012), Agricultural Organization of Markazi Province.

2-Asadi, H., Soltani, Gh., & torkamaani, J.(2007). Irrigation water pricing in Iran: A case study on land downstream of Taleghan dam. Agricultural Economic and Development (Special agricultural policy), 15: 58.

3- Buysse, J., Van Huylenbroeck, G., & Lauwers, L. (2007). Normative, positive and econometric mathematical programming as tools for incorporation of

multi-functionality in agricultural policy modelling. Agriculture, Ecosystems and Environment, 120, 70-81.
4- Chang, H., & Mishra. P.L. (2008). Impact off-farm labor supply on food expenditures of the farm household. Journal of Food Policy, 51, 619-624.
5- Dinar, A., & Maria-Saleth, R. (2005). Issues in Water Pricing Reforms: From Getting Correct Prices to Setting Appropriate Institutions. In Folmer, H. and Tietenberg, T. (Eds.). International Yearbook of Environmental and Resource Economics New Horizons in Environmental Economics, Edward Elgar Publishing, Cheltenham, UK.
6- Dinar, A., & Mody, J. (2004). Irrigation water management policies: Pricing and allocation principles and implementation experiences. Natural Resources Forum, 28, 112-122.
7- Fragoso, R., & Marques, C. (2009). Avaliação económica de tarifas de água no uso agrícola: um estudo de caso no Sul de Portugal. Revista de Economia e Sociologia Rural, 47(3), 699-718.
8- Frija, A., Chebil, A., Speelman, S., & Van Huylenbroeck, G. (2008). Effect of changes in the institutional structure of irrigation water property rights on the willingness to pay of farmers for water: case of Tunisia. XIIth Congress of European Association of Agricultural Economics. EAAE, Ghent, Belgium.
9- Golan, A., Judge, G., & Miller, D. (1996). Maximum Entropy Econometrics, Wiley, Chichester United Kingdom.
10- Heckelei, T., & Britz, W. (2005). Models based on positive mathematical programming: state of art and further extensions. In: Arfini, F. (ed.) Modelling Agricultural Policies: State of Art, New Challenges. Monte Universita Parma, Parma, 48-74.
11- Heckelei, T., & Wolff, H. (2003). Estimation of constrained optimization models for agricultural supply analysis based on generalized maximum entropy. European Review of Agricultural Economics, 30(1), 27-50.
12- Henry de Frahan, Buysse, B., Polomé, J., Fernagut, B., Harmignie, O., Lauwers, L., Van Huylenbroeck, G., & Van Meensel, J. (2007). Positive mathematical programming for agriculture and environmental policy analysis: review and practice. In: Weintraub, A., Bjorndal, T., Epstein, R., & Romero, C. (eds.), Management of Natural Resources: A Handbook of Operations Research Models, Algorithms and Implementations. Kluwer Academic Publihers.
13- Herrera, P.A., van Huylenbroeck, G., & Espinel, R.L. (2004). An application of the contingent valuation method to assess the efficiency of the institutional structure of irrigation property rights: the case of the peninsula of Santa Elena, Ecuador. International Journal of Water Resources Development, 20: 537-551.
14- Howitt, R. (1995). Positive mathematical programming. American Journal of Agricultural Economics, 77(2), 329-342.
15- Hussein zad, J. (2004). The role of agricultural water demand management policy of the precious, Iranian Water Resources Management Conference, University of Tabriz.
16- Jafari, M. (2013). Investigating the effect of drought and increasing price of agricultural inputs on cropping patterns of Khomein plain. The Dissertation of M.Sc. in Agricultural Economics Engineering, University of Sistan and Baluchestan.
17- Johansson, R.C., Tsur, Y., Roe, T.L., Doukkali, R., & Dinar, A. (2002). Pricing irrigation water: a review of theory and practice. Water Police, 4, 173-199.
18- Karbasi, A., & Falsafizade, N. (2010). Factors affecting the diversity of agricultural crops, Journal of Agricultural Economics Research, Vol 4.
19- Liao, Y.S., Giordano, M., & Fraiture, C. (2007). An empirical analysis of the impacts of irrigation pricing reforms in China. Water Policy, 9, 45-60.
20- Mills, G. (1984). Optimization in Economic Analysis. George Allen & Unwin Ltd, London.
21- Mosayebi, M., & Maliki, M. (2012). Evaluate the status of water resources in terms of droughts and wet periods (Case Study: Watershed Khomeini turret - Central Province), First National Conference on Wilderness, Wilderness International Research Center, Tehran University.
22- Ray, I. (2007). Get the prices right: a model of water prices and irrigation efficiency in Maharashtra, India. In Molle, F., Berkoff, J. (eds.), Irrigation water pricing: The gap between theory and practice. CAB International, Oxfordshire, UK.
23- Roe, T., Dinar, A., Tsur, Y., & Diao, X. (2005). Feedback links between economy wide and farm-level policies: With application to irrigation water management in Morocco. Journal of Policy Modeling, 27, 905-928.
24- Rogers, P., Bhatia, R., & Huber, A. (1998). Water as a social and economic good: How to put the principle into practice. Global Water Partnership/ Swedish International Development Cooperation Agency, Stockholm, Sweden.
25- Rogers, P., Silva, R., & Bhatia, R. (2002). Water

is an economic good: How to use prices to promote equity, efficiency, and sustainability. Water Policy, 4, 1-17.

26- Speelman, S., Farolfi, S., Frija, A., D'Haese, M., & D'haese, L. (2010a). The impact of the water rights system on smallholder irrigators' willingness to pay water in Limpopo province, South Africa. Environment and Development Economics, 15, 465-483.

27- Speelman, S., Frija, A., Buysse, J., & Van-Huylenbroeck, G. (2011). The importance of irrigation water rights: lessons from South Africa and Tunisia. Water Policy, 13(5), 1-14.

28- Tiwari, D., & Dinar, A. (2002). Balancing future food demand and water supply: The role of economic incentives in irrigated agriculture. Quarterly Journal of International Agriculture, 41(1-2), 77-97.

29- Tsur, Y., & Dinar, A. (1997). On the relative efficiency of alternative methods for pricing irrigation water and their implementation. World Bank Economic Review, 11, 243-262.

30- Tsur, Y., Roe, T.L., Doukkali, R.M., & Dinar, A. (2004). Pricing irrigation water: Principles and cases from developing countries. RFF Press, Washington, DC, USA

31- Tsure, Y. (2005). Economic aspects of irrigation water pricing. Canadian Water Resources Journal, 30(1), 31-46.

32- Veettil, P.C., Speelman, S., Frija, A., Buysse, J., &Van-Huylenbroeck, G. (2011). Complementarity between water pricing, water rights and local water governance: A Bayesian analysis of choice behaviour of farmers in the Krishna river basin, India. Ecological Economics, 70, 1756-1766.

33- Ward, F.A. (2009). Economics in integrated water management. Environmental Modelling and Software, 24, 948-958.

Research Performance of Agriculture Faculty Members: A Comparative Study at West Part of Iran

Nematollah Shiri [1], Nader Naderi [2] and Ahmad Rezvanfar [3]*

Abstract

Based on personal and professional characteristics, the present study compares the research performance among faculty members of agricultural colleges in west part of Iran. The statistical population of this study consisted of all faculty members in the agricultural colleges of universities of Ilam, Razi and Kurdistan at Iran, which 116 faculty members were selected as the sample using the proportionate stratified random sampling method. The main instrument in this study was questionnaire which its validity was confirmed by the panel of experts. The data was analyzed using descriptive and inferential statistics with SPSSWin20 software. Results showed that the present status of research performance among faculty members of agricultural colleges in west part of Iran was weak. Results of mean comparisons showed that there was significant difference between research performance based on age, work experience, academic degree, educational group and gender variables. Findings of this study can pave the way for formulating sound programs in higher agricultural education system to promote research performance among faculty members of agricultural colleges.

Keywords:
Research Performance, Personal and Professional Characteristics, Agriculture Faculty Members

[1] *Ph.D. Student, Department of Agricultural Extension & Education, Razi University, Kermanshah, Iran.*
[2] *Assistant Professor, Department of Agricultural Extension & Education, Razi University, Kermanshah, Iran.*
[3] *Professor, Department of Agricultural Extension & Education, University of Tehran, Karaj, Iran.*
* *Corresponding author's email: nshiri82@gmail.com*

INTRODUCTION

Scientific and technical capability in the production of knowledge and its application in practice can be considered as the most obvious indicator of development in any country (Najafipour *et al.*, 2009). As such, in the twenty-first century perspective, the promotion of academic research excellence is considered as one of the overriding goals of the university (Tien, 2007). Because, in higher education system, the research performance plays an important role in promotion, tenure and salary and also measured as the main indicator of success in universities (Blonedell, 2001; Kotrlik *et al.*, 2002; Wichian *et al.*, 2009). Due to increasing changes in response to various fields of agricultural science, the higher agricultural education system needs to maintain and enhance the research quality (FAO, 1997). FAO notes that, consistent with other research, the agricultural sciences have undergone several changes. Therefore, aligning the universities' scientific members and agricultural higher education centers with new paradigms and exchange of ideas, scientific meetings and using research findings can significantly develop agricultural higher education institutions and centers (Movahedi *et al.*, 2012). Research performance in universities and higher education institutions is a multidimensional concept that includes several indicators (Tien, 2007). Research performance is the one of the main aspects of the academic performance that plays an important role in the academic ranking of universities (Jung, 2012; Shin and Cumming, 2010). In a broad definition, research performance can include refereed publications, library and field articles, book chapters and monographies (Ransdell, 2001). In the other definitions, research performance also can covers categories such as: research reports published in national and international journals, presentations, patents, citations of articles and rewards (Zainab, 2000).

Researchers mainly measure research performance with calculating and combining the indicators derived from the sum of the number of all completed research reports, the number of published research reports and used research reports (Wichian *et al.*, 2009). In turn, most studies have used the number of categories such as, books, articles, conferences and research projects to assess the research performance among faculty members in universities (Jung, 2012; Hedjazi and Behravan, 2011; Shin and Cummings, 2010; Wichian, 2009; Law and Chon, 2007; Zhao and Ritchie, 2006; Bowen, 2005; Sax *et al.*, 2002; Changsrisang, 2002; Bouden and Cilliers, 2001; Taylor, 2001). In the present study, therefore, we also evaluate the research performance among faculty members of agricultural colleges in west part of Iran, with using most important and basic indicators of research performance, i.e., books, articles, conferences and research projects.

The importance of research on the growth and development of communities is critically important. As societies develop, they must improve its position more than anything else with deepening their research and development (Karimian *et al.*, 2011). In Iran, more than 70 percent of the research capability of researchers is concentrated in the universities and research institutions (Hosseinpour, 2011). According to statistics, the number of documents indexed in 2008 at Iran, was 13,568 cases and shows that on average, every four faculty members have produced a document (Saburi, 2009). The same ratio is 40 at Thai public universities (Wichian *et al.*, 2009). Although, in the recent years, we can observe that there has been relatively suitable growth of research activities at Iran, but on a global scale, comparative comparison of the research indicators suggest that the utility of these indicators are still not enough (Karimian *et al.*, 2011). Turkey has a considerable distance from Iran, yielded first rank, when compared to other countries in the region (Saburi, 2009). Therefore, because universities and higher education centers have required resources, specialists, research facilities and also have important mission of knowledge production, they are more responsive to current gap than other parties (Toreghi, 2005). Moreover, due to universities have important mission toward realization of the national aspirations, they are more inclined to increasingly improve their dynamic production of the science and research (Karimian *et al.*, 2011). In this regard, researchers believe that Iran, however, has the capacity, talent and important intellectual capitals and the field is ready for a huge scientific leaps, but now more than ever the question is that why the growth of academic research in Iran is not enough? Researchers addressed this question form different perspectives and found that knowledge production is influenced by personal and professional characteristics (Callcut *et al.*, 2004; Castill and Cano, 2004; Smeby and Try, 2005; Ouimet, 2005; Lertputtarak, 2008; Wichian *et al.*, 2009; Jung, 2012).

Despite the importance of personal and professional characteristics in the research performance, to date, there is no study to profoundly study the effects of these characteristics on the research performance among faculty members of Iran's universities and higher education instructions. Already, it is generally accepted that in the light of an effective and efficient system of higher education, holistic development is as possible as other fields. Given the importance of the research problem and extant literature, in this study, the research performance among faculty members of agricultural colleges in west part of Iran were studied according to their personal and professional characteristics. Therefore, with focus on personal and professional characteristics, the present study aimed to compare the research performance among faculty members of agricultural colleges in west part of Iran. Also, the derived specific objectives of the study are as follow:

1- Investigate the faculty members' personal and professional characteristics;

2- Investigate the current status of research performance among faculty members;

3-Compare the research performance among faculty members based on their personal and professional characteristics.

MATERIALS AND METHODS

This study categorizes in applied and descriptive-survey studies and used quantitative research paradigm. The statistical population consisted of all agricultural faculty members of universities, Ilam (31), Razi (59) and Kurdistan (47) at Iran (N=137). Using the sampling table (Patten, 2002), 116 (26 Ilam University, 51 Razi University, 39 Kurdistan University), were selected via the proportionate stratified random sampling method (n=116). The main research instrument for data collection was a questionnaire consisted of two parts, which first section includes personal and professional characteristics. Through a systematic review of the literature, in the second section, we applied four indicators (i.e., article, conference, research project and book) to measure research performance. The data concerning the research performance of faculty members in 2011 and 2012 was extracted from personal and research files in the form of a documentary study. Validity of the questionnaire was assessed through panel of expert in department of agricultural extension and education faculty members and education and psychology of university of Tehran. SPSSWin20 software was used to analyze the data in two parts of descriptive (Frequency, percentage, mean and standard deviation) and inferential (Tests of mean comparison) statistics.

RESULTS

Personal and professional characteristics

Based on the findings, the average age of faculty members was 40.5 years (SD=8.19) and with the age range 29 to 67 years, which most of them (45.7 %) categorized in the age stratum 39 to 48 years. Also, the average work experience of the faculty members was 10.16 years (SD= 7.47) and with the age range 1 to 30 years, which most of them (60.3%) categorized in the work experience stratum 10 years and less than 10 years. Furthermore, based on the findings, 25 percent of the faculty members (29 cases) were working in the Department of Agronomy and Plant Breeding and 2.6 percent of them (n=3) were working in the Department of Science and Food Industry. Other personal and professional characteristics of faculty members were shown in Table 1.

Research performance

In order to assess the research performance of the faculty members, four indicators of articles, conferences, research projects, and books were used. The results of the prioritization of indicators to measure the research performance is presented in Table 2.

Based on the findings presented in Table 2, among the four indicators of measuring research performance, conference is located at top priority, while book is last priority. Overall, the av-

Table1: Descriptive statistics of respondents regarding their personal and professional characteristics

Variable	Category	Frequency	Percent (%)
Gender	Male	108	93.1
	Female	8	6.9
Marital status	Married	96	82.8
	Single	20	17.2
Academic degree	Assistant	102	87.9
	Associate	8	6.9
	Professor	6	5.2
University	Ilam	26	22.4
	Razi	51	44.0
	Kurdistan	39	33.6

erage research performance of faculty members was 2.71(lower than mean=5.28) with a coefficient of variation of 1.87. These findings suggest that the research performance among faculty members of agricultural colleges in west part of Iran is weak.

Comparison of the research performance based on personal and professional characteristics

In order to compare the research performance of faculty members based on age, work experience, university, academic degree and educational group variables, we applied Kruskal - Wallis test (Table 3). As findings show, there is significant difference in the research performance of faculty members based on age, work experience, academic degree and educational group. According to ranking mean, faculty members who located in the age class of 59 years and more, have more research performance than other faculty members. Faculty members with work experience class of 11 to 20 years, show higher research performance than other faculty members. Faculty members, who possess an academic degree of associate professor, are more likely to show research performance than other faculty members. Finally, faculty members who were working in the department of agricultural extension and education, have more research performance than other their counterparts.

To compare the research performance of faculty members based on gender, marital status, and using sabbatical variables, we applied Mann-Whitney Test (Table 4). Surprisingly, our findings indicate that there is no significant difference in the research performance of faculty members based on their marital status and using sabbatical. However, there was significant difference between faculty members on their gender with higher performance of male faculty members than their counterparts.

Finally, we include graduate university as independent variable into independent t-test in order to compare the research performance of faculty members (Table 5). The results presented in Table 5, indicate that there is no significant difference in the research performance of faculty members based on the grouping variable of graduate university.

DISCUSSION

Faculty members of Iranian higher agricultural education system have crucial role of accelerating the development process through knowledge

Table 2: Prioritization of indicators assessing research performancefessional characteristics

Indicators	Mean	C.V.	Min	Max	Priority
Article	2.37	2.22	0.00	13.50	2
Conference	6.32	4.60	0.00	19.00	1
Research project	1.65	1.56	0.00	9.00	3
Book	0.43	0.60	0.00	3.00	4
Research performance (Total)	2.71	1.87	0.50	10.56	-

Table 3: Comparison of research performance of respondents related to their personal and professional characteristics

Independent variable	Category	Frequency	Ranking Mean	Kruskal-Wallis Test	Significant Level
Age	up to 38	49	47.58	13.628**	0.003
	39-48	53	62.00		
	49-58	7	80.00		
	more than 59	7	86.93		
Work experience	up to 10	70	51.25	8.266*	0.016
	11-20	35	70.13		
	more than 21	11	67.64		
University	Ilam	26	66.35		0.251
	Razi	51	59.23		
	Kurdistan	39	52.32	2.761	
Academic degree	Assistant	102	54.62		0.003
	Associate	8	90.31		
	Professor	6	82.08	11.491**	
Educational group	Agri. Extension	7	92.71		0.001
	Agronomy	29	69.48		
	Plant Protection	11	68.18	27.268**	
	Irrigation	13	49.58		
	Animal Science	23	64.26		
	Agri. Mechanics	8	41.75		
	Soil Science	10	61.30		
	Horticulture	7	12.14		
	Agri. Economics	5	53.00		
	Food Science	3	50.33		

** $P < 0.05$, * $P < 0.10$

production and its transfer to their clients. Therefore, understanding the factors that affect the academic success and performance are critically important. In this regard, the present study conducted to compare the research performance among faculty members of agricultural colleges in west part of Iran, based on their personal and professional characteristics. Findings of this study could increase our understanding of the personal and professional factors affecting the research performance and would help planners of universities to develop coherent programs for promoting research performance.

Findings showed that the age, as a personal factor, has a major role in the research performance of faculty members. Older faculty members were more likely to show higher levels of research performance. This finding is corresponds with previous studies, such as Callcut *et al.,* (2004), Castill and Cano (2004) and Smeby and Try (2005). Hence, as our findings show, we argue that one of the reasons for the poor research performance among faculty members of agricultural colleges in west part of Iran is the effect of age on the research performance. Therefore, this study encourages planners of Iranian agricultural higher education system to develop a systematic program in which younger faculty members can benefit from the experiences of older faculty members.

Results showed that the work experience, also, plays an important role in the research performance of faculty members, so that faculty members with more work experience have more research performance than their counterparts. This finding can be dovetailed with of the studies such as, Callcut *et al.*, (2004), Castill and Cano (2004) and Jung (2012). Hence, we can state that one of the other reasons for the poor performance among faculty members of agricultural colleges in west part of Iran is their weak work experience and, in this regard, we suggest that the participatory culture should be encouraged, in which, all experienced and less experienced faculty members will have more opportunities to work together and use and exchange their expe-

Table 4. Comparison of research performance of respondents related to their gender, marital status, and using sabbatical.

Independent variable	Category	Frequency	Ranking Mean	Kruskal-Wallis Test	Significant Level
Gender	Male	108	61.08	153.000***	0.002
	Female	8	23.63		
Marital status	Married	96	60.14	803.000	0.251
	Single	20	50.65		
Using studying opportunities	Yes	17	69.56	653.500	0.142
	No	99	56.60		

** P < 0.01.

Table 5. Comparison of research performance of respondents related to their graduate university.

Independent Variable	Category	Frequency	Mean	SD	t	Significant level
Graduate University	Interior	77	9.96	7.416	-1.685	0.095
	Abroad	39	12.44	7.610		

** P < 0.01.

riences.

Our findings indicated that the academic degree, as a professional factor, can significantly contribute to research performance. Faculty members with an academic degree of associate professor were as being more research performance than their counterparts. This finding is consistent with Smeby and Try (2005). Hence, given that most faculty members among agricultural colleges in west part of Iran were placed in assistant professor degree, we can say that one of the factors contributing to their poor performance is low academic degree (Assistant) in the majority of them. In this regard, having professor academic degree could be the most important incentive for faculty members with academic degree of associate professor in order to proceed to further promotions, however, faculty members with academic degree of assistant professor have a great distance from those faculty members, already, possesses an academic degree of professor, and, in turn, have no incentive to conduct research in order to be upgraded. Therefore, the planners of Iranian agricultural higher education system can take measures such as increase in salary and welfare facilities, if we would expect that assistant faculty members should be more active in the field of research and knowledge production.

Results showed that the educational group of faculty members can have a major role in their research performance. Faculty members who were working in the department of agricultural extension and education show more research performance than their counterparts. This finding could be due to the nature of farming fields and the conditions and facilities for research that they are primarily needed. Accordingly, researchers who are working in the field of agricultural extension and education are more active and productive in poor laboratory facilities, due to they are often interested to the farming conditions of social, cultural and economic, and are mainly applied a non-experimental design. Therefore, it is recommended that planners of agricultural higher education system can improve the research performance of all faculty members with providing research equipment and facilities required for the knowledge production.

Finally, our results showed that the gender can significantly affect in the research performance, in that, male faculty members were more likely to show higher research performance than their counterparts, which is congruent with findings of Castill and Cano (2004), Jung (2012) and Ouimet (2005). This finding may be because of the female faculty members at Iran have been faced with two major obstacles for scientific work in universities, i.e., being as busy because of probably much work in the home and other motivational and cultural restrictions for work in social environments. Therefore, we suggest that planners of Iranian agricultural higher education system take necessary actions to eliminate the motivational and cultural barriers affecting the participation of female's faculty members in academic research activities.

REFRENCES

1- Boaden, R.F., & Cilliers, F.F. (2001). Quality and the research assessment exercise. Quality Assurance in Education, 9(1), 5–13.

2- Bloedel, J.R. (2001). Judging research productivity on entrepreneurial campuses. Evaluation Research Productivity. 105. Available: http:// merrill. ku. edu/ publications/2001whitepaper/bloedel.html

3- Bowen, J.T. (2005). Managing a research career. International Journal of Contemporary Hospitality Management, 17(7), 633–637.

4- Callcut, R.A., Rikkers, L., Lewis, B., & Chen, H. (2004). Dose academic advancement impact teaching performance of surgical faculty? Surgery, 136(2), 277-281.

5- Castillo, J.X., & Cano, J. (2004) Factors explaining job satisfaction among faculty. Journal of Agricultural Education, 45(3), 65-74.

6- Changsrisang, A. (2002). Factors that influence research productivity of faculties at nursing colleges of the Ministry of Defense and the National Police Bureau. Bangkok: Naval Nursing Colleges.

7- FAO. (1997). Issues and opportunities for agricultural education and training in the 1990s and beyond. FAO publications, Rome, Italy.

8- Hosseinpour, M. (2011). A study of debilitating factors of research from the viewpoint of faculty members in human sciences. New Findings in psychological, 6 (19), 79- 95.

9- Hedjazi, Y. & Behravan, J. (2011). Study of factors influencing research productivity of agriculture faculty members in Iran. High Educ, 62, 635–647.

10- Jung, J. (2012). Faculty research productivity in Hong Kong across academic discipline. Higher education studies, 2(4), 1- 13.

11- Karimian, Z., Sabbaghian, Z., & Sadaghpor, B.S. (2011). Investigation of barriers and challenges of research and product of scientific in medical sciences university. Iranian higher education, 3(4), 35- 63.

12- Kotrlik, J.W., Bartlett, J.E., Higgins, C.C., & Williams, H.A. (2002). Factors associated with research productivity of agricultural education faculty. Journal of Agricultural Education. 43(3). Available: http://pubs.aged.tamu.edu/jae/pdf/Vol43/43-03-01. pdf

13- Law, R., & Chon, K. (2007). Evaluating research performance in tourism and hospitality: The perspective of university program heads. Tourism Management, 28, 1203–1211.

14- Leetputtarak, S. (2008). An investigation of factors related to research productivity in a public university in Thailand: a case study. A dissertation submitted in partial fulfillment of the requirements for the Degree of Doctor of Education, School of Education, Faculty of Arts, Education and Human Development, Victoria University, Melbourne, Australia.

15- Movahedi, R., Asgari, N., & Chizari, M. (2012). Factors affecting teaching quality and research performance of faculty members: The case of the agricultural faculty, BuAli University. Iranian agricultural Extension and Education Journal, 7 (2), 63- 74.

16- Najafipour, H., Darvish Moghadam, S., Azmandian, J., & Hosseini, H. (2009). The effect of providing Education-Research services by founding a clinical research development center on the research Performance of clinical faculty members of Kerman University of Medical Sciences. Strides in Development of Medical Education, 6 (2), 157- 164.

17- Ouimet, M. (2005). Factors associated with research productivity and knowledge transfer in Canadian Medical Schools: A cross-sectional survey. Paper to be presented at the triplex helix conference, the capitalization of knowledge: Cognitive, economic, social and cultural aspects, Turin, Italy, 18-21 May.

18- Patten, M.L. (2002). Proposing Empirical Research. Los Angeles: Pyrczak Publishing.

19- Ransdell, L.B. (2001). Using the precede-proceed model to increase productivity in health education faculty. The International Electronic Journal of Health Education, 4, 276–282. Online at: http:// www.iejhe.org.

20- Sax, L.J., Hagedorn, L.S., Arredondo, M., & DiCrisi F.A. (2002). Faculty research productivity: exploring the role of gender and family-related factors. Res Higher Educ. 43, 423–446.

21- Saburi, A. A. (2009). Product of Iran sciences in 2008. Quarterly of Approach, No. 43, 21- 31.

22- Smeby, J.C., & Try, S. (2005). Departmental Contexts and Faculty Research Activity in Norway. Research in Higher Education, 46(6), 593–619.

23- Shin, J.C., & Cummings, W.K. (2010). Multilevel analysis of academic publishing across disci-

pline: Research performance, collaboration, and time on research. Scientometrics, 85(2), 582-594.
24- Taylor, J. (2001). The impact of performance indicators on the work of university academics: Evidence from Australian Universities. Higher Education Quarterly, 55(1), 42–61.
25- Toreghi, J. (2005). Improvement the University mangers for Tomorrow University. Esfahan: Kankash press.
26- Tien, F.F. (2007). Faculty research behavior and career incentives: the case of Taiwan. International Journal of Educational Development. 27, 4- 17.
27- Wichian, S.N., Wongwanich, S., & Bowarnkitiwong, S. (2009). Factors affecting research productivity of faculty members in government universities: Lisrel and Neural Network Analyses. Journal of the Association of American Medical Colleges, 30, 67- 78.
28- Zainab, A.N. (2000). Publication productivity, focus on institutional, collaborative and communicational correlates: A review of the literature. Malaysian Journal of Library & Information Science, 5, 53-94.
29- Zhao, W., & Ritchie, J. R. B. (2006). A supplementary investigation of academic leadership in tourism research. Tourism Management, 1985–2004.

26

Socio-Economic Factors Influencing Adoption of Energy–Saving Technologies among Smallholder Farmers: The Case of West Pokot County, Kenya

Andiema Chesang Everlyne [1], Nkurumwa Oywaya Agnes [2] and Amudavi Mulama David [3]

Abstract

Fuel wood provides the main source of energy for cooking and space heating for over 80 percent of households living in Kenya. The heavy reliance on the biomass energy has exerted an imbalance in demand and supply consequently resulting in adverse environmental effects in Kenya. As part of innovation efforts, several energy-conserving technologies have been developed. A unique cook stove named Maendeleo was developed and promoted in Kenya and more so, West Pokot County, northern of Kenya, with the goal of reducing the quantity of wood households use for energy, and ultimately reduce pressure on local forests. However, despite the demonstrated technological multiple benefits and the institutional promotional efforts of the Maendeleo stove technology; the adoption level of this innovation has remained low. An important question investigated in this study was what makes potential users not utilize such valuable innovations? Socio-cultural, economic, political and institutional barriers are considered to contribute to low uptake of such innovations. This study therefore, sought to assess socio-economic factors influencing the adoption of the Maendeleo stove in the rural setting of Kapenguria Division. A survey research methodology with ex-post facto design was employed. The results showed that the age of the respondents had the highest influence on the non-adoption of the Maendeleo stove. Given the relatively low adoption level of Maendeleo stove in the county, and the projected increase in the number of people relying on biomass, this study recommends that the government and development partners put in place a programme for the promotion and dissemination of Maendeleo stove. There should be further investigation into the adoption behaviour of the respondents on the reasons for non-adoption and discontinuance of use of the Maendeleo stove.

Keywords:
Adoption, Smallholder, Energy-saving technologies

[1] Graduate Student, Department of Agricultural Education & Extension, Egerton University, Kenya.
[2] Lecturer, Department of Agricultural Education & Extension, Egerton University, Kenya.
[3] Director, Biovision Africa Trust, C/O icipe-African Insect Science for Food and Health, Nairobi, Kenya.
* Corresponding author's email: andiemaeve@yahoo.com

INTRODUCTION

In most developing countries, biomass-based energy accounts for more than 90% of all household energy consumption (FAO, 2010; Field, 2010). The dependence of the world's population on biomass fuels for domestic energy consumption has a negative impact on the social well-being of the users and the rural environment (United Nations, 2009; Inayat, 2011). The use of biomass in inefficient ways increases fuel wood demand of a household, yet energy needs of the developing world have to be met in a sustainable manner (Ndung'u, 2009). The invention and diffusion of improved cook stoves in developing countries is one of the strategies perceived as instrumental in combating the negative effects related to the use of traditional hearths (Rwiza, 2009). There have been several improved stove programs facilitated and implemented for communities in different parts of the world with support from governments, organizations, scientific institutions and funding agencies, for more than five decades (Reddy, 2008).

The adoption and continued use of improved cook stoves in the developing countries is of social, economic and environmental concern (Inayat, 2011). In particular such innovations have potential for delivering triple dividends: household health, local environmental quality, and regional climate benefits (Lewis and Pattanayak, 2012). Although the social, economic, and environmental benefits of the improved stove programmes seem to be rather clear, the rate of adoption of technologies promoted is not as fast as initially anticipated (FAO, 2010). However, there may be a range of personal, social, cultural and economic factors, as well as the characteristics of the innovation itself that prevent the technology adoption, notwithstanding the argument by scholars that the poor economic conditions in developing countries should encourage innovations in new technologies (Lundvall, 2007).

A number of socio-economic factors such as lack of knowledge about the costs and benefits of improved cooking technology (Muneer and Mohamed, 2003; Bikram, 2008), income level of the household, and lack of proper monitoring systems of the stove programs are responsible for slow adoption rates. Maendeleo stove is a stove that was developed in Kenya in the 1980's as one of the strategies to reduce fuel wood consumption. The basic component is the stove liner, (made of a pottery cylinder fired in a kiln), which incorporates a door for fuel and air intake and pot supports on the stove liner, built into a mud and stone surrounded stove in the kitchen or in a metal cladding (Mandeleo portable). Research findings by Ndung'u, (2009) show the stove could provide fuel wood savings of up to 43 per cent and producing up to 60% less smoke compared to a three-stone fire commonly used. With proper use; a fuel wood saving efficiency of up to 50% is achievable. Despite the fact that the Maendeleo stove technology has been promoted in Kenya for nearly twenty years and has been produced on a more commercial basis, the stove has remained at a low level of use within rural communities-only 4% of the targeted populations is using this stove (Ingwe, 2007). A number of farmers' characteristics, attributes of the technology and institutional factors are hypothesized as influencing adoption of these technologies.

Much emphasis has been placed on several socio-economic factors as being the most important factors influencing adoption of the Maendeleo stove technology– namely age of the target respondents, level of their education, household in, household size and farm size. These are some of the variables that have varying degrees of influence over the adoption of influence over the adoption of practices or technologies being promoted for uptake. Karin *et al.* (2007) observed that the socio-economic factors Were positively correlated with the adoption of the energy-saving cook stoves. Demographic factors such as age, education, household income, household size and land size, all have varying degrees of influence over the adoption of changed practices e,g (Jeanette *et al.,* 2010).

The age of a potential adopter has been found, with mixed results to influence adoption of innovations. A farmer's age is expected to increase Maendeleo stove technology adoption in the sense that older farmers over time have gained knowledge of the Maendeleo stove technology and experience in the use of the stove and are better able to evaluate technology information than would do younger farmers. An analysis of this study shows that age of the respondent is associated with the adoption of the Maendeleo stove technology. However, there is conflicting evidence on this relationship, with some researchers finding no significant relationship between age and adoption rates of improved stove technologies (for example; Cary *et al.,* 2001, 2002; Lockie and Rockloff, 2004). Other researchers have found a more direct relationship between adoption rates and an adopter's age. The younger the farmer, the more likely he is to adopt innovations early in his life cycle (Diederen *et al.,* 2003).

Education helps the transformation of information (processed data) to knowledge (information that is modeled to be useful) which in turn influences adoption. Farmers who have adequate information about knowledge of technology use are likely to adopt it (Abebaw and Belay, 2001; Rogers, 2003). Traditionally, educated people were expected to understand the benefit of the innovation in question at a faster speed than the uneducated (Makame, 2007; Rollins, 2009). Educational status is assumed to influence the adoption decision of many technologies because with higher level of education the farmer would be in a position to technically and economically assess the new technology to clear doubts and uncertainties associated with it, and enhance its adoption (Aneani *et al.,* 2012). The more aware (educated) respondents were, the more likely they were to use efficient cooking technologies (Inayat, 2011). In the present study, level of education was hypothesized as a proxy for more awareness about the pros and cons of using the Maendeleo stove technology. Although it is not necessary that more education equates to greater awareness, it is assumed that more educated people have more knowledge about benefits of Maendeleo stove technology than do un-educated people. Awareness of the Maendeleo stove technology in which the respondents had been introduced in the study area was high, the main sources of information being family members, farmer to farmer contact, community group and extension staff. In this context, awareness has been identified as a major factor impacting the adoption of the Maendeleo stove technology. Technical information and the frequency of exposure to this information are important in influencing adoption of Maendeleo stove technology and other technologies (Fernandez-Cornejo *et al.,* 2007). In the study area, the more educated respondents may be assumed to be more aware of the environmental and health effects of using biomass fuels, and therefore, formal education may increase the speed of adoption. There was a significant difference in the years of schooling of the respondents among the young and the older respondents, with the former being more educated. There was no significant difference observed between respondents with formal education and non-formal education in terms of gap between adopters and non-adopters. Adopters had significantly higher contact with extension visits than did non-adopters.

Household income is an indicator of prosperity and may be expected to have positive effect on adoption of technologies as wealthier households may have higher probability of investing in and using improved stoves (Inayat, 2011). Household income has a unidirectional, linear relationship with the household's type of fuel used as energy source and type of cook stove (Rwiza, 2009). Household income of a household was measured as total sum of money in Kenya shillings the household earns per year. the present study focused on the cost of the Maendeleo stove technology; this refered to the economic purchasing power and installation ability of the stove of the user. Cost of adopting a new technology remains a very important factor influencing the decision to adopt the technology (Huh and Kim, 2008). It follows that if a stove is too expensive, adoption decisions will be negatively affected. Rogers (2003) assumed that economic motivation is one of the main thrusts for adopting an innovation, especially if the idea is expensive in both the initial and running costs. The study revealed that cost of the Maendeleo stove was affordable as none of the respondents found the stove to be expensive. Technology with low initial cost is more likely to be adopted than would be technology with high initial cost. Low initial cost has a positive influence on the rate and speed of technology adoption. This perceived cost therefore, may be expected to increase adoption of the Maendeleo stove technology unless other attributes of the technology or other extraneous variables negatively influenced utilization of the technology.

Household size was expected to have a positive influence on the adoption of the Maendeleo stove technology. Family size is expected to have a positive influence on the model of the stove used (Inayat, 2011). It is assumed that larger households will cook more food for the household members requiring use of large pans and more fuel wood hence will be more inclined to adopt the Maendeleo stove technology. It is expected, therefore, that a larger household size will affect positively the decision of adopting the Maendeleo stove technology.

Land size is one of the first and most widely used factors on which the empirical adoption literature has focused. Most studies find a positive relationship between farm size and adoption. Farmers with larger farms are more likely to adopt relatively new innovations (Diederen, *et al.*, 2003). Thus, it can be assumed that a household with large landholding will be more likely to adopt improved cook stoves than would households with small landholdings (Field, 2010). The importance of including this variable in the model was the fact that acquisition of land is an important determinant of socio-economic status in West Pokot County. It was expected that a household with large landholding will be more likely to adopt Maendeleo stove technology. In the present study, farm size may be expected to influence a farmer's ability to set aside a portion of the land for wood production which in turn may affect availability of fuel and hence the ability to use energy saving Maendeleo stoves. It was expected that a household with large landholding would be more likely to adopt improved cook stove technology.

Although the social, economic, and environmental benefits of the improved stove programmes seem to be rather clear, the rate of adoption is not as fast as hoped and anticipated. A literature review on the diffusion of innovations reveals possible explanation about the slow diffusion process of technologies. Factors that may have influence on the adoption of Maendeleo stove include: socio-economic factors (i.e. age, education, household income, household size, land size); stove-related factors (cost, size, perceived benefits, biomass flexibility, operatability and quality) and, finally institutional factors (access to extension services, land tenure, membership to groups). These factors were used to examine issues in the context of the Maendeleo stoves in the study area. The purpose of this study was to determine the influence of socio-economic characteristics on adoption of the Maendeleo stove among small-scale farmers in Kapenguria Division of West Pokot County. The null hypothesis was: H_{01}: farmers' socio-economic characteristics have no statistically significant influence on adoption of Maendeleo stove.

MATERIALS AND METHODS

The study employed an ex- post facto survey research design. This design was found to be suitable for this study as the dissemination of Maendeleo stoves had already occurred and is on-going and the factors influencing adoption could only be studied retrospectively.

The survey was conducted in West Pokot County which lies along Kenya's western border with Uganda between latitudes 240 40'N and 107'N and between longitudes 34037'E and 35049'E. It covers an area of 2317.5sq km, which is approximately 5% of the area of Rift Valley Province. The county varies greatly in geographical features. The southeast part falls within the mountainous Cherangani Hills, which reaches up to 2550 metres above sea level, the lowest is 1550metres above sea level. The county has a bimodal rainfall pattern which is normally unevenly distributed ranging from 700mm to 1600mm per annum. The temperatures range from a minimum of 15 degrees centigrade to a maximum of 34 degrees centigrade. The County has four divisions namely: Kapenguria, Kongelai, Chepareria and Sook; 23 locations with 82 sub-locations.

Kongelai and Sook's residents are purely pastoralists; Chepareria residents are agro pastoralists and those of the larger part of Kapenguria practice mixed farming.

The study was conducted in 2 locations namely; Kaibos and Talau (with 2 sub-locations each) Kaibos, Kipkorinya, Chepkoti and Talau respectively; of Kapenguria Division, West Pokot sub-county. The four sub-locations had many similarities in terms of their farming system and the socio-cultural environment. The agricultural sector in the region is dominated by smallholder farmers. The division covers an area of 335.6 sq km with 9 locations and 28 sub-locations. The accessible population of this study comprised of 82,057 people with a total of 16,131 households. The division is a cosmopolitan, rural-urban setting with a rapid increase in population because it is a high potential area agriculturally and also due to influx of new comers from the drier areas of the district. The fast growing population has accelerated the rate of destruction of trees because there is a high demand for fire wood to use for cooking with highly inefficient open fire stoves.

The sampling frame consisted of a list of households which was generated from the four sub-locations with the help of the Agricultural Extension staff and the local leaders. From the sampling list 160 households were selected through simple random sampling using computer randomizer program. A structured questionnaire was administered to adult female respondents who are in charge of cooking for the household because cooking and taking care of children is almost done entirely by women. Studies indicate that women have a central stake when it comes to the adoption and use of improved stoves.

The distribution of the respondents in the study areas is depicted in table 1.

The analytical tools employed in this study were both the descriptive and inferential statistics. The descriptive statistics used were frequency counts, percentages and means, while the inferential statistical tool used was the logit regression model. The logit model is a standard method for understanding the association between explanatory variables and a binary dependent variable (Greene, 2008; Hosmer and Lemeshow, 2000). The objective of the research was to understand the degree and the trend of the relationship between dependent and independent variables in terms of adoption of Maendeleo stove technology. Since the adoption of Maendeleo stove technology is a dichotomous or binary dependent variable, with the option of either adoption or non-adoption, the binary logistic regression model was applied as the most appropriate tool to investigate how each independent variable affects the probability of the occurrence of events (Long and Freese, 2006). The logistic regression model explores the socio-economic factors influencing the adoption of Maendeleo stove technology. Thus it helps to explore the degree and direction of relationship between dependent and independent variables in the adoption of Maendeleo stove technology at the household level. Accordingly, Maendeleo stove technology in the study area is influenced by a set of independent variables and is specified as follows:

$$Y_i / 1- Y_i = \beta_0 + \beta_1 x_{1i} + \beta_2 x_{2i} + \beta_3 x_{3i} + ... + \beta_k x_{ki}.$$

Where the subscript i means the ith observation in the sample. Y is the probability that a household adopts the Maendeleo stove technology and (1-Y) is the probability that a household does not adopt Maendeleo stove technology. $\beta 0$ is the intercept term and $\beta 1, \beta 2, ..., \beta k$ are the coefficients of the independent variables X1, X2,........., Xk.

In the analysis of the hypothesis of this study adoption of Maendeleo stove was considered in relation to the five explanatory variables in the socio-economic characteristics; namely age, level of education, household income, household size and farm size.

The model can then be expressed as follows: $Yi = \beta_0 + \beta_1$ (x_1= Age (in years); $+\beta_2$ (x_2= Education (level of education attained); $+\beta_3$ (x_3= Household income (shillings); $+\beta_4$ (x_4= Farm size (in acres)

Table 1: Distribution of respondents by location and sub-location.

Location	Sub-location	No. of Respondents
Kaibos Location	Kaibos Sub-location	39
	Kipkorinya Sub-location	41
Talau Location	Chepkoti Sub-location	43
	Talau Sub-location	37
Total		160

+ β_5 (x_5 = Household size (persons).

Age was measured in years. Highest education level attained by the household was coded as 1 = non-formal education 2 = primary 3 = secondary, 4 = tertiary. Four dummy variables representing the level of education attained by the farmer were used, where non-formal education was used as a reference category variable. Household income is a continuous variable measured in Kenya shillings based on estimates of the last 12 months. Household size is a continuous variable measured by the number of persons living in the household and land size is a continuous variable measured in acres.

RESULTS AND DISCUSSION

Socio-economic characteristics of the respondents

Age of the respondents

The respondents were asked to indicate their age in years. Table 2 summarizes the results with regards to age of the respondents.

Findings revealed that the minimum age of the respondents was 19 years while the maximum was 69 years with a mean of 42 (±0.903) years. The respondents were grouped into six age categories as shown in Table 2. The majority (83%) of the respondents were in the age category of 31-60 years while the rest (17) were in the age group of 30 years and below. Observations from the study revealed that the respondents in the age category of 31–60 years adopted the Maendeleo stove technology.

The observed results indicate that the older respondents had prior exposure to the benefits of the Maendeleo stove technology having been promoted in the county in the last 30 years through the extension services. These findings are consistent with research work done by Mignouna *et al.,* (2011) which indicated that a farmer's age is expected to increase technology adoption in the sense that older farmers over time have gained farming knowledge and experience and are better able to evaluate technology information than younger farmers. Okunade (2007) in a study of Nigerian women farmers found a significant positive relationship between age and adoption of farm technologies. He concluded that the older the farmers were, the more their years of farming experience and hence the better the decision the farmer would make in adopting new technologies.

The younger generations on the other hand were aware of the Maendeleo stove technology but, lacked information on the benefits of the Maendeleo stove technology, due to the limited technical information about the Maendeleo stove technology by the agricultural extension

Table 2: Frequency distribution of the respondents by age.

Age categories (Years)	Frequency	Percent	Cumulative Percent
<21	5	3.1	3.1
21 – 30	23	14.4	17.5
31 – 40	47	29.4	46.9
41 – 50	43	26.9	73.9
51 – 60	33	20.6	94.4
>60	9	5.6	100.0
Total	**160**	**100.0**	

Mean 42.31, se 0.903, median 42, mode 49, std dev 11.43, minimum 19 maximum 69.

Table 3: Level of education attained by the respondents.

Level of formal Education	Frequency	Percent	Cumulative Percent
No formal education	11	6.9	6.9
Primary	118	73.8	80.6
Secondary	17	10.6	91.3
Tertiary	14	8.8	100.0
Total	**160**	**100.0**	

Mean 42.31, se 0.903, median 42, mode 49, std dev 11.43, minimum 19 maximum 69.

agents. This was accelerated by the un-availability of the Maendeleo stoves in the study area and lack of confidence that the stove is durable as the respondents feared the stove would not work or would break quickly. As a result; the respondents opted for the alternative improved stoves like the chepkube brooder, rocket stoves etc.

Level of education of the respondents

The respondents were asked to indicate the highest level of formal education that they had attained.

The level of education of the respondents ranged from no formal education at all, primary education, secondary education and tertiary education, the findings were as indicated in Table 3. The findings indicated that only a minority (6.9%) of respondents had never received formal education while a higher percentage (73.8%) of the respondents had attained primary level education. Educational status is assumed to influence the adoption decision of many technologies because with higher level of education the farmer would be in a position to technically and economically assess the new technology to clear doubts and uncertainties associated with it and enhance its adoption (Aneani *et al.,* 2012). In the study area, the respondents with formal education may be assumed to be more aware of the environmental and health effects of using biomass fuels, and therefore, formal education may increase the speed of adoption. The respondents with no formal education however, were aware of the benefits of the Maendeleo stove technology having gotten the technical information through the agricultural extension pathways.Technical information about the energy-saving Maendeleo stove technology are disseminated through the agricultural extension pathways such as farmers' field days, demonstrations, farmers' trainings among others, by the Home Economics Officers in the State Department of Agriculture. Access to information through extension services reduces the uncertainty about a technology's performance hence may change individual's assessment from purely subjective to objective over time thereby facilitating adoption. Technical information through extension services is critical in promoting adoption of modern agricultural production technologies because it can counter balance the negative effect of lack of years of formal education in the overall decision to adopt some technologies (Kubok, 2007). However, the study findings indicated that the level of education is not a major determinant of adoption of the Maendeleo stove technology since households who adopted the Maen-

Table 4: Frequency distribution of the household size.

Number of Members	Frequency	Percent	Cumulative Percent
1-3 members	10	6.3	6.3
4-6 members	55	34.4	40.7
7-9 members	55	34.4	75.1
10-12 members	31	19.4	94.5
Above 13	9	5.5	100.0
Total	**160**	**100.0**	

Mean 7.5; SE 0.239; median 7, mode 8, std dev 3.033, minimum 1 and maximum 18.

Table 5: Frequency distribution of the land size. (n=160).

Size of Land(in acres)	Frequency	Percent	Cumulative Percent
<1.0	64	40	40
1.0-5.0	88	55	95
5.1-10.0	6	3.8	98.8
10.1-15	2	1.2	100.0
Total	**160**	**100.0**	

Mean 4.0 ±0.297, median 3, mode 1, minimum 0, and maximum 17.

deleo stove technology cut across all the levels of education.

Distribution of household size

The household size referred to the number of all the people living within that household. The respondents were asked to indicate the number of people living in their homes and their responses were summarized in Table 4. The study findings indicated that household sizes ranged between 1 and 18 members. The average number of people living within the households was eight. The majority (75.1%) of households had nine or fewer members while the rest (24.9%) had ten or more members in the households. In the study area, household size was expected to have a positive influence on the adoption of the Maendeleo stove technology as it may be assumed that families with a large number of members have a higher demand for energy fuel and may be more inclined to adopt improved cook stoves.

Fuel savings are roughly proportional to baseline fuel use. Thus, it was also expected that adoption of energy-saving cook stove would be higher for those with higher baseline fuel use and expenditures and with larger families (Ostrom, 2010). The study findings indicated that households with more household members did not use the Maendeleo stove technology though they retained it, instead returned to their former inefficient stove design. Their reason was that the Maendeleo stove technology had pot holders of specific diameter. They could not use the Maendeleo stove technology when they wanted to use a pot whose diameter was larger than the holder's. Another reason why some respondents used both the traditional and Maendeleo stove technology was because of the inflexibility nature of Maendeleo stove technology with one pot holder. This makes respondents who want to cook two pots of food simultaneously to make fire on the traditional stove as well.

To solve the problem of holder-pot incompatibility the respondents re-deployed the traditional three-stone fires but, the few households that had adopted the Maendeleo stove technology had either modified the pot rests or used pot holders while cooking ugali which they had difficulty cooking using the Maendeleo stove technology. Therefore incompatibility between a stove and utensils may lead some users to keep their traditional stove as it does not have this problem. The acceptance or abandonment (return to former stove design) was dictated by the number and sizes of pots the stove could accommodate (Rwiza, 2009).

Distribution of farm size

The respondents were asked to indicate the size of land they owned as a household. Table 10 shows the average size of land owned by the respondents.

An analysis of the data showed that the mean land area owned by the households was four acres and ranged between zero and seventeen acres. The majority (95%) of the households had farms that were below five acres while a minority (5%) owned land that was more than five acres. Farm size may be expected to influence a farmer's ability to set aside a portion of the land for wood production which in turn may affect availability of fuel and hence the ability to use energy saving stoves (Makame, 2007).

It was expected that households with small landholding would be more likely to adopt the Maendeleo stove technology because of the high cost of buying or gathering fuel and the in-

Table 6: Annual household income. (n=160).

Income categories (KShs.)	Frequency	Percent	Cumulative Percent
>50,000	58	36.3	36.3
50,001-100,000	50	31.2	67.5
100,001-150,000	21	13.1	80.6
150,001-200,000	13	8.2	88.8
>200,000	18	11.2	100.0
Total	**160**	**100.0**	

Mean 116,438 ± 8,801; median 100,000; mode 50,000; std dev 111,330; minimum 5,000 and maximum 18,630,000.

Table 7: Logistic regression analysis of socio-economic factors influencing adoption of Maendeleo stove (dependent variable 1=adopter of MS; 0=non-adopter).

Variable name	β	S.E.	Wald	Sig.	Exp(B)
Age of respondent	0.046	0.017	7.022	0.008	1.047
Primary education dummy	0.572	0.704	0.660	0.417	1.771
Secondary education dummy	0.881	0.889	0.983	0.321	2.414
Tertiary dummy	1.385	0.974	2.022	0.155	3.994
Income dummy	0.132	0.152	0.752	0.386	1.141
Household size	-0.032	0.063	0.251	0.616	.969
Farm size	0.113	0.094	1.449	0.229	1.120
Constant	-2.733	1.149	5.652	0.017	0.065
Number of observations	160				
Overall percentage prediction	51.9				
-2 Log likelihood	205.211				
Cox & Snell R Square	0.097				
Nagelkerke R Square	0.130				
Chi-square	16.371				
Significance	0.22				

Mean 116,438 ± 8,801; median 100,000; mode 50,000; std dev 111,330; minimum 5,000 and maximum 18,630,000.

adequate land to establish woodlots for fuel wood. Despite the fact that they had later realized the benefits of the Maendeleo stove technology and accepted it, they could not have the stove because of unavailability of the Maendeleo stoves. Households with large landholdings likewise were expected to adopt the Maendeleo stove technology since they had adequate space for establishing wood lots for fuel wood compared to their counterparts but, to the contrary because they still felt they had no fuel wood shortage.

Household income

Household income was measured as total sum of money in Kenya shillings as earned by all members of the household per year. Data on household incomes was categorized and presented as given in table 6.

The average annual income for the households was KShs. 116,438 (± 8,801), with a minimum of KShs. 5,000 and a maximum of KShs. 18,630,000. The majority (80.6 %) of the households earned less than KShs. 150,000 while the rest (19.4%) had an income of more than KShs. 150,000. Household income is an indicator of prosperity and may be expected to have positive effect on adoption of technologies as wealthier households may have higher probability of investing in and using improved stoves (Inayat, 2011). Although income is the most important pointer of the economic status of a farmer, it is difficult to collect reliable information on income from the respondents as most consider it confidential. The study revealed that all the respondents irrespective of their income level found the Maendeleo stove technology less costly and affordable. Technology with low initial cost is more likely to be adopted than that with technology with high initial cost (Rogers, 2003). Low initial cost has a positive influence on the rate and speed of technology adoption. This perceived cost therefore, may be expected to increase adoption of the Maendeleo stove technology unless other attributes of the technology or other extraneous variables negatively influenced utilization of the technology.

RESULTS

The study investigated the hypothesis that farmers' socio-economic characteristics have no significant influence on adoption of Maendeleo stove. The results are indicated in table 7. The -2log likelihood estimate of 205.211 with a statistically significant chi-square of 16.371 ($P < 0.05$) indicated that the independent variables jointly determined the adoption decision of the small-scale farmers.

The model as a whole explained between 9.7

percent (Cox and Snell R square) and the pseudo R-squared was estimated to be 0.130 implying that about 13.0 per cent of the variation in the dichotomous dependent variable (adoption of Maendeleo stove) was explained jointly by the predictors. The results further demonstrated that 51.9 per cent of the cases were correctly predicted by the model. The intercept of the model was significant ($P < 0.05$).

The socio-economic model indicated that the significant determinant of adoption was age of the respondent ($P < 0.05$). This means that age influenced the likelihood of the respondent adopting the Maendeleo stove. As the respondent got older by one year, the log of the odds ratio increased by 0.046 which led to an increase in the odds ratio by 1.047 times.

The positive and significant contribution of age suggests that adoption of Maendeleo stove was higher among older respondents than younger ones. The observed results indicate that the older respondents in this study may have had prior exposure to the benefits of the Maendeleo stove technology having been brought up in the area more than twenty years ago unlike the younger respondents who were not sensitized on the same. These findings are consistent with research work done by Mignouna *et al.* (2011) which indicated that a farmer's age is expected to increase technology adoption in the sense that older farmers over time have gained farming knowledge and experience and are better able to evaluate technology information than younger farmers. Okunade (2007) in a study of Nigerian women farmers found a significant positive relationship between age and adoption of farm technologies. He concluded that the older the farmers were, the more their years of farming experience and hence the better the decision the farmer would make in adopting technologies.

These findings differ from those of Wambugu (2006) in his study of fodder shrubs farmers who reported no relationship between age of farmers and their adoption behavior. However, the studies by Aneani *et al.* (2012) on adoption of some cocoa production technologies by cocoa farmers in Ghana, reported a statistically negative relationship between age and adoption. As the age of the farmer increases the physical strength declines thereby reducing the farmer's ability to use new technology also older farmers may be more conservative, less flexible and more skeptical about the benefits of Maendeleo stove. Akudugu *et al.* (2012) and Marchionni and Ritchie (2007) concluded that the adopter of a new technology is typically younger as younger people are more likely to adopt improved technological practices as they are risk takers and that since they are still accumulating economic resources they would opt to adopt more technologies.

According to the findings of the analysis, the level of education, household income, farm size and household size tended to be less probable in influencing the decision of a farmer to adopt the Maendeleo stove. Though from research findings by Inayat (2011), household size is expected to have a positive influence on the model of the stove used; this is contrary to the results from the study. Though it may be assumed that families with a large household will be more inclined to adopt the improved stoves to minimize on the fuel use, this was not the case since larger households use bigger utensils to cook which are not able to fit well on the stoves. While education is critical in enabling a technology user to assess the usefulness of a new technology (Makame, 2007), this is inconsistent with the this study findings since use of the Maendeleo stove is dependent on meeting a basic need of cooking fuel by the households.

Much empirical adoption literature focuses on farm size as the first and probably the most important determinant of adoption of different agricultural innovations and technologies (See for instance Doss and Morris, 2001; and Daku, 2002). This is because farm size can affect and in turn be affected by the other factors influencing adoption. The effect of farm size on adoption could be positive, negative or neutral. There are mixed findings in the literature on the influence of landholding size on households' decisions whether or not to adopt new technologies as shown by inconsistent results in the studies by Kassie *et al.* (2009); Waithaka *et al.* (2007). Fernandez-Cornejo, (1996) and Kasenge (1998) found farm size to be positively related to adoption.

However, findings from the current study indicate that farm size had a negative relationship with the adoption of the Maendeleo stove technology since cooking fuel to meet the energy demand for the household can be met using a small portion of the farm for the supply of required fuel wood. These findings are consistent with findings by Yaron *et al.* (1992); and Harper *et al.* (1990) who found negative relationship between adoption and farm size. Household income, from the findings, show that it is not a factor influencing use of the stoves as they are affordable to the potential users and the income level may not limit affordability.

CONCLUSION

The level of adoption of the Maendeleo stove among the small-scale farmers surveyed was low because about half of the respondents who had adopted the cook stove had abandoned using it and returned to their former stove design. The respondents quoted various reasons that prompted them to abandon the stove and among the reasons given were: the stove could not accommodate a large number and sizes of pots, the stove design could not allow for other non-cooking attributes (heat, light, insect repellent, etc.), and lack of confidence that the stove is durable. The rest of the respondents had no Maendeleo stove technology at all although they were aware of the stove due to lack of technical information on the benefits of the Maendeleo stove technology and non-availability of the stoves.

Agricultural extension services provided by the Ministry of Agriculture are the major source of agricultural information in the study area. Access to extensions services therefore creates the platform for acquisition of the relevant information that promotes technology adoption. In the last 30 years many organizations such as the Ministry's of Agriculture and other stakeholders have been disseminating and promoting the Maendeleo stove technology. Delivery of extension services in Kenya by the main extension service providers in the government and NGOs has been declining over the years. However, technology transfer through the extension service appears to have been slow and inefficient. The implication of these findings is that extension visits are important to technology adoption. The greater the degree of contact of farmers with extension personnel, the greater is the possibilities of farmers being influenced to adopt agricultural technologies. Frequent visits to the farmer by the extension agent would provide the farmer with necessary information about the availability of needed resources, market and prices as well as the profitability status of the Maendeleo stove technology to clear any doubts and uncertainties concerning it to increase the probability of its adoption.

The focus of the study was to determine the socio-economic factors influencing adoption of the energy-saving Maendeleo stove technology among smallholder farmers. Among the socio-economic characteristics considered only the respondent's age, had a significant association with adoption of the Maendeleo stove technology. Other socio-economic characteristics (level of education, household income, household size, and farm size) had no significant influence on adoption of the Maendeleo stove technology. The findings are important as they show that none of these features prevents smallholder farmers from adopting the Maendeleo stove technology. Rich and poor, educated and uneducated, young and old; all appear to be potential adopters.

ACKNOWLEDGEMENTS

I am grateful to all the academic staff of the Department of Agricultural Education and Extension Egerton University and my colleagues for their moral, advice and for creating a harmonious environment during my stay at the University. My profound acknowledgement and thanks are due to the Ministry of Agriculture staff West Pokot County for their direct and indirect contributions to this study. Special thanks to the respondents for their time, patience and cooperation during the period of field data collection. The author would also like to acknowledge the helpful comments provided by the editor and anonymous reviewers which helped improve the analysis in this paper.

REFERENCES

1- Abebaw, D., & Belay, K. (2001). Factors influencing adoption of high yielding maize varieties in Southwestern Ethiopia: An application of logit. Quarterly Journal of International Agriculture, 40

(2), 149-167.

2- Akudugu, M.A., Guo, E., & Dadzie, S.K. (2012). Adoption of modern agricultural production technologies by farm households in Ghana: What factors influence their decisions. Journal of Biology, Agriculture and Healthcare, 2(3). Available online at http//www.iiste.org/journals/index.php/JBAH/article/download/.../1454.

3- Aneani, F., Anchirinah, V. M., Owusu-Ansah, F., & Asamoah, M. (2012). Adoption of Some Cocoa Production Technologies by Cocoa Farmers in Ghana. Sustainable Agriculture Research, 1(1), p103.

4- Bikram, M.T. (2008). Revising the need of improved stoves: estimating health, time, and carbon benefits. South Asia Network for Development and Environmental Economics (SANDEE), Nepal. Working Paper No. 44-09.

5- Cary, J., Webb, T., & Barr, N. (2001). The adoption of sustainable practices: Some new insights. Land and Water report available on: www. lwa. gov. au/download/final_reports/BRR19. pdf Accessed, 12(12).

6- Cary J, Webb, T., & Barr, N. (2002). 'Understanding Landholders' Capacity to Change to Sustainable Practices: Insights about practice adoption and capacity to change.' Agriculture, Fisheries and Forestry Australia, Canberra.

7- Daku, L. (2002). Assessing farm-level and aggregate economic impacts of olive integrated pest management programs in Albania. Ph.D Dissertation, Virginia Polytechnic Institute and State University, David, Lynne Riener Publishers.

8- Diederen, P.J.M., van Meijl, J.C.M., & Wolters, A.M. (2003). Modernisation in agriculture: what makes a farmer adopt an innovation? International Journal of Agricultural Resources, Governance and Ecology, 2(3), 328-342.

9- Doss, C.R., & Morris, M.L. (2001). How does gender affect the adoption of agricultural innovation? The case of improved maize technologies in Ghana. Journal of Agricultural Economics, 25, 27-39.

10- Fernandez-Cornejo, J. (1996). The microeconomic impact of IPM adoption: Theory and application. Agricultural and Resource Economic Review, 25, 149-160.

11- Fernandez-Cornejo, J., Mishra, A., Nehring, R., Hendricks, C., Southern, M., & Gregory, A. (2007). Off- Farm Income, Technology Adoption, and Farm Economic Performance. Economic Research Report No. 36. Economic Research Service. United States Department of Agriculture.

12- Field, T. (2010). Wood-Starved and Footsore- Global Fuelwood Shortage-World Forests. Retrieved June, 4, 2010, from http:// blogs. bnet. com / favicon. ico.

13- Food & Agriculture Organization of the United Nations (2010). Corporate Document Repository; World Energy; Wood for Energy; Problems and Promises. Retrieved September 6, 2010, from http://www.fao.org/docrep/Q4960E/q4960e03,htm# Top of Page.

14- Greene, W.H. (2008). Econometric Analysis, 6th Edition, Upper Saddle River, New Jersey, Prentice-Hall, New York University.

15- Harper, J. K., Rister, M. E., Mjelde, J. W., Drees, B. M., & Way, M.O. (1990). Factors influencing the adoption of insect management technology. American Journal of Agricultural Economics, 72(4), 997-1005.

16- Hosmer, D.W., & Lemeshow, S. (2000). Applied Logistic Regression. Second Edition. A Wiley-Inter science Publication. New York.

17- Huh, Y.E., & Kim, S.H. (2008). Do early adopters upgrade early? Role of post-adoption behavior in the purchase of next-generation products. Journal of Business Research, 61(1), 40-46.

18- Inayat, J. (2011). What makes people adopt improved cook stoves? Empirical evidence from rural north west Pakistan. Working Paper 012. The Governance of Clean Development Working Paper Series. School of International Development, University of East Anglia UK

19- Ingwe, A. (2007). Rocket Mud Stoves in Kenya. Boiling Point issue No. 53, Household Energy Network (HEDON).

20- Jeanette, S., Beth, C., & Ray, B. (2010). Factors influencing the adoption of NRM innovations or changed practices. Working Paper No. 1. Understanding Land Manager Constraints to the Adoption of Changed Practices or Technological Innovations: Literature Review. Queensland Department of Natural Resources, Mines and Water, Charles Stuart University. Queensland.

21- Karin, T., Alicia, C., Omar, M., & Leticia, M. (2007). Social perceptions about a technological innovation for fuelwood cooking: Case study in rural Mexico. Universidad Nacional Autónoma de México, Mexico.

22- Kasenge, V. (1998). Socio-economic factors influencing the level of soil management practices on fragile land. In proceedings of the 16th Conference of Soil Science Society of East Africa (Eds.: Shayo-

Ngowi, A.J., G. Ley and F.B.R Rwehumbiza), 13th-19th, Tanga, Tanzania pp.102-112.

23- Kassie, M., Zikhali, P., Manjur, K., & Edwards, S. (2009). Adoption of Sustainable Agriculture Practices: Evidence from a Semi-Arid Region of Ethiopia. Natural Resources Forum. 33, 189-198. doi: 10.1111/j.1477-8947.2009.01224.x, http: // dx. doi.org/10.1111/j.1477-8947.2009.01224.x.

24- Kubok, L.K. (2007). Factors influencing adoption of furrow irrigation technologies among women farmers in Chepareria Division of West Pokot District, Kenya. Unpublished M.Sc. Thesis, Njoro, Kenya: Egerton University.

25- Lewis, J.J., & Pattanayak, S.K. (2012). Who adopts improved fuels and cookstoves? A Systematic Review. Environmental Health Perspectives, 120(5), 637-645.

26- Lockie, S. & Rockloff, S. (2004). Landowner attitudes to wetlands and wetland conservation programs and incentives. Draft report for Coastal CRC. Australia.

27- Long, S.T., & Freese, J. (2006). Regression model for categorical dependent variables using stata. Texas: A Stata Press Publication, College Station.

28- Lundvall, B.A. (2007). National Innovation Systems. Industry and Innovation, 14 (1), 95-119.

29- Makame, O.M. (2007). Adoption of improved stoves and deforestation in Zanzibar Management of Environmental Quality. An International Journal, 18 (3), 353 – 365.

30- Marchionni, C., & Ritchie, J. (2007). Organizational factors that support the implementation of a nursing. Best Practice Guideline. Journal of Nursing Management, 16 (3), 74-266.

31- Mignouna, D.B., Manyong, V.M., Rusike, J., Mutabazi, K.D.S., & Senkondo, E.M. (2011). Determinants of adopting imazapyr-resistant maize technologies and its impact on household income in Western Kenya. AgBioForum, 14(3), 158-163. Available on the World Wide Web: http://www.agbioforum.org.

32- Muneer, S.T., & Mohamed, W.M. (2003). Adoption of biomass improved cookstoves in patriarchal society: an example from Sudan. The Science of the Total Environment, 307(1-3), 259-266.

33- Ndung'u, M. (2009). Ministry of Agriculture: Home Economics Technical Update. No.3, August, 2009. Nairobi, Kenya. Neuman, W.L. (1997). Social Research Methods: Qualitative and Quantitative Approaches, 3rd ed. Boston: Allyn and Bacon.

34- Okunade, E.O. (2007). Effectiveness of extension teaching methods in acquiring knowledge, skill and attitude by women farmers in Osun State. Journal of Applied Sciences Research, 3(4), 282-286.

35- Ostrom, T. (2010). Considering Sustainability Factors in the Development Project Life-Cycle: A Framework for Increasing Successful Adoption of Improved Stoves. Unpublished Master's thesis - Michigan Technological University- 2010.

36- Reddy, N.S. (2008). Factors influencing community adoption of good stoves. Proceedings of the PCIA 3rd Forum, Bangalore, India.

37- Rogers, E.M. (2003). Diffusion of innovations, 5th ed, Free Press, New York.

38- Rollins, T. (2009). Using the innovation adoption diffusion model to target educational programming. Journal of Agricultural Education Vol. 34.

39- Rwiza, M. (2009). Innovations and sustainability: The case of improved stoves' adoption and use in Tanzania. Unpublished Masters Thesis. Lund University, Sweden.

40- United Nations, (2009). DSD: Areas of Work: Energy for Sustainable Development [Internet]. Available from: <http: // www. un. org /esa /dsd / dsd _ aofw _ene/ene_index.shtml [Accessed 23 May 2009].

41- Wambugu, (2006). Factors influencing the effectiveness of farmers as disseminators of fodder shrubs in the Central Kenya Highlands: Research project report, Nairobi, Kenya.

42- Waithaka, M. M., Thornton, P. K., Shepherd, K. D., & Ndiwa, N.N. (2007). Factors Affecting the Use of Fertilizers and Manure by Smallholders: the Case of Vihiga, Western Kenya. Nutr Cycl Agroecosyst. 78, 211-224. doi: 10.1007/s10705-006-9087-x, http: //dx.doi.org/ 10.1007/s10705-006-9087-x.

43- Yaron, D., Voet, H., & Dinar, A. (1992). Innovations on family farms: the Nazareth region in Israel. American Journal of Agricultural Economics, 74(2), 361-370.

Permissions

All chapters in this book were first published in IJAMAD, by Islamic Azad University; hereby published with permission under the Creative Commons Attribution License or equivalent. Every chapter published in this book has been scrutinized by our experts. Their significance has been extensively debated. The topics covered herein carry significant findings which will fuel the growth of the discipline. They may even be implemented as practical applications or may be referred to as a beginning point for another development.

The contributors of this book come from diverse backgrounds, making this book a truly international effort. This book will bring forth new frontiers with its revolutionizing research information and detailed analysis of the nascent developments around the world.

We would like to thank all the contributing authors for lending their expertise to make the book truly unique. They have played a crucial role in the development of this book. Without their invaluable contributions this book wouldn't have been possible. They have made vital efforts to compile up to date information on the varied aspects of this subject to make this book a valuable addition to the collection of many professionals and students.

This book was conceptualized with the vision of imparting up-to-date information and advanced data in this field. To ensure the same, a matchless editorial board was set up. Every individual on the board went through rigorous rounds of assessment to prove their worth. After which they invested a large part of their time researching and compiling the most relevant data for our readers.

The editorial board has been involved in producing this book since its inception. They have spent rigorous hours researching and exploring the diverse topics which have resulted in the successful publishing of this book. They have passed on their knowledge of decades through this book. To expedite this challenging task, the publisher supported the team at every step. A small team of assistant editors was also appointed to further simplify the editing procedure and attain best results for the readers.

Apart from the editorial board, the designing team has also invested a significant amount of their time in understanding the subject and creating the most relevant covers. They scrutinized every image to scout for the most suitable representation of the subject and create an appropriate cover for the book.

The publishing team has been an ardent support to the editorial, designing and production team. Their endless efforts to recruit the best for this project, has resulted in the accomplishment of this book. They are a veteran in the field of academics and their pool of knowledge is as vast as their experience in printing. Their expertise and guidance has proved useful at every step. Their uncompromising quality standards have made this book an exceptional effort. Their encouragement from time to time has been an inspiration for everyone.

The publisher and the editorial board hope that this book will prove to be a valuable piece of knowledge for researchers, students, practitioners and scholars across the globe.

List of Contributors

Nasiru Suleman and Solomon Sarpong
Department of Statistics, Faculty of Mathematical Sciences, University for Development Studies, P. O. Box 24 Navrongo, Ghana, West Africa

Rasheed Gbolagade Adeola
Department of Agricultural Extension and Rural Development, Ladoke Akintola University of Technology, P.M.B. 4000 Ogbomoso, Nigeria

Ali Bagherzadeh
Assistant Professor of Economics, Khoy Branch, Islamic Azad University, Iran

Javad Mahmoodi Karamjavan
Assistant Professor, Department of Agricultural Extension and Education, Tabriz Branch, Islamic Azad University, Tabriz, Iran

Reza Movahedi
Assistant Prof. Agricultural Extension and Education Department., Bu-Ali- Sina University, Hamedan, Iran M.Sc. Graduated on Rural Development, Bu-Ali Sina University, Hamedan, Iran

Hadi Fathi and Somaye Latifi
M.Sc. Graduated on rural development, Bu-Ali Sina University, Hamedan, Iran

Maduabuchi Mkpado
Department of Agricultural Economics and Extension Federal University of Oye-Ekiti, Nigeria

C.J. Arene
Department of Agricultural Economics University of Nigeria, Nsukka

Hamed Kashfi
M.Sc. Animal Husbandry Management, Gorgan University of Agricultural Sciences and Natural Resources

Ahmad Reza Yazdani and Farhad Shirani Bidabadi
Members of the Scientific Board of Gorgan University of Agricultural Sciences and Natural Resources

Mehrdad Latifi
Shahroud's Central Veterinary Medicine Labratoary

Fakhrodin Ghasemi Sahebi
M.Sc., Irrigation and Drainage, Razi University, Kermanshah, Iran

Masoud Hekmat
Department of Agricultural Management, Shoushtar Branch, Islamic Azad University, Shoushtar, Iran

Eman Pourkhiz
Young Researchers Club, Ghaemshahr Branch, Islamic Azad University, Iran

Zabiholah Gholami
Department of Economics, Islamic Azad University, Qaemshahr Branch

Javad Shahraki
Assistant Professor of Agricultural Economics, University of Sistan and Baluchestan

Neda AliAhmadi
Master of Agricultural Economics in MA great of the University of Sistan and Baluchestan

Hoda Manafian
M.Sc Department of Rural Development and Management, Faculty of Agricultural Economic and Development, College of Agriculture and Natural Resources, University of Tehran, Iran

Zhila Daneshvar Ameri
Assistant Professor of Rural Development, Department of Agricultural Economic and Development, College of Agriculture and Natural Resources, University of Tehran, Iran

Saba Salehi Esfahani
Department of Recreation, Parks & Tourism Management, Kent State University, United States of America (USA)

Abdulazeez, Muhammad-Lawal and Adekunle, Dayo
Department of Agricultural Economics and Farm Management, University of Ilorin, P.M.B. 1515, Ilorin, Nigeria

Omotesho, Kemi Funmi
Department of Agricultural Extension and Rural Development, University of Ilorin, P.M.B. 1515, Ilorin, Nigeria

Adekola, Olabisi Fatimat
Department of Agronomy, University of Ilorin, P.M.B. 1515, Ilorin, Nigeria

Ahmad Reza Ommani
Assistant Professor, Department of Agricultural Extension and Education, Shoushtar Branch, Islamic Azad University, Shoushtar, Iran

Raju Ghimire and Rudra Bahadur Shrestha
Doctoral Student, Department of Tropical Agriculture and International Cooperation, National Pingtung University of Science and Technology, Pingtung 91201, Taiwan

Wen-Chi Huang
Associate Professor, Department of Agribusiness Management, National Pingtung University of Science and Technology, Pingtung 91201, Taiwan

Ronald O. Nyamwamu
Extension Division, Ministry of Agriculture, P.O. Box 30028, Nairobi, Kenya

Justus M. Ombati and John G. Mwangi
Department of Agricultural Education & Extension, Egerton University, PO Box 536-20115 Egerton, Kenya

Rasool Lavaei Adaryani, Fatemeh Adel and Ali Amiri
M.Sc. Students in Rural Development, University of Tehran

Mohammad Reza Akbari
M.Sc. Student in Agricultural Extension and Education, University of Tehran

Behrouz Eskandarpour
Faculty Member of Payam Noor University, Iran

Parisa Khaligh Khiyavi
Lecturer of Economics, Rasht Branch, Islamic Azad University, Rasht, Iran

Mohammad Kavoosi Kalashami
Assistant Professor, Department of Agricultural Economics, Faculty of Agricultural Sciences, University of Guilan, Iran

Ademoye Emmanuel Akinboboye, Fadipe , A.A., Adigun and Jubril Olayinka, Animashaun
Department of Agricultural Economics and Farm Management, University of Ilorin, PMB 1515, Ilorin, Nigeria

Amir Hossein Chizari
Assistant Professor, Department of Agricultural Economics, Faculty of Agricultural Economics and Development, Tehran University

Masoud Fehresti Sani
Ph.D Student of Department of Agricultural Economics, Faculty of Agricultural Economics and Development, Tehran University

Mohammad Kavoosi Kalashami
Assistant Professor, Department of Agricultural Economics, Faculty of Agriculture, Guilan University

Mojtaba Sookhtanlo, Hesamedin Gholami and Seyyed Reza Es'haghi
Department of Agricultural Extension and Education, University of Tehran, Karaj, Iran

Lounis Semara
Ph.D Student of Animal Production, Department of Agriculture and Animal Science, Setif 1 University, Algeria

Charefeddine Mouffok
Assistant Professor, Department of Agriculture and Animal Science, Setif 1 University, Algeria

Toufik Madani
Professor, Department of Agriculture and Animal Science, Setif 1 University, Algeria

Ubon Asuquo Essien, Chukwuemeka John Arene and Noble Jackson Nweze
Department of Agricultural Economics, University of Nigeria, Nsukka

Mostafa Baniasadi
MSc Graduated of Agricultural Economics, Department of Agricultural Economics, University of Tehran

Saeed Yazdani and Habib Allah Salami
Professors of Agricultural Economics, Department of Agricultural Economics, University of Tehran

Gholamreza Zamanian
Assistant Professor in Economics, University of Sistan and Baluchestan, Iran

Mehdi Jafari
Ph.D Student of Agricultural Economics, University of Sistan and Baluchestan, Iran

Shahram Saeedian
Graduate student of Agricultural Economics, University of Sistan and Baluchestan, Iran

Nematollah Shiri
Ph.D. Student, Department of Agricultural Extension & Education, Razi University, Kermanshah, Iran

Nader Naderi
Assistant Professor, Department of Agricultural Extension & Education, Razi University, Kermanshah, Iran

Ahmad Rezvanfar
Professor, Department of Agricultural Extension & Education, University of Tehran, Karaj, Iran

Andiema Chesang Everlyne
Graduate Student, Department of Agricultural Education & Extension, Egerton University, Kenya

Nkurumwa Oywaya Agnes
Lecturer, Department of Agricultural Education & Extension, Egerton University, Kenya

Amudavi Mulama David
Director, Biovision Africa Trust, C/O icipe-African Insect Science for Food and Health, Nairobi, Kenya